AF371258

Computational Analytical Methods for Buildings and Cities: Space Syntax and Shape Grammar

Computational Analytical Methods for Buildings and Cities: Space Syntax and Shape Grammar

Editors

Michael J. Ostwald

Ju Hyun Lee

Basel • Beijing • Wuhan • Barcelona • Belgrade • Novi Sad • Cluj • Manchester

Editors
Michael J. Ostwald
The University of New South
Wales
Sydney
Australia

Ju Hyun Lee
The University of New South
Wales
Sydney
Australia

Editorial Office
MDPI
St. Alban-Anlage 66
4052 Basel, Switzerland

This is a reprint of articles from the Special Issue published online in the open access journal *Buildings* (ISSN 2075-5309) (available at: https://www.mdpi.com/journal/buildings/special_issues/computational_buildings).

For citation purposes, cite each article independently as indicated on the article page online and as indicated below:

Lastname, A.A.; Lastname, B.B. Article Title. *Journal Name* **Year**, *Volume Number*, Page Range.

ISBN 978-3-0365-9694-5 (Hbk)
ISBN 978-3-0365-9695-2 (PDF)
doi.org/10.3390/books978-3-0365-9695-2

Cover image courtesy of Michael J. Ostwald and Ju Hyun Lee

Contents

About the Editors

Michael J. Ostwald

Michael J. Ostwald is a Professor of Architectural Analytics at UNSW, Sydney. He has a PhD in architectural history and theory and a DSc in design mathematics and computing. Michael has previously been a Professorial Research Fellow at Victoria University Wellington, a visiting Professor and Research Fellow at RMIT University, an ARC Future Fellow at Newcastle, and a visiting fellow at ANU, MIT, HKU, and UCLA. He completed postdoctoral research on geometry at UCLA (Calif.), CCA (McGill, Montreal), and Harvard (Mass.). Michael is the Editor-in-Chief of the *Nexus Network Journal: Architecture and Mathematics* (Springer) and on the editorial boards of *ARQ* (Cambridge) and *Architectural Theory Review* (Taylor and Francis).

Ju Hyun Lee

Ju Hyun Lee is an Associate Professor (Scientia Academic) of Architecture and Computational Design at UNSW, Sydney. He has made significant research contributions in the fields of architectural computing and design cognition and has held multiple academic roles in Australia and South Korea. He completed postdoctoral research at the University of Newcastle and was a senior research fellow at University of South Australia. He is Chair of AKA.N and Co-Director of Advanced Architectural Analytics lab at UNSW. Ju Hyun is the co-author of *Grammatical and Syntactical Approaches in Architecture* (IGI Global) and *Design Thinking: Creativity, Collaboration and Culture* (Springer). He has been awarded eleven competitive research grants, including two ARC Discovery projects.

Preface

This Special Issue is about Space Syntax and Shape Grammar, which are respectively concerned with *space* and *form*. Both computational theories and relevant techniques have been extensively developed and tested in architecture and urban design research over the last four decades. Despite acknowledging that buildings and cities develop from the complex interplay between both *space* and *form*, computational analyses and generative approaches often tend to focus on either one or the other, rarely integrating both seamlessly. In this context, this publication aims to provide a comprehensive understanding of both the syntactical and grammatical properties of architectural and urban environments, thereby opening the door to integration. This preface introduces two theories and elucidates their significance for both architectural practitioners and academics. It then provides an overview of the scope of this Special Issue.

Space Syntax theory underscores the social impact of *space*, particularly spatial organization or structure, and employs a variety of techniques to analyze the two-dimensional spatial characteristics of architectural and urban environments. Drawing from graph theory in mathematics, the syntactic properties of architectural or urban space are abstracted into a graph comprised of nodes and edges. This graph, once created, undergoes mathematical analysis, and the results obtained are then correlated back to the original spatial properties. Through this process, Space Syntax methods can be used to uncover socio-spatial relationships in built environments.

Shape Grammar, on the other hand, adheres to a rigorous *formal* logic, resembling a mathematical procedure, to generate a shape or form. Essentially, it serves as a production model or system, encompassing various variations, such as graphic, functional, parallel, parametric, and set grammars. Shape Grammar for the architectural design language has been a common research focus in the past. In many instances, it is utilized for design analysis and generation. Recent studies on Shape Grammar leverages generative algorithms not only to automatically generate design variations, but also to optimize and evaluate them.

This Special Issue consists of fourteen research articles, with twelve focusing on Space Syntax and two on Shape Grammar. The research articles using Space Syntax methods explore a myriad of built environments, e.g., historic courtyard houses, laboratories, art museums, mega-shelters, public libraries, subway stations, historic streets, rural settlements, and urban districts. In contrast, two pioneering grammatical studies address the generation of architectural and spatial layouts. Collectively, this Special Issue stands as a valuable reference for architectural and urban researchers, providing fresh insights into the analysis, evaluation, and generation of *space* and *form* in built environments.

Forty-one researchers from multiple countries and anonymous reviewers of *Buildings* significantly contribute to this Special Issue and the future of architectural and urban research, fostering the creative and systematic development of built environments. This book is tailored to architects, urban planners, researchers, and students interested in understanding and examining computational approaches in architecture and urban design. We invite readers to embark on this journey of exploration and discovery in the language of architecture and urban design.

Michael J. Ostwald and Ju Hyun Lee
Editors

 *buildings*

Editorial

Computational Analytical Methods for Buildings and Cities: Space Syntax and Shape Grammar

Michael J. Ostwald and Ju Hyun Lee *

School of Built Environment, Faculty of Arts, Design & Architecture, The University of New South Wales, Sydney, NSW 2052, Australia; m.ostwald@unsw.edu.au
* Correspondence: juhyun.lee@unsw.edu.au

Citation: Ostwald, M.J.; Lee, J.H. Computational Analytical Methods for Buildings and Cities: Space Syntax and Shape Grammar. *Buildings* **2023**, *13*, 1613. https:// doi.org/10.3390/buildings13071613

Received: 20 June 2023
Accepted: 20 June 2023
Published: 26 June 2023

1. Introduction

During the first century BC, the famous Roman architect, Vitruvius, defined architecture as encompassing three essential properties: *firmitas* (firmness), *utilitas* (utility), and *venustas* (attractiveness or beauty). These principles have been debated and promulgated in the field of architecture for many centuries, and more recently, they have been reconceptualised by architectural computing researchers as 'grammar', 'syntax' and 'style', respectively [1]. Grammar refers to 'form', representing the physical presence and stability ('firmness') of a building. Syntax refers to the interface between people and space, which has parallels to the types of functional needs encapsulated under 'utility'. The last property, 'style', pertains to the distinctive way architecture and urbanism employ language to express or communicate formally and spatially. The classic computational design approaches utilised to study and define the first two properties are Shape Grammar and Space Syntax, the combination of which has also been linked to the creation and generation of style [1].

Space Syntax, theorised by Bill Hillier and Julienne Hanson [2], examines the correlation between spatial logic and social behaviour by identifying the syntactic attributes of spatial configurations within the constructed environment. Introduced by pioneers such as George Stiny and James Gips [3], Shape Grammar is a computational and generative method for identifying and understanding the design principles that define the formal properties of a building or style. In this regard, 'syntactical' methods focus on the spatial or topological configurations of a building, while 'grammatical' methods address the shape rules responsible for a building's formal or geometric properties. Collectively, this Special Issue, "*Computational Analytical Methods for Buildings and Cities: Space Syntax and Shape Grammar*", presents advanced research findings using two widely recognized computational methods to analyse and model architectural and urban spaces.

2. Contributions

This Special Issue comprises 14 articles, with 11 original research articles and one systematic review on the applications of Space Syntax methods and 2 on Shape Grammar research. The 14 articles were authored by 41 international scholars, with a research focus on buildings and cities in Australia, Brazil, China, Denmark, Greece, Germany, Hungary, France, Sweden, South Africa, and the United States. The eleven original research articles on Space Syntax explore a range of built environments—courtyard houses [4], laboratories in history [5], art museums [6], mega-shelters [7], public libraries and their surrounding areas [8], a subway station [9], historic streets [10], rural settlements [11], and urban districts [12–14].

Zolfagharkhani and Ostwald [4] present a quantitative and computational study of the spatial topology of 37 historic courtyard houses in Yazd, Iran, using a conventional justified plan graph (JPG) method. By demonstrating how the JPG can rigorously examine socio-spatial characteristics of architecture over time, they provide illustrations and arguments to challenge traditional assumptions surrounding the Yazd courtyard house. In addition to a comprehensive overview of the historical, cultural, and climatic context of this iconic architectural type, their research offers insights into its spatial logic and diversity, which can inform future conservation and adaptation plans.

Zhang and Cui [5] studied the diversity and complexity of scientific laboratories in history, identifying the spatial topology of their JPGs. By reflecting changes in scientific paradigms, they observe the evolution of spatial genotypes in historic laboratories. In addition to conventional properties such as *integration*, new syntactic measures are presented in the study, such as *distributedness index* and *space link ratio*, which can be applied to examine the topological network and flexibility of building plan layouts.

Lee and Kim [6] present an in-depth syntactic analysis of four international art museums to explore the ways their auxiliary paths influence visitors' experiences and behaviours. Developed using convex graph identification, spatial sequences and possible trails are identified and compared with syntactic and isovist properties. The combination of JPG and visibility graph analysis (VGA) is used to reveal a hidden socio-spatial logic shaping human perception and engagement in architecture.

Kim et al. [7] conducted a comparative analysis of the spatial configuration of four mega-shelters for disaster refugees in Australia, Japan, and the U.S., using a comprehensive combination of axial line analysis (ALA), VGA and JPG. Space syntax is valuable in conducting a rigorous analysis of significant but sensitive environments. Their research highlights spatial cognitive abilities and communications to develop guidelines for the planning and management of disaster shelters.

Zhao and Hong [8] examine the correlation between the accessibility and utilization of public libraries in Seoul, analysing the syntactic and behavioural (usage) data from 783 public libraries and their surrounding areas. In terms of spatial analysis, two accessibility indicators are measured using axial maps and topological connections between public libraries and subway stations. The degree of utilization activation is then calculated based on the number of visitors over five consecutive years, which allowed the quantitative examination of the socio-spatial relationships using correlations and regression results.

Kim and Kim [9] statistically reveal pedestrian movements in a subway station predicted using agent-based modelling (or agent-based simulation or ABS) drawing on Space Syntax theory. It is evident in this research that a moderate correlation exists between simulated movement and specific pedestrian patterns in the station area.

Xu, Rollo, and Esteban [10] measure the experiential qualities of historical streets in Nanxun, China, which is a Beijing–Hangzhou Grand Canal town. They use a combined method of ALA, VGA, and ABS and derive four factors, complexity, coherence, mystery, and legibility, from environmental psychology and urban design theories, the knowledge of which is intended to help designers and policymakers understand and improve cognitive experiences in historical streets.

Zhao et al. [11] explore the characteristics of 18 typical rural settlements along the Grand Canal in Tianjin, China, using field investigation and GIS analysis. Their research uses a special Space Syntax modelling tool, called "spatial design network analysis (sDNA)", measuring *proximity* and *through-movement degree*. In addition, the categorisation of historical rural settlements and on-site phenotypes is valuable for the conservation, development, and management of this cultural heritage type.

Wang et al. [14] also use the GIS-sDNA method to examine the liveability and vitality of Qingdao's historical city in China, providing a detailed spatial interpretation of the city. The authors specifically developed a composite measure of 16 built environment variables, which they call a "JANE index" (derived from Jacobs' theory of urban vitality), and then analysed the spatial distribution and correlation of urban vitality. Their framework can be applied to systematic research on the liveability and vitality of different areas and contexts.

Pan et al. [12] investigate the correlation between public participation and spatial form in urban space in Nanjing Xinjiekou, using observations and syntactic measures such as *choice* and *integration*. This empirical study assists in the development of a morphological understanding of location-specific spaces and their future spatial planning.

Guo, Hu, and Tang [13] propose a new method, a 3D Node-Landmark Grid Analysis Model (3D NL GAM), to quantify the structural salience of urban landmarks and evaluate the intelligibility of urban environments in Changsha, China. The 3D NL GAM integrates 3D visibility measurement and Space Syntax principles to calculate the structural salience of landmarks based on their visibility, spatial properties, and links between wayfinding decision nodes and landmarks. This can be applied to improve the intelligibility and attractiveness of new or existing urban districts.

Lee, Ostwald, and Zhou [15] provide a comprehensive and systematic review of 38 articles on socio-spatial experience, using the PRISMA (Preferred Reporting Items for Systematic Reviews and Meta-Analyses) framework. This is the first PRISMA-compliant scoping review of Space Syntax studies that contributes to identifying key socio-spatial phenomena in architectural, medical, and urban spaces, including significant research problems and directions for future research.

In contrast, two grammatical studies focus on the generation of architectural and spatial layouts. Paulino, Ligler and Napolitano [16] develop a shape grammar for *Sobrado* buildings in São Luís, Brazil, transforming the spatial configuration of the historic architecture into contemporary social housing (e.g., for a multi-family apartment). Their research contributes to the adaptive reuse of traditional architecture by offering a comprehensive explanation of the *reviver* grammar and its application.

Lastly, Muslimin [17] addresses a combination of Space Syntax and Shape Grammar approaches by incorporating graph and shape computation to generate schematic plan layouts. Two case studies on hospitality and retail sectors demonstrate the shape-to-graph and graph-to-shape transformations, leading to sequential rules for functions and spatial arrangements. This integrated method is useful for analysing the functional evolution of architecture and generating alternative building plans.

3. Discussion and Conclusions

This Special Issue offers valuable references for architectural and urban research, presenting new insights into analysing, evaluating, and generating spatial configurations in buildings and urban patterns in cities. While new or combined measures, such as urban vitality and 3D visibility, are introduced in this collection, three basic Space Syntax techniques—ALA, JPG, and VGA—are consistently employed to examine a variety of architectural and urban spaces. An important message from this collection is that syntactic properties can be regarded as either independent or dependent variables in such design information models, enabling the systematic measurement and development of the built environment.

In this Special Issue, two research articles discuss the use of Shape Grammar methods for generating spatial layouts, highlighting the incorporation of syntactic information in spatial configurations as part of the generative process. This research confirms that architectural design languages can be developed from a combined understanding of Space Syntax and Shape Grammar approaches.

The collective efforts of the 41 researchers whose articles are featured in this Special Issue contribute to future architectural and urban, promoting creative and systematic development of the built environment.

Author Contributions: All authors equally contributed to this editorial. Conceptualization, M.J.O. and J.H.L.; formal analysis, J.H.L.; writing—original draft preparation, M.J.O. and J.H.L.; writing—review and editing, M.J.O. All authors have read and agreed to the published version of the manuscript.

Conflicts of Interest: The authors declare no conflict of interest.

References

1. Lee, J.H.; Ostwald, M.J. *Grammatical and Syntactical Approaches in Architecture: Emerging Research and Opportunities*; IGI Global: Hershey, PA, USA, 2020.
2. Hillier, B.; Hanson, J. *The Social Logic of Space*; Cambridge University Press: Cambridge, MA, USA, 1984.

3. Stiny, G.; Gips, J. Shape grammars and the generative specification of painting and sculpture. In *Information Processing 71*; Freiman, C.V., Ed.; North Holland: Amsterdam, The Netherlands, 1972; pp. 1460–1465.

4. Zolfagharkhani, M.; Ostwald, M.J. The Spatial Structure of Yazd Courtyard Houses: A Space Syntax Analysis of the Topological Characteristics of the Courtyard. *Buildings* **2021**, *11*, 262. [CrossRef]

5. Zhang, X.; Cui, T. Evolution Process of Scientific Space: Spatial Analysis of Three Groups of Laboratories in History (16th–20th Century). *Buildings* **2022**, *12*, 1909. [CrossRef]

6. Lee, J.H.; Kim, Y.S. Rethinking Art Museum Spaces and Investigating How Auxiliary Paths Work Differently. *Buildings* **2022**, *12*, 248. [CrossRef]

7. Kim, Y.O.; Kim, J.Y.; Yum, H.Y.; Lee, J.K. A Study on Mega-Shelter Layout Planning Based on User Behavior. *Buildings* **2022**, *12*, 1630. [CrossRef]

8. Zhao, X.; Hong, K. Basic Analysis of the Correlation between the Accessibility and Utilization Activation of Public Libraries in Seoul: Focusing on Location and Subway Factors. *Buildings* **2023**, *13*, 600. [CrossRef]

9. Kim, J.Y.; Kim, Y.O. Analysis of Pedestrian Behaviors in Subway Station Using Agent-Based Model: Case of Gangnam Station, Seoul, Korea. *Buildings* **2023**, *13*, 537. [CrossRef]

10. Xu, Y.; Rollo, J.; Esteban, Y. Evaluating Experiential Qualities of Historical Streets in Nanxun Canal Town through a Space Syntax Approach. *Buildings* **2021**, *11*, 544. [CrossRef]

11. Zhao, Y.; Yan, J.-W.; Li, Y.; Bian, G.-M.; Du, Y.-Z. In-Site Phenotype of the Settlement Space along China's Grand Canal Tianjin Section: GIS-sDNA-Based Model Analysis. *Buildings* **2022**, *12*, 394. [CrossRef]

12. Pan, M.; Shen, Y.; Jiang, Q.; Zhou, Q.; Li, Y. Reshaping Publicness: Research on Correlation between Public Participation and Spatial Form in Urban Space Based on Space Synta: A Case Study on Nanjing Xinjiekou. *Buildings* **2022**, *12*, 1492. [CrossRef]

13. Guo, Y.; Hu, X.; Tang, J. Structural Landmark Salience Computation in Compact Urban Districts with 3D Node-Landmark Grid Analysis Model: A Case Study on Two Sample Districts in Changsha, China. *Buildings* **2023**, *13*, 1024. [CrossRef]

14. Wang, S.; Deng, Q.; Jin, S.; Wang, G. Re-Examining Urban Vitality through Jane Jacobs' Criteria Using GIS-sDNA: The Case of Qingdao, China. *Buildings* **2022**, *12*, 1586. [CrossRef]

15. Lee, J.H.; Ostwald, M.J.; Zhou, L. Socio-Spatial Experience in Space Syntax Research: A PRISMA-Compliant Review. *Buildings* **2023**, *13*, 644. [CrossRef]

16. Paulino, D.M.S.; Ligler, H.; Napolitano, R. A Grammar-Based Approach for Generating Spatial Layout Solutions for the Adaptive Reuse of Sobrado Buildings. *Buildings* **2023**, *13*, 722. [CrossRef]

17. Muslimin, R. Experience Grammar: Creative Space Planning with Generative Graph and Shape for Early Design Stage. *Buildings* **2023**, *13*, 869. [CrossRef]

Article

The Spatial Structure of Yazd Courtyard Houses: A Space Syntax Analysis of the Topological Characteristics of the Courtyard

Mina Zolfagharkhani [1,*] and Michael J. Ostwald [2]

[1] Faculty of Art and Architecture, Science and Research Branch, Islamic Azad University, Semnan 36167_43154, Iran

[2] UNSW Built Environment, University of New South Wales, Sydney, NSW 2052, Australia; m.ostwald@unsw.edu.au

* Correspondence: minazolfagharkhani@gmail.com

Abstract: An important "architectural type" in Iranian history is the Yazd courtyard house. This historic building type features a walled boundary that contains a complex pattern of open (to the sky), semi-enclosed and enclosed spaces. The planning of the courtyard in these houses has typically been interpreted as either a response to changing socio-cultural values or to local climatic conditions. Such theories about the planning of these houses are based on a series of assumptions about (i) the numbers of courtyards and rooms they contain, (ii) their unchanging nature over time and (iii) a topological pattern existing in the relationship between the courtyard and the rest of the plan. Yet, these assumptions, all of which have an impact on the socio-cultural or climatic interpretation of this famous architectural type, have never been tested. In response, this paper uses a computational and mathematical method drawn from Space Syntax to measure the spatial topology of 37 plans of Yazd's most significant courtyard houses. These houses, which are classified by the Yazd Cultural Heritage Organization, were constructed between the 11th and 20th CE centuries and are all exemplars of this type. This paper develops three hypotheses around the assumptions found in past research about the characteristic planning of the Yazd courtyard house. Then, using quantitative measures derived from plan graph analysis, the paper develops a series of longitudinal trends to test the hypotheses and explore changes that have occurred in this architectural type over time.

Keywords: Yazd courtyard house; Persian architecture; space syntax; graph theory; heritage architecture

Citation: Zolfagharkhani, M.; Ostwald, M.J. The Spatial Structure of Yazd Courtyard Houses: A Space Syntax Analysis of the Topological Characteristics of the Courtyard. *Buildings* **2021**, *11*, 262. https://doi.org/10.3390/buildings11060262

Academic Editors: Morten Gjerde and David Arditi

Received: 10 May 2021
Accepted: 16 June 2021
Published: 19 June 2021

Publisher's Note: MDPI stays neutral with regard to jurisdictional claims in published maps and institutional affiliations.

1. Introduction

The city of Yazd is the capital of the Yazd Province in central Iran. While its origins are typically traced to the fifth century BCE (third century AH) and the rule of Sassanian King Yazdegerd I, the region was inhabited for several thousand years before this time [1]. Today, the city of Yazd is famous for its historic wind towers (*bâdgir*), cisterns (*ab anbars*), underground water channels (*qanats*), Zoroastrian fire temple (*dar-e mehr*) and central courtyard houses. Collectively, these elements are responsible for Yazd's UNESCO World Heritage status. The last of these architectural types, the courtyard houses, are the focus of the present paper. They are typically described as possessing a highly homogeneous formal and spatial structure in their planning that has remained largely unchanged for over a thousand years [2].

Past research into the planning of Yazd courtyard houses has typically argued that the courtyard is either an environmental response to the arid, desert environment, or it has socio-cultural significance, being shaped by Zoroastrian or Islamic faiths in the region. For example, researchers have repeatedly measured the orientation, size and proportions of the courtyard in Yazd houses, as part of studies using climatic and environmental modelling [3,4]. Conversely, researchers have used the courtyard's location in the plan to explain a pattern of social, phenomenal and symbolic relations in these houses [5–7]. While such research has proposed multiple interpretations of the planning and origins of these houses, there are several presuppositions implicit in them that have never been established [3–7]. Three of these relate to untested ideas about the architectural planning of these houses.

(i)   Past research typically focusses on the role of *the* (singular) courtyard in a Yazd house, and in parallel it is often assumed that these houses are relatively consistent in terms of total size (number of rooms). To confirm if this is reasonable, it is necessary to measure the number of courtyards, perimeter rooms and other rooms in the most significant examples of this type.

(ii)   Past research treats the planning of the Yazd courtyard house as largely unchanged over time, despite being constructed from the 11th to the 20th century. To determine if this is a legitimate supposition, a longitudinal assessment of the characteristics of the plan, and especially of the location and role of the courtyard, is required.

(iii)   A common view in most research that seeks to generalize the properties of this architectural type is that they possess a pattern of spatial organization around their courtyards. This is not a straightforward proposition to make or assess. The properties of plans are complex, and without measuring the networks of social and functional relations they create, this view cannot be validated.

These three gaps in our understanding of the architectural planning of Yazd courtyard houses are the focus of the present paper. None of them are directly about socio-cultural or environmental factors, rather, they are all concerned with understanding the spatial properties of architectural plans that have been interpreted in different ways in the past. As such, this paper addresses foundational assumptions in past research about the topological properties of architectural plans. That is, the ways rooms are interconnected, and courtyards are located, relative to other spaces in a plan.

To address these three knowledge gaps, this paper uses a variation of a standard architectural analytical method to measure the topological properties of the planning of the Yazd courtyard house. Plan graph analysis, a Space Syntax technique originally developed in the 1980s, uses graph theory to measure various topological properties of plans, which can then be correlated to social, perceptual and cognitive factors [8–10]. These results can be normalized for comparison of different sized buildings and changing planning characteristics can be charted over time [10]. The three gaps in our knowledge of the Yazd courtyard house can all begin to be addressed using this method.

In this paper, plan graph analysis is used to measure the topological properties of 37 architectural plans, spanning from the Seljuk-Atabakan (11th century CE) to the Pahlavid eras (20th century CE). This process involves the development of seven mathematical measures for connectivity relations between 618 rooms in these 37 houses, leading to 4326 raw data points which are compared using standard statistical methods. These 37 cases are a purposive set rather than a population sample. They have been independently identified by heritage architects and historians as exemplars of this building type and they also encompass the works that are most commonly studied in past research.

In addition to filling three gaps in our knowledge about this famous building type, this paper has several original dimensions. First, it is one of only a small group of papers on Yazd architectural planning to draw on both English and Iranian language sources. Second, research about Yazd courtyard houses has often relied on very small sample sizes or has only considered small houses to draw conclusions. The sample size for the present paper is one of the biggest of this type, it includes both large and small houses and the volume of data it develops is robust enough to draw conclusions about the planning of the works studied. Third, this paper constructs its mathematical analysis around the courtyard's topological location in the plan, whereas past Space Syntax research uses the exterior as the primary point of reference for developing results.

Notwithstanding these features, this paper has several methodological limitations. As the focus is on the topology or connectivity of the plan, it does not consider cross-sectional variations in the architecture, orientation, contextual issues or environmental factors. This analysis is limited to the spatial relations defined by the planning of these buildings. It is not concerned with tectonics or materials, measuring energy and climate, or the history of the people who inhabited these houses. Indeed, this paper cannot offer any new conclusions or direct evidence about either environmental or socio-cultural conditions. None of the measures developed in this paper are climatic, although several can be extrapolated to comment on social or cognitive patterns in design. Instead, the focus of this paper is solely on the mathematical properties of architectural plans, which are central to many theories and interpretations of the origins and significance of the Yazd courtyard house.

This paper commences with a background of the Yazd courtyard house and an overview of past research about its history, planning and development. Thereafter, an introduction to the Space Syntax plan graph analysis method is provided, along with a description of the particular way it is applied in this paper. The three knowledge gaps are then reframed as hypotheses with measurable conditions. The results are presented and discussed, before the paper concludes with a reflection on the hypotheses, and what the results imply for the standard interpretations of this important heritage building type.

2. Background

The development of domestic Iranian architectural planning throughout history is often explained as a product of three factors. First, religious beliefs and practices emphasized the importance of unity, introversion and a connection to nature [11]. The Qur'an, for example, explicitly calls for balance, privacy and comfort, which scholars argue is reflected in the creation of inwardly focused, self-contained buildings with only a single entrance [12,13]. Rather than possessing exterior windows, these houses had internal courtyards and rooftops to provide a physical, phenomenological and spiritual connection to the world. Second, whereas in rural and agrarian society, houses were at least partially designed to meet the practical needs of animal husbandry and agricultural storage, in cities, domestic planning reflected socio-cultural or religious values and practices along with familial structures [14]. Thus, the extent to which an extended family included multiple generations, regularly received visitors or had family servants or retainers was a major influence on the planning of urban houses. Third, architectural planning was adapted to respond to local climatic conditions, which often led to the creation of multi-purpose rooms, that would be used in different ways during different seasons, a practice called "seasonal relocation" [15]. Researchers have argued that these three factors, either individually or collectively, are responsible for the evolution of the Iranian house throughout history [16,17].

At a macroscale, the city of Yazd was first defined by a wall enclosing its major buildings in the 11th century CE, and over time, the city expanded with the addition of further walls enclosing residential neighborhoods [1]. The house plans in the city were rarely orthogonal, their walls being shaped by historic street networks and ownership patterns [18]. Yazd's historic domestic architecture is often divided into three types—courtyard houses, platform house (*chahar soffeh*) and garden houses—of which the first are the most common [6] (Figure 1). In reality, though, these types are not mutually exclusive. For example, one of the most famous types of platform houses, pre-dating the Seljuk and Timurid eras, has a four-platform plan with a central courtyard. It has been argued that this plan was originally Zoroastrian and was later adapted in post-Islamic Iran to provide a clear hierarchy of spaces around a central courtyard, with a perimeter of semi-open spaces and then a series of fully enclosed spaces [19]. As such, many platform and garden houses feature courtyards, although they do not necessarily conform to all characteristics of courtyard houses. In the present paper, one of the 37 cases examined (Vizier's House) conforms to both the courtyard house and platform (*chahar soffeh*) house types and its inclusion reflects this dual categorization.

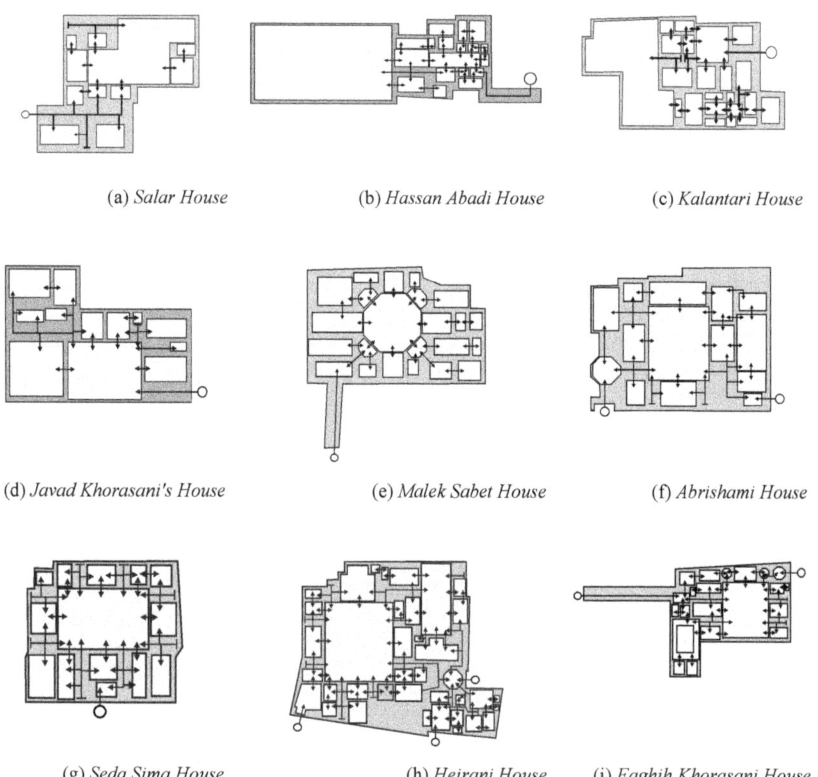

(a) Salar House   (b) Hassan Abadi House   (c) Kalantari House

(d) Javad Khorasani's House   (e) Malek Sabet House   (f) Abrishami House

(g) Seda Sima House   (h) Heirani House   (i) Faghih Khorasani House

Figure 1. Nine example plans of Yazd courtyard houses depicting habitable spaces (white), both enclosed and open to the sky, and trafficable connections between spaces (arrows).

Past research has tended to explain the spatial planning of the Yazd courtyard house using either climatic responsiveness or socio-religious reasoning, and it is clear that both factors partially explain the planning of these buildings. For example, Adeli and Abbasi Harofteh [20] argue that the combination of open yards and enclosed rooms is an appropriate response to the hot and dry climate of Yazd. The cruciform pattern with a central courtyard is ideal for "seasonal relocation", as it has summer-living (northeast) and winter-living (southwest) zones [21]. Moreover, the courtyards are often angled so that half the interior spaces capture the sun in the colder parts of the year and are shaded in the hotter months. In a study of three Yazd houses, for example, Soflaei, Shokouhian and Soflaei [22] identify a consistent orientation and proportion for the courtyards to achieve this outcome. In a study of ten Yazd courtyard houses, Khajehzadeh, Vale and Yavari (p. 481, [3]) conclude that the courtyards were designed "on the basis of receiving sunshine for some internal facades on cold winter days and sufficient shade on others for hot summer days". Conversely, historians have argued that residential planning is designed primarily to achieve privacy and to emphasize spatial hierarchy through separation of spaces by courtyards and corridors. The courtyard symbolized the centrality and unity of the reception and living spaces, with axial family spaces and perimeter service spaces reinforcing the hierarchical structure [12].

Architectural planning is both a reflection of the needs of the society that created it, and a mechanism which shapes the way a society functions [23]. In addition to the dominant socio-cultural and environmental paradigms for explaining the planning of the Yazd courtyard house, there are also multiple pragmatic factors involved [1,6,18].

These houses were multi-generational residences, with different zones for the family's oldest members, their children and spouses and finally their grand-children. As such, the Yazd courtyard house often accommodated multiple extended families. In addition, large families often had servants, creating the need for more extensive networks of rooms and spaces for work and living. The multiplicity of courtyards and perimeter rooms accommodated a similarly diverse range of domestic, agricultural and even industrial practices, providing food, materials and products for the household to use, sell or trade. Some rooms and courtyards would have functioned as stables for horses or livestock, and a room was often set aside for cotton and silk weaving. Furthermore, in traditional Iranian culture, houses of this type served to ensure privacy and safety for both families and their guests. In the latter case, the vestibule at the entrance of the house was a zone of separation, between the public and private realms. The vestibule commonly contained links to guest quarters, providing a welcoming space, and also delineating their rooms from family and service areas [5].

While rooms in these houses were not "functionally defined" (in the sense that their use changed to adapt to climate and the needs of the extended family), there were some consistent factors. Many main courtyards were paved and used for family celebrations, rituals and events. Some courtyards were designed for a single function and were located in the plan for a specific purpose, say, adjacent to a guest room, or alternatively as a space for animals. Others would have contained vegetable gardens and ponds. As these houses had few conventional windows, the courtyards were the primary means of accessing light, sun, air and views. The rooms that surrounded the courtyards would have had indirect access to natural light and ventilation through the roof opening. These spaces would have included both semi-public (guests and family) and semi-private (family) functions, depending on the seasons. In the next layer of rooms, which are two steps removed from natural light and air, there would have been kitchens, vestibules, entrances, stables and bathrooms. Because the rooms in these houses often had multiple doors, they would have functioned in part as habitable spaces and also as passageways to navigate around the house. This spatial separation, which required people to pass through a matrix of rooms, would have been the main way of shaping and controlling the relationships between, and conflicting spatial needs of, children, family members and guests [2].

Throughout the latter part of the 20th CE century, much of this changed. Growing industrialization led to a rapid urban growth, overcrowding and pollution in many historic cities, including Yazd. In response, the affluent and middle-class of Yazdi society gradually abandoned the historical core of the city and relocated to the suburbs. Today, some of the larger historic courtyard houses have been converted into government offices, and medium-sized houses have become cultural centers, restaurants and hotels. The smallest houses are often still residences today, where they are typically inhabited by low-to-middle income families.

This background to the Yazd courtyard house highlights common interpretations of this building type and practical dimensions. Within this large topic area, the present paper is concerned with three gaps in our knowledge which recur throughout much of the past research. First, despite evidence to the contrary, it would be possible to read past research without knowing that houses of this type typically have multiple courtyards, not all of which have the same orientation or proportion [19–21]. They also vary significantly in scale (number of rooms) and complexity (spatial hierarchy) [12–14]. Second, there is a tendency in past research to treat the planning of these houses as largely unchanging over time [20,21]. Third, past research overwhelmingly describes these buildings as having a relatively consistent planning pattern [2,4]. This paper begins the process of understanding these three gaps in our knowledge, which in turn have an impact on the dominant environmental and socio-cultural interpretations of this historic building type.

3. Method and Measures

First developed in the 1980s, Space Syntax is a set of theories, methods and techniques for mathematically measuring and comparing the spatial or spatio-visual properties of a building [10,23]. This approach is grounded in the view that the spatial properties of architectural plans are significant because they both reflect and reinforce the dominant social relationships, hierarchies and controls of a building type, society or era [8,24,25]. Logically, Hillier argues, "a spatial layout can reflect and embody a social pattern" and in parallel, "space can also shape a social pattern" (p. 104, [8]). Ostwald and Dawes propose that "such a pattern serves to enshrine the collective social structures and values of a group in the spatial configuration of buildings which have been designed to accommodate them" (p. 24, [9]).

With a focus solely on spatial properties, Space Syntax was the catalyst for a paradigm shift in architectural analysis, rejecting the standard conventions which sought to understand architecture in terms of formal properties (shape), style (aesthetics) and geography (orientation and context). Instead, Space Syntax reasoning proposed an understanding of architecture that is inherently spatial, relational and topological. Conceptually, Space Syntax suggests that for understanding the ways people inhabit, cognitively process and behave in space, plans are more significant than elevations.

A further important aspect of Space Syntax theory is that it maintains that, logically and mathematically, the topologically central spaces in a plan are more likely to be experienced than those at the periphery of a plan. To frame this another way, "central" spaces offer greater opportunity for social interaction and co-presence than peripheral ones. This is not saying that heightened social interaction will only occur in central spaces, or that peripheral spaces will be empty. It does suggest, however, that the activities that happen over time, and which multiple people engage in, will tend to occur in the more spatially integrated rooms. Thus, Space Syntax models the statistical or mathematical properties of spaces, which have been shown to correlate to patterns in human behavior [6–9].

One of the most well-known Space Syntax techniques is Convex Space analysis, which is also known as Justified Plan Graph analysis [26]. This method commences by identifying the main visually or functionally defined spaces in a plan and the connections between them. The "permeability"—capacity to move—relationships between these spaces are then analyzed using graph theory [10,27]. This requires converting the spaces in the plan into the nodes of a graph, and connections between these spaces into edges in the graph (Figure 2). The base of the graph or "carrier" is then chosen to reflect the property being examined. In most cases, this is the exterior, as it is significant for understanding or studying privacy, access and control.

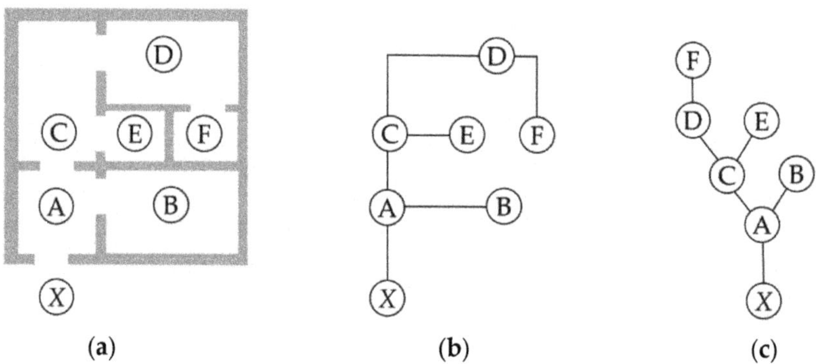

Figure 2. Example plan graph abstraction process: (**a**) an architectural plan with six rooms, annotated A–F and with the exterior (X) as carrier; (**b**) diagram of connections between spaces; (**c**) justified graph of the topology of the plan, relative to the exterior.

Once the graph for each plan is developed, then seven primary topological measures can be derived from it for each node (or room), generating measures for the complete graph (or building) and two additional measures can then be developed from each graph. Furthermore, using these nine measures, comparisons can be undertaken between sets of houses to reveal their planning structures and how they vary by region or over time [10,27]. Detailed descriptions, formulas and worked examples for all of these standard measures are widely available [10,26,28,29] and the key mathematical and interpretive principles are summarized hereafter.

3.1. Total Depth (TD)

TD is the sum of the connections between a given node in the graph and every other node, weighted by depth. This is, in practical terms, an indication of the distance between a room and every other room in a plan.

$$TD = (0 \times n_x) + (1 \times n_x) + (2 \times n_x) + \cdots + (X \times n_x) \tag{1}$$

3.2. Mean Depth (MD)

MD is the average depth of a node in a graph, where *K* is the complete number of nodes (the denominator is $K{-}1$, because it excludes the node being measured). A room that is higher than the mean for a plan is more spatially isolated than one which is lower than the mean.

$$MD = \frac{TD}{(K-1)} \tag{2}$$

3.3. Relative Asymmetry (RA) and Integration (i_{RA})

RA is a measure of the relative isolation of a node within the graph, and its inverse is $i_{(RA)}$, which is the level of integration of a node. Simplistically, more integrated spaces are more public and people will be more likely to meet one another in these spaces. Less integrated spaces are more private or have reduced likelihood of being sites of random encounters. *RA* is also important because it allows a researcher to compare measures derived from similar size graphs by normalizing *MD* to a range between 0.0 and 1.0.

$$RA = \frac{2(MD-1)}{(K-2)} \tag{3}$$

$$i = \frac{1}{RA} \tag{4}$$

3.4. Real Relative Asymmetry (RRA) and Integration (i_{RRA})

Because graph values are relative to the number of nodes and their distribution, to undertake a comparison between different sized graphs (building plans with different numbers of rooms), *RRA* normalizes the measures relative to an idealized diamond-graph *D*, for *K* number of spaces [23,24]. Once *RRA* is determined, a normalized integration (i_{RRA}) measure can also be developed for comparing properties of different sized graphs. It must be noted that there is disagreement in past research about the relative merits of *RRA* verses *RA*, with both being normalized, albeit in different ways. For this reason, some past research reports one of the other pair of results, whereas other researchers report both and use them as complimentary data sets, which may reinforce a key message [26,28]. In this paper, both are reported as part of a comprehensive review of the results. For a detailed review of the merits of both approaches and their interpretation, see [26].

$$RRA = \frac{RA}{D_k} \tag{5}$$

$$i_{RRA} = \frac{1}{RRA} \tag{6}$$

3.5. Control Value (CV)

CV is a measure of the extent to which one node influences all others. It has also been described by Klarqvist as "a dynamic local measure" that defines "the degree to which a space controls access to its immediate neighbours" (p. 11, [28]). In practical terms, it can be used to identify the spaces in a plan that will tend to be "sites of attraction" or have "pulling potential or capacity" (p. 63, [8]). The CV of a node *a* (and where *Val(b)* is the number of connections to node *b*) is determined using the following formula.

$$CV_a = \sum_{D(a,\,b)=1} \frac{1}{Val(b)} \tag{7}$$

3.6. Unrelativized Difference Factor (H) and Relativized Difference Factor (H)*

Once the seven topological measures are developed for each room, the mean properties of an entire plan can be determined. Two important properties are *H* and *H**. The logic for these measures is derived from Shannon's *H*-Measure and resultant entropy calculations, and they are used to interpret whether a plan is ordered or deliberate, as opposed to random or disordered [10,28].

$$H = -\sum \left[\frac{a}{t} In\left(\frac{a}{t} \right) \right] + \left[\frac{b}{t} In\left(\frac{b}{t} \right) \right] + \left[\frac{c}{t} In\left(\frac{c}{t} \right) \right] \tag{8}$$

$$H^* = \frac{(H - In2)}{(In3 - In2)} \tag{9}$$

Hanson describes *H* as measuring the "spread or degree of configurational differentiation across a set of integration values" (p. 30, [9]). *H* is calculated using the maximum (*a*), mean (*b*) and minimum (*c*) *RAs* in the graph, the sum of which is (*t*). For *H**, the formula uses natural logarithm to base *e (ln)*.

4. Application

In this paper, plan graph analysis is used to measure the topological properties of the plans of 37 historic houses, from six eras (Seljuk-Atabakan, Ilkhanid-Al-Muzaffar, Safavieh, Zandieh and Afshari, Qajar and Pahlavi). This section describes the case selection approach and the methodological application.

4.1. Case Selection

Past research about the socio-cultural and environmental properties of Yazd courtyard houses has typically been based on an assessment of just one or two cases [21], and it is rare to see as many as 10 cases in such a study [3]. While different approaches and methods call for different numbers of cases, past research emphasizes that it is not possible to develop a statistically significant "sample" of Yazd courtyard houses, relative to the total "population" of these cases. There are multiple reasons for this, including a relative paucity of cases that have not been extensively modified in the past, and the fact that these houses are, for the most part, privately owned and inaccessible to scholars.

Knowing that a statistically significant sample cannot be established, the conventional approach used by scholars is a variation of the "snow-balling" strategy, wherein the cases studied in past research are selected as being significant for future research. In contrast, the present paper adopts a purposive approach, using the archives and surveys of the Yazd Cultural Heritage Organization (YCHO) as a starting point. The buildings identified by YCHO, and which make up the set studied in this paper, have four significant properties.

1. They are largely in original condition, not having been extensively adapted and modified over time.
2. Their urban contexts are not significantly degraded by modern development or other factors such as natural disasters.
3. They have been pre-classified by experts into different planning types: four-row, platform, garden-villa, central courtyard.
4. They are exemplars of the era in which they were designed and built.

Using the YCHO's parameters for determining significant works, the number of cases available across the time period and in the particular planning type ("courtyard house"), is 37 (Figure 3, Table 1). Focusing just on the set of courtyard houses documented by the YCHO, the cases span from the Seljuk-Atabakan to the Pahlavi eras. There are few documented courtyard houses which pre-date the earlier of these eras and none which match the criteria. The Pahlavi period extends across the first half of the 20th century, when relatively traditional methods and approaches were still in use. By the latter part of the 20th century, however, this changed, and cases could not be legitimately regarded as reflecting the traditional building type.

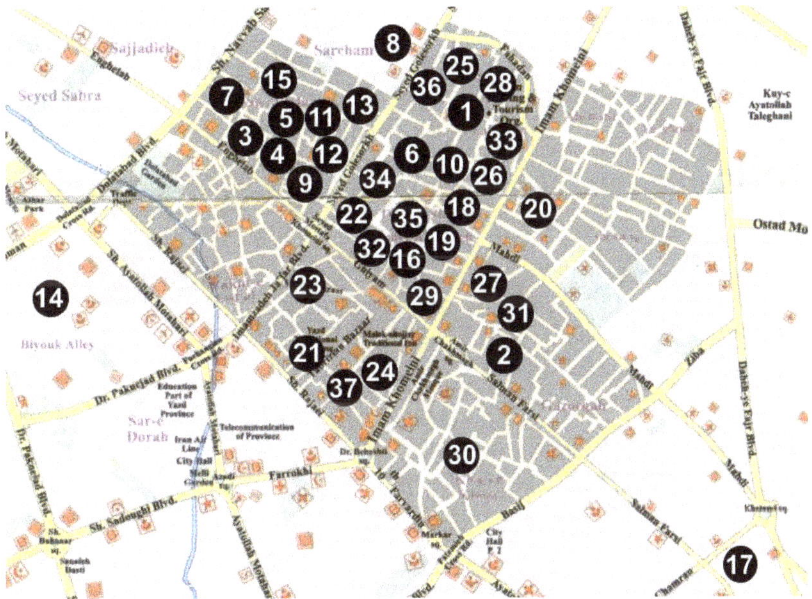

Figure 3. Plan of the historic city of Yazd (Iran), showing the location of the 37 cases measured and compared in this paper (see Table 1 for key).

Table 1. Complete list of cases by ERA, and Key to Figure 3.

Era	House	Key
Seljuk-Atabakan	Tagh bolandha	1
Ilkha-nid-Muzaffarid	Moravej 1	2
	Shahneyi	3
	Barjini	4
	Karimi	5
	Tavana	6
	Javad Khorasa-ni's	7
	Seyyed Go-le-sorkh's	8
	Mohammad-ali Gol's	9
	Salar	10
	Hassan Abadi	11
	Najib	12
	Momtaz	13
	Luck Zadeh	14
	Kalantari	15
Safavid	Malek Sabet	16
	Moravej	17
	Vizier's	18
	Mashruteh	19
	Shah Yahya	20
	Nadeb	21
Zand-Afsharid	Abrishami	22
Qajar	Shafi Pour	23
	Sadoughi	24
	Seda Sima	25
	Shokouhi	26
	Sigari	27
	Heirani	28
	Kolahdouz	29
	Golshan	30
Pahlavid	Farokhi Yazdi	31
	Nekouie	32
	Kashefi	33
	Valizade	34
	Ali Vaziri	35
	Shakeri	36
	Faghih Khorasa-ni	37

4.2. Methodological Approach

It is not surprising, given the widespread application of Space Syntax methods, that they have previously been used to investigate the properties of Iranian housing [30,31], including selected aspects of Yazd courtyard housing, which have often been used to support claims about environmental and socio-cultural properties [6,7]. All of these past applications of syntactical methods have used the entrance as carrier, which produces data for examining overarching public–private or visitor–inhabitant relationships. For the present paper, in order to isolate and measure the properties of the plans relative to their courtyards, the courtyards are used as the carriers.

Six steps were followed to develop the data for this paper.

1. The floor plans for each of the 37 YCHO-classified houses are coded and annotated to differentiate "open" (to the sky) and "closed" (enclosed by walls and roofed) spaces.
2. Spaces are "functionally" defined, rather than in accordance with formal convex space conventions. The connections between spaces (doors, non-habitable annexes and short corridors) are annotated as lines with arrows.

3. A grading map is developed, identifying the topological "step-distance" between rooms and yards (courtyards = 0, one step removed from a yard = 1, two steps removed from a yard = 2, etc.). Thus, courtyards are grade 0, perimeter spaces are grade 1, with deeper spaces graded 2 or higher. Multiple spaces at the same level of depth are numbered for identification purposes (2.1, 2.2, 2.3, etc.). This grading is not only a topological property, but it reflects the value and status of rooms. Historically, valuable living spaces were assumed to be closest to the courtyards and conversely, service areas such as kitchens, stables and bathrooms were more distant. This also reflects the "privacy gradient theory" of spatial organization [9]. Importantly, for the majority of environmental performance-related interpretations of these houses, depth of rooms from courtyards is a core consideration.

4. For each room in each house, seven measures—TD, MD, RA, $i_{(RA)}$, RRA, $i_{(RRA)}$, CV— are developed using Agraph software [32].

5. For each house, four additional measures—H, H^*, means based on the number of spaces in the house (K) and the number of spaces at each grade—are developed.

6. Longitudinal graphs are produced of the mean measures and mean grades over time, and trendlines are generated from these ("least-boxes" method) to support interpretation of the data.

Agraph [32], from the Norwegian University of Science and Technology (Norges teknisk-naturvitenskapelige universitet), is a common software package for plan graph analysis in architecture. It does differ from DepthMap, the other common plan graph analysis software, in several ways, including its calculations for some metrics and the way it handles rounding-errors, irrational numbers and reporting. For example, one reporting issue arises from compounded rounding errors using averages of mean results. Issues of this type are common in software programs for this purpose, and they do result in anomalies in the data. Such variations and issues are discussed in [26]. For example, a "methodological problem occurs when the total depth of a node equals the number of rooms minus one" (p. 468, [26]), because it is not possible to calculate the reciprocal of an inverse number. This mathematical problem would not occur in hierarchical or arborescent plans, but could occur in shallow, rhizomorphous plans. While Yazd courtyard houses are not shallow in a conventional planning sense, when measured relative to the topology of the courtyard, they can be extremely shallow, resulting in this irrational number problem.

4.3. Methodological Example

Examples of the application of these six steps are provided for the Tagh Bolandha House (Figure 4a–c and Table 2) from the Seljuk-Atabakan period and the Golshan House (Figure 5a–c and Table 3) from the Qajar period. For each house, every room is converted into a graph node. For each node, seven measures are produced: TD, MD, RA, i, RRA, i_{RRA} and CV. For example, consider node 1.9 in the Tagh Bolandha House, a large space to the center-right of the biggest courtyard. This node's $TD = 68$, $MD = 2.13$, $RA = 0.07$, $RRA = 0.42$, and $CV = 0.99$. Although considering the properties of a single room in this plan would not be a standard way of reading the results, it can be undertaken by comparing these values with the means for the entire plan. For example, the large room designated as node 1.9 is less deep than average ($\mu\,TD = 79.39$), and it is therefore more accessible than other rooms. Its topological "step distance" from a courtyard is close to average ($\mu\,MD = 2.48$), and it is slightly more integrated ($\mu\,i = 12.09$) than an average room in the house. A close review of the plans reveals that this room is not only directly connected to one courtyard, but it requires only one additional connection to access two more courtyards (by way of nodes 1.10 and 1.19). The existence of such a room poses a challenge for two of the hypotheses that are tested in this paper, although it would only be a significant challenge if multiple rooms with similar properties existed in other plans of this architectural type.

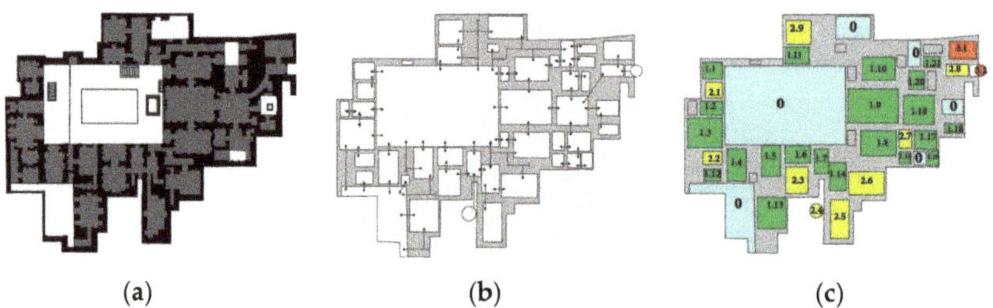

Figure 4. Tagh Bolandha House: (**a**) floor plans, (**b**) space access and connection plan and (**c**) space grading.

Table 2. Tagh Bolandha House results.

Grade/Identify	TD	MD	RA	i	RRA	$i_{(RRA)}$	CV
0	43	1.34	0.02	45.09	0.13	7.71	13.64
1.1	72	2.25	0.08	12.40	0.47	2.12	1.04
1.2	72	2.25	0.08	12.40	0.47	2.12	0.37
1.3	71	2.22	0.08	12.72	0.46	2.17	1.54
1.4	74	2.31	0.08	11.81	0.50	2.02	0.04
1.5	74	2.31	0.08	11.81	0.50	2.02	0.04
1.6	72	2.25	0.08	12.40	0.47	2.12	0.54
1.7	68	2.13	0.07	13.78	0.42	2.36	1.74
1.8	70	2.19	0.08	13.05	0.45	2.23	0.82
1.9	68	2.13	0.07	13.78	0.42	2.36	0.99
1.10	73	2.28	0.08	12.10	0.48	2.07	0.29
1.11	73	2.28	0.08	12.10	0.48	2.07	0.54
1.12	74	2.31	0.08	11.81	0.50	2.02	0.04
1.13	74	2.31	0.08	11.81	0.50	2.02	0.04
1.14	67	2.09	0.07	14.17	0.41	2.42	1.87
1.15	74	2.31	0.08	11.81	0.50	2.02	0.04
1.16	74	2.31	0.08	11.81	0.50	2.02	0.04
1.17	68	2.13	0.07	13.78	0.42	2.36	1.24
1.18	74	2.31	0.08	11.81	0.50	2.02	0.04
1.19	64	2.00	0.06	15.50	0.38	2.65	1.45
1.20	70	2.19	0.08	13.05	0.45	2.23	0.24
1.21	74	2.31	0.08	11.81	0.50	2.02	0.04
2.1	103	3.22	0.14	6.99	0.84	1.19	0.50
2.2	102	3.19	0.14	7.09	0.83	1.21	0.33
2.3	97	3.03	0.13	7.63	0.77	1.30	0.75
2.4	99	3.09	0.14	7.40	0.79	1.27	0.25
2.5	98	3.06	0.13	7.52	0.78	1.29	0.20
2.6	71	2.22	0.08	12.72	0.46	2.17	0.49
2.7	99	3.09	0.14	7.40	0.79	1.27	0.33
2.8	91	2.84	0.12	8.41	0.70	1.44	2.20
2.9	73	2.28	0.08	12.1	0.48	2.07	0.54
3.1	122	3.81	0.18	5.51	1.06	0.94	0.33
3.2	122	3.81	0.18	5.51	1.06	0.94	0.33
Mean	79.39	2.48	0.09	12.09	0.56	2.07	1.00

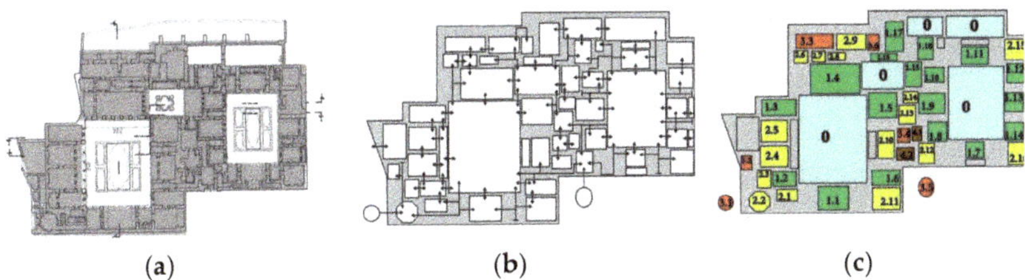

(a)   (b)   (c)

Figure 5. Golshan House: (**a**) floor plans, (**b**) space access and connection plan and (**c**) space grading.

Table 3. Golshan House results.

Grade/Identify	TD	MD	RA	i	RRA	$i_{(RRA)}$	CV
0	58	1.38	0.02	53.81	0.13	7.86	10.67
1.1	93	2.21	0.06	16.88	0.41	2.46	0.77
1.2	93	2.21	0.06	16.88	0.41	2.46	1.28
1.3	96	2.29	0.06	15.94	0.43	2.33	0.53
1.4	95	2.26	0.06	16.25	0.42	2.37	0.69
1.5	95	2.26	0.06	16.25	0.42	2.37	0.23
1.6	95	2.26	0.06	16.25	0.42	2.37	0.83
1.7	96	2.29	0.06	15.94	0.43	2.33	0.67
1.8	93	2.21	0.06	16.88	0.41	2.46	0.37
1.9	98	2.33	0.07	15.38	0.45	2.24	0.23
1.10	94	2.24	0.06	16.56	0.41	2.42	1.44
1.11	97	2.31	0.06	15.65	0.44	2.29	0.56
1.12	95	2.26	0.06	16.25	0.42	2.37	1.03
1.13	98	2.33	0.07	15.38	0.45	2.24	0.53
1.14	98	2.33	0.07	15.38	0.45	2.24	0.53
1.15	94	2.24	0.06	16.56	0.41	2.42	1.23
1.16	99	2.36	0.07	15.11	0.45	2.21	0.03
1.17	99	2.36	0.07	15.11	0.45	2.21	0.03
1.18	97	2.31	0.06	15.65	0.44	2.29	0.56
2.1	92	2.19	0.06	17.22	0.40	2.51	0.69
2.2	91	2.17	0.06	17.57	0.39	2.57	2.28
2.3	127	3.02	0.10	10.13	0.68	1.48	0.25
2.4	131	3.12	0.10	9.67	0.71	1.41	0.75
2.5	133	3.17	0.11	9.46	0.72	1.38	1.00
2.6	95	2.26	0.06	16.25	0.42	2.37	0.69
2.7	131	3.12	0.10	9.67	0.71	1.41	1.16
2.8	96	2.29	0.06	15.94	0.43	2.33	0.28
2.9	93	2.21	0.06	16.88	0.41	2.46	1.03
2.10	96	2.29	0.06	15.94	0.43	2.33	0.42
2.11	97	2.31	0.06	15.65	0.44	2.29	0.53
2.12	89	2.12	0.05	18.32	0.37	2.67	2.47
2.13	89	2.12	0.05	18.32	0.37	2.67	1.33
2.14	135	3.21	0.11	9.26	0.74	1.35	0.25
2.15	136	3.24	0.11	9.16	0.75	1.34	0.33
2.16	98	2.33	0.07	15.38	0.45	2.24	0.36
3.1	132	3.14	0.10	9.57	0.72	1.40	0.16
3.2	132	3.14	0.10	9.57	0.72	1.40	0.16
3.3	130	3.10	0.10	9.78	0.70	1.43	0.58
3.4	126	3.00	0.10	10.25	0.67	1.50	2.20
3.5	130	3.10	0.10	9.78	0.70	1.43	0.14
3.6	134	3.19	0.11	9.36	0.73	1.37	0.25
4.1	167	3.98	0.15	6.89	0.99	1.01	0.33
4.2	167	3.98	0.15	6.89	0.99	1.01	0.33
Mean	107.67	2.56	0.08	14.86	0.52	2.17	0.93

A detailed mixed-method, qualitative and quantitative analysis of the properties of the entire plan of the Tagh Bolandha House can be undertaken in this way, although it is more common to develop rank-order lists of different room types (known as an "inequality genotype") to capture larger patterns in the planning. With the complete set of data for each house available, two further measures can be derived for the Tagh Bolandha House: $H = 0.86$ and $H^* = 0.41$. These results indicate that the plan has broadly homogenized properties relative to the courtyard locations, with a tight cluster of integration values [8]. Superficially at least, this implies that there is only a very loose spatial matrix of rhizomorphous structure to the plan, without a high level of control or social hierarchy, relative to the location of the courtyards [24,27].

5. Hypotheses

The three knowledge gaps identified in the introduction and literature review are reframed in this section as hypotheses, which are then aligned to testing conditions. These are not hypotheses in a scientific sense, rather, they are conversions of implicit assumptions into mathematically testable forms. These hypotheses are as follows

Hypothesis 1 (H1). *Yazd courtyard houses typically have a single courtyard and have similar total numbers of spaces.*

The first part of this hypothesis would be supported by the data if $\mu\,K_{G0} < 1.3$. The 1.3 value, while artificial, would allow a small number of houses to have a secondary courtyard, which would not impact the results of past research. If, however, $\mu\,K_{G0} > 1.3$, then assumptions about the role of the (singular) courtyard in this planning type would need to be revisited. For the second part of this hypothesis, if SD of K is $> 25\%$ of $\mu\,K$, then assumptions in past research about the consistent size of courtyard houses may need to be revised. Because no absolute metric is available, the 25% is an artificial indicator to accommodate some variance in the data around an otherwise clustered result.

Hypothesis 2 (H2). *The planning of Yazd courtyard houses is largely unchanged over time.*

This would be supported if longitudinal trendlines generated from TD, MD, RA and CV are flat/level and the SDs of these measures are less than 25% of the mean in each case.

Hypothesis 3 (H3). *Yazd courtyard houses exhibit a clear pattern of planning relative to the courtyard.*

There are three possible ways to test this claim. First, this type of consistency could be evident in the ratio of $K_{G0}/K_{G1}/K_{G2\text{-}4}$ (courtyards to perimeter rooms to higher gradient rooms). Second, trendlines for RRA (which is normalized to accommodate different size graphs) and CV could reveal a pattern that operates regardless of scale. Third, H can be used to determine if there is any evidence of intent in the planning. If $\mu\,H < 0.5$, the planning is highly deliberate or distinctive.

6. Results and Discussion

The number of rooms (K) and the number of courtyards (K_{G0}) in each house are charted chronologically (Figure 6), as too are the number of rooms at different gradients of depths from the courtyards ($K_{G1\text{-}4}$) (Figure 7). The mean number of enclosed rooms in each house in the set is 16.70, the mean number of courtyards is 2.05 and the standard deviation for K is 7.99 (47% of μK). The largest house, in terms of room numbers, is the Golshan House ($K = 42$) and the smallest is Javad Khorasani's House ($K = 8$). The number of courtyards is highest in the Tagh bolandha House ($K_{G0} = 6$) and only six houses have a single courtyard (Salar, Javad Khorasani's, Malek Sabet, Vizier's, Seda Sima and Valizade). When the data are analyzed chronologically, a slight rise in K over time is visible (largely due to the data for the Qajar period), while the trendline for number of courtyards (K_{G0}) is level (Figure 6). Trendlines for perimeter rooms (K_{G1}) also rise slightly over time, whereas for deeper rooms (K_{G2-4}) the trend is largely flat (Figure 7).

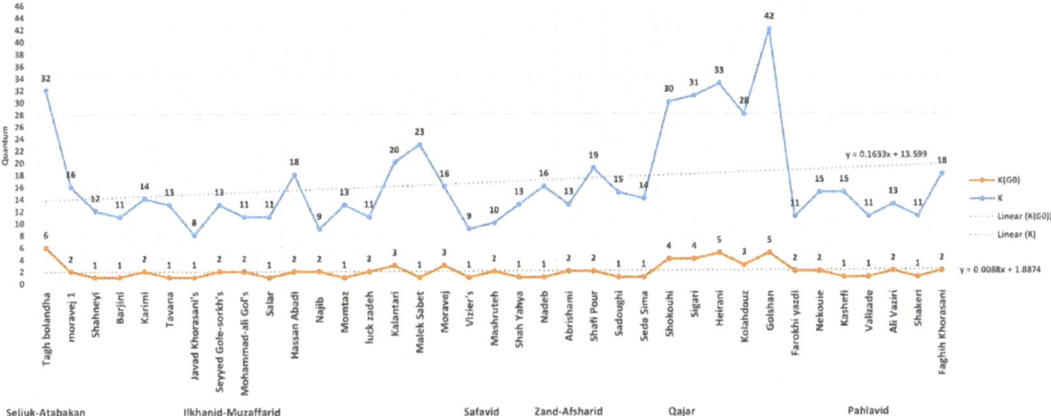

Figure 6. Number of enclosed rooms (K), number of courtyards (K_{G0}) and trendlines for the 37 houses, presented chronologically by era.

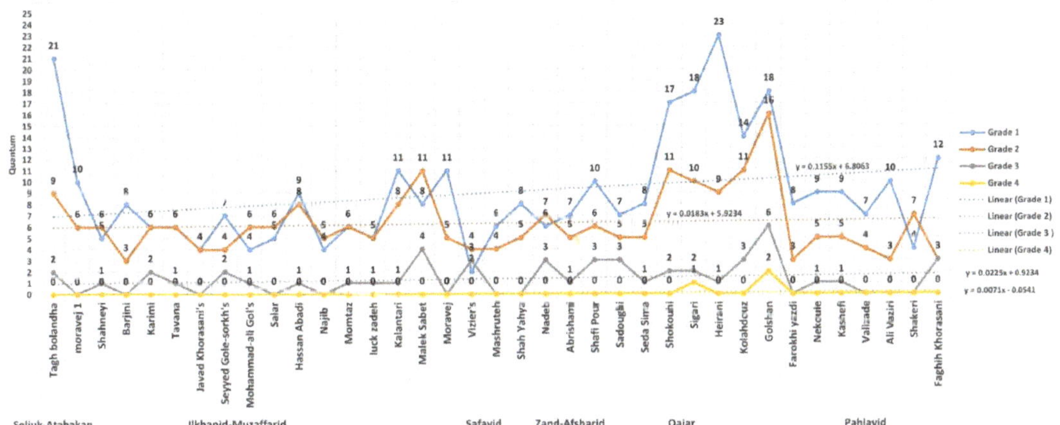

Figure 7. Number of gradient 1+ rooms (K_{G1-4}) and trendlines for the 37 houses, presented chronologically by era.

The ratio of courtyards ($\mu = 2.05$), to perimeter rooms ($\mu = 9.00$), to second gradient rooms ($\mu = 6.27$), is 1/4.30/3.05. Despite these mean figures, the ratio of courtyards to the total number of rooms is less constant. For example, in both the Tagh Bolandha House and the Moravej House, this ratio is 1/5.3, even though the two are very different sizes. In the Golshan House the ratio is 1/8.4 and in the Salar House it is 1/11, confirming that, despite some clustering in the data, the ratio may be evolving over time.

The syntactical results for each house are presented in Table 4, and the mean results for each period are in Table 5. The topological depth data indicate a marginal rise in *TD* (Figure 8) and a decrease in *MD* (Figure 9) over time. As *K* is rising (Figure 6), it might be assumed that a parallel rise in *TD* would occur, but this is not the case. Logically, the inclusion of additional courtyards, coupled with planning that clusters rooms around these yards, could be responsible for the trends in both *TD* and *MD*. This interpretation is supported by the gradient data, where the number of K_{G3} and K_{G4} rooms remains very low regardless of the total house size (Figure 7).

Table 4. Results per house: mean for *TD, MD, RA, i, RRA, i(RRA)* and *CV* and holistic result for *H* and *H**.

Era	House	*TD*	*MD*	*RA*	*i(RA)*	*RRA*	*i(RRA)*	*CV*	*H*	*H**
Seljuk-Atabakan	Tagh Bolandha	79.39	2.48	0.10	12.09	0.56	2.07	1.00	0.86	0.41
Ilkha-nid-Muzaffarid	Moravej 1	35.18	2.20	0.16	7.71	0.65	1.88	1.23	0.83	0.34
	Shahneyi	28.46	2.37	0.25	4.62	0.90	1.28	1.00	0.94	0.61
	Barjini	24.17	2.20	0.24	5.37	0.84	1.53	1.00	0.86	0.42
	Karimi	33.73	2.41	0.22	5.53	0.84	1.43	1.00	0.85	0.39
	Tavana	30.86	2.37	0.23	5.21	0.86	1.39	1.00	0.88	0.46
	Javad Khorasa-ni's	19.11	2.39	0.4	3.03	1.25	0.96	1.00	0.93	0.58
	Seyyed Go-le-sorkh's	35.86	2.76	0.29	4.00	1.10	1.07	1.00	0.90	0.51
	Mohammad-ali Gol's	29.33	2.67	0.33	3.47	1.17	0.99	1.00	0.95	0.63
	Salar	24.67	2.24	0.25	4.81	0.87	1.37	0.99	0.93	0.58
	Hassan Abadi	48.11	2.67	0.20	5.83	0.85	1.35	0.98	0.90	0.51
	Najib	19.00	2.11	0.28	4.30	0.91	1.31	1.00	0.94	0.61
	Momtaz	32.29	2.48	0.25	4.51	0.93	1.21	1.00	0.94	0.61
	Luck Zadeh	27.83	2.53	0.31	3.83	1.07	1.09	1.00	0.90	0.52
	Kalantari	47.62	2.38	0.15	8.09	0.66	1.78	0.98	0.85	0.39
Safavid	Malek Sabet	74.00	3.22	0.20	5.52	0.98	1.13	1.00	0.93	0.58
	Moravej	37.76	2.36	0.18	6.70	0.74	1.64	1.06	0.88	0.46
	Vizier's	19.80	2.20	0.30	4.18	0.98	1.28	1.00	0.89	0.49
	Mashruteh	16.18	1.62	0.14	6.48	0.47	1.91	1.00	0.66	−0.07
	Shah Yahya	25.00	1.92	0.15	11.40	0.58	3.05	1.00	0.71	0.04
	Nadeb	31.29	1.96	0.13	10.01	0.52	2.44	0.96	0.83	0.34
Zand-Afsharid	Abrishami	23.29	1.79	0.13	12.50	0.49	3.34	0.99	0.70	0.02
Qajar	Shafi Pour	39.20	2.06	0.12	11.02	0.53	2.48	1.01	0.80	0.26
	Sadoughi	33.38	2.23	0.18	6.87	0.70	1.72	1.03	0.85	0.39
	Seda Sima	26.00	1.86	0.13	12.96	0.51	3.36	0.99	0.71	0.04
	Shokouhi	64.90	2.16	0.08	15.22	0.45	2.71	1.00	0.76	0.17
	Sigari	84.19	2.72	0.11	10.09	0.66	1.76	1.00	0.81	0.29
	Heirani	64.59	1.96	0.06	24.03	0.36	4.04	0.96	0.62	−0.17
	Kolahdouz	65.17	2.33	0.10	12.06	0.53	2.22	0.97	0.85	0.39
	Golshan	107.67	2.56	0.08	14.86	0.52	2.17	0.93	0.86	0.41
Pahlavid	Farokhi Yazdi	22.17	2.02	0.20	6.77	0.71	1.93	1.00	0.81	0.29
	Nekouie	31.25	2.08	0.15	8.30	0.62	2.08	0.99	0.82	0.31
	Kashefi	32.75	2.18	0.17	7.03	0.67	1.77	1.00	0.84	0.37
	Valizade	19.83	1.80	0.16	10.27	0.56	2.93	1.00	0.77	0.19
	Ali Vaziri	23.00	1.77	0.13	6.74	0.48	1.80	0.99	0.68	−0.02
	Shakeri	23.00	2.09	0.22	5.74	0.77	1.64	0.99	0.88	0.46
	Faghih Khorasa-ni	41.16	2.29	0.15	7.87	0.66	1.82	0.99	0.79	0.24
Mean (*μ*)		38.41	2.26	0.19	8.08	0.73	1.89	1.00	0.84	0.35
Stand. Dev. (*SD*)		21.31	0.33	0.08	4.27	0.22	0.73	0.04	0.09	0.21
SD/μ × 100 (%)		55	15	42	52	30	38	4	10	60

Table 5. Mean results for six eras. Note that the Seljuk-Atabakan and Zand-Afsharid eras each have only one case.

Era	*TD*	*MD*	*RA*	*i*	*RRA*	*i(RRA)*	*CV*	*H*	*H**
Seljuk-Atabakan	79.39	2.48	0.10	12.09	0.56	2.07	1.00	0.86	0.41
Ilkhanid-Muzaffarid	31.16	2.41	0.25	5.02	0.92	1.33	1.01	0.90	0.51
Safavid	34.01	2.21	0.18	7.38	0.71	1.91	1.00	0.82	0.31
Zand-Afsharid	23.29	1.79	0.13	12.50	0.49	3.34	0.99	0.70	0.02
Qajar	60.64	2.24	0.11	13.39	0.53	2.56	0.99	0.78	0.22
Pahlavid	27.59	2.03	0.17	7.53	0.64	2.00	0.99	0.80	0.26
Mean	42.68	2.19	0.16	9.65	0.64	2.20	1.00	0.81	0.29

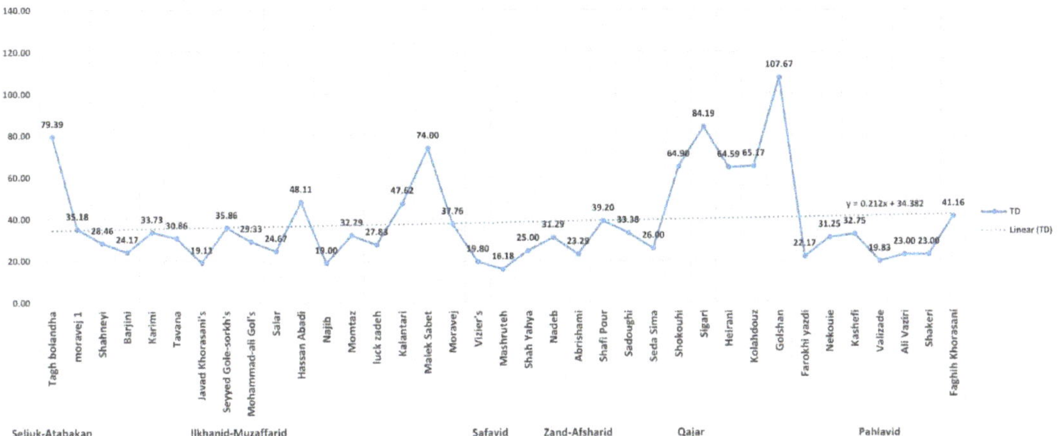

Figure 8. Total Depth (*TD*) results and trendlines for the 37 houses, presented chronologically by era.

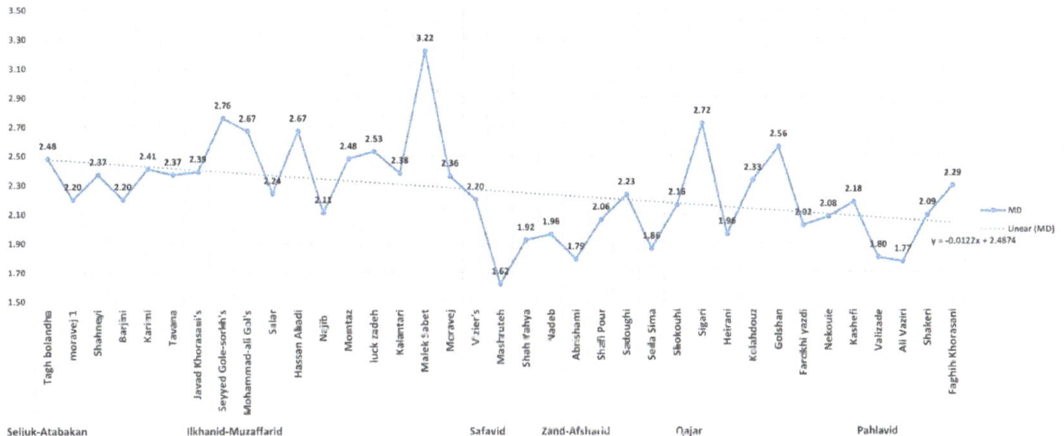

Figure 9. Mean Depth (*MD*) results and trendlines for the 37 houses, presented chronologically by era.

The highest *RA* and *RRA* (Figure 10) values are in the data for Javad Khorasani's House (Ilkhanid period) at 1.25 and 0.4, respectively, and the lowest are in Heirani House (Qajar period) at 0.06 and 0.36, respectively. Interestingly, the normalized *RRA* results for the two are similar (0.4 verses 0.36), which may reflect the similar ratio of courtyards to rooms identified previously (K_{G0}/K). The difference in results between the smallest and largest values is relatively pronounced, and with the increase in scale, rooms become less private, because the addition of further yards produces more perimeter spaces, which are higher integration zones. On average, over time, spaces have become more integrated around the courtyards and arguably, privacy has been reduced because of this. This might reflect a simplification of environmental design strategy over time, as well, with deeper rooms being less important.

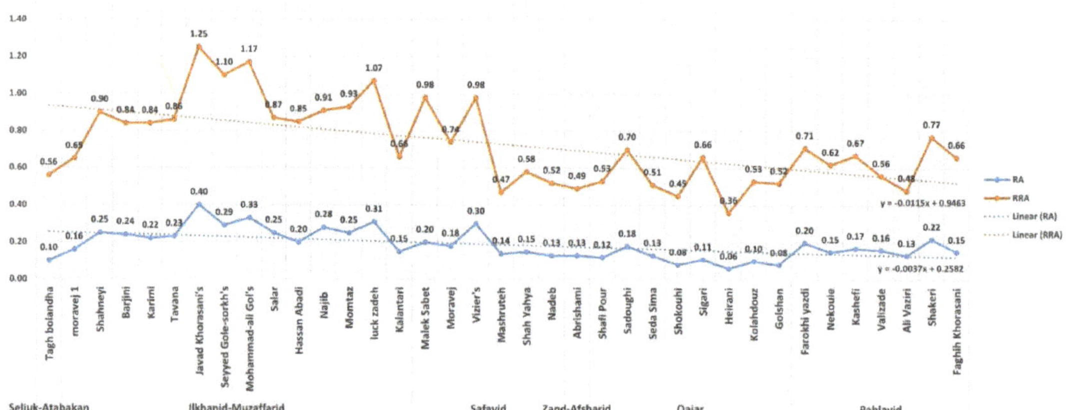

Figure 10. Relative Asymmetry (*RA*) and Real Relative Asymmetry (*RRA*) results and trendlines for the 37 houses, presented chronologically by era.

Results for mean *CV* for each house may indicate that a pattern exists in their planning. Not only is the trendline relatively flat, but the $SD/\mu \times 100$ for CV is 4%, which indicates a tightly clustered set of data, with only a few statistical outliers (Figure 11). The primary outliers are the Moravej 1 House, Moravej House and the Golshan House.

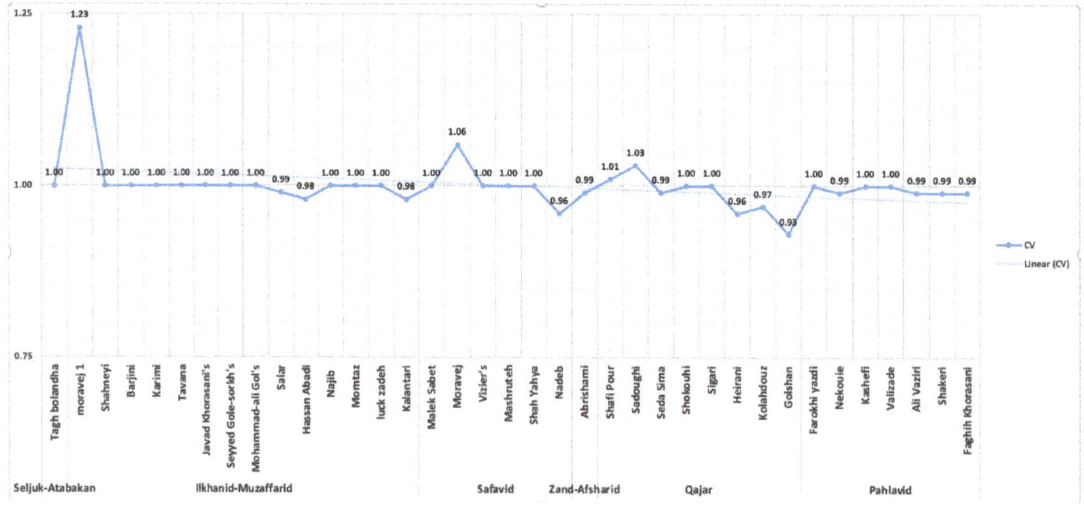

Figure 11. Control Value (*CV*) results and trendlines for the 37 houses, presented chronologically by era.

Difference Factor (*H*) provides an indication of the degree to which the spatial distribution of rooms in a plan, relative to the position of the yard, is more or less deliberate or distinctive. An *H* result close to 0 would infer a high level of specificity or hierarchy in a plan, whereas a result closer to 1 might suggest a more generic and undifferentiated set of spatial relations in the plan. The mean for *H* across the set is 0.84 (Table 4) and the trendline marginally reduces over time (Figure 12), which confirms that there is a degree of homogeneity in the planning.

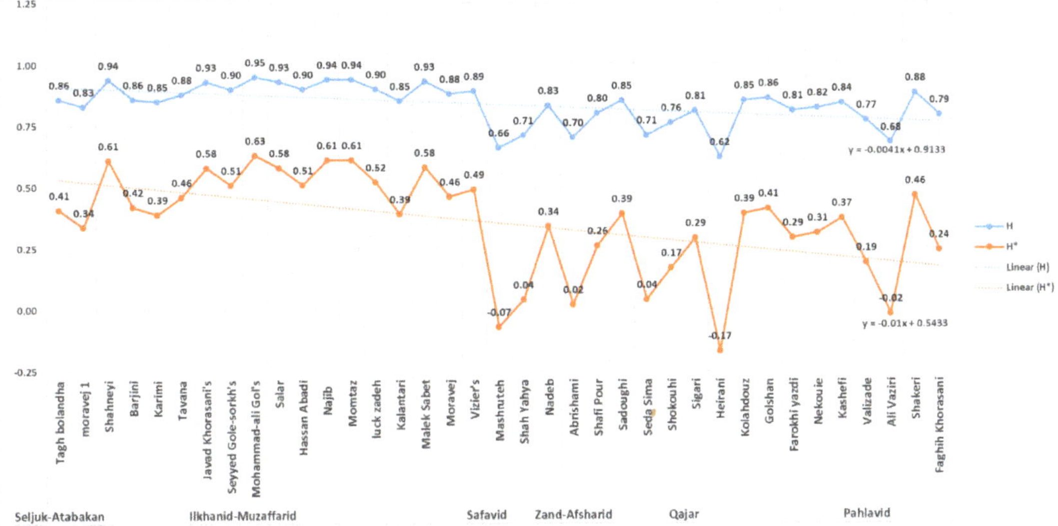

Figure 12. *H* and *H** results and trendlines for the 37 houses, presented chronologically by era.

7. Conclusions

Using a variation of the Space Syntax plan graph analysis method, justified by courtyard location, this paper set out to examine three hypotheses about the planning of the Yazd courtyard house. Before reviewing the results, it is important to reiterate two limitations. The method chosen for this paper can only be used to directly comment on the topological properties of the architectural plan, which in turn may have indirect consequences for socio-cultural or climatic interpretations of the function of these houses. Furthermore, the particular software application resulted in some anomalous results and compounded rounding errors, which do occur when dealing with highly connected, matrix-style plans, and cannot easily be resolved. The use of the courtyard as carrier for the graph, while both original and appropriate, was responsible for some of these anomalies, and a future study will use a more conventional external carrier for a parallel analysis. The results for each hypothesis are as follows.

H1. *Yazd courtyard houses typically have a single courtyard and similar total numbers of spaces.* The data do not support the first part of this hypothesis, as $\mu\,K_{G0} = 2.05$. On average, the exemplar Yazd courtyard houses have two or more courtyards, with a maximum of six. For the second part, the data do not support this hypothesis, as the standard deviation for K is 7.99 (47% of μK) and the houses range from 8 to 42 rooms. Notwithstanding this diversity, their size is, on average, increasing over time.

H2. *The planning of Yazd courtyard houses is largely unchanged over time.* Despite trendlines for *TD* being relatively level, SD/μ for *TD* is 55%, confirming a high level of variation. The trendlines for *MD* and *RA* are both falling and standard deviations emphasize the diversity in some (SD/μ for RA = 42%), but not all (SD/μ for MD = 15%) aspects of the data. Nevertheless, the data suggest that spatial hierarchy and depth (*TD* and *MD*), and integration and separation (public verses private, *i* and *RA*) have all gradually changed over time. In addition, despite the plans increasing in both size and complexity, the houses have become less private or spatially isolated.

H3. *Yazd courtyard houses exhibit a clear pattern of planning relative to the courtyard.* This hypothesis is about patterns in the frequency and topological location of the courtyard in the architectural plan. The result for this hypothesis is partially supported, although there is a need for future research to confirm this. Three sets of data are used for this analysis: (i) the ratio $K_{G0}/K_{G1}/K_{G2\text{-}4}$, (ii) trendlines for *RRA* and CV, and (iii) $\mu\,H$. The

$K_{G0}/K_{G1}/K_{G2}$- ratio in the data is $1/4.30/3.05$ and, despite variations, there are multiple cases that cluster around this ratio. One implication of this result might be that changes in the number of rooms in a house were accommodated in the plan through the addition of further courtyards. Alternatively, over time, the practical importance of having a single, central courtyard may have been reduced, possibly as a reflection of economic changes, increased migration and urbanization rather than changing environmental conditions. The trendline for RRA reinforces this reading. Thus, despite a growth in K over time, and a parallel reduction in normalized levels of isolation of a node within the graph (RRA), the normalized integration is relatively stable. In terms of spatio-structural complexity, a loose pattern may well exist around the ratio of yards to rooms. The final indicator, $\mu H = 0.84$, suggests that these 37 plans only possess a low level of spatial differentiation or deliberation. Such a finding would make sense for a building type that has a completely flexible or rhizomorphous spatial arrangement to accommodate seasonal adaption. Indeed, this approach was also common in some Renaissance European villas, and seasonal adaption has recently been tested for the first time using plan graph analysis of Palladian villas [33]. The μH results for Palladian villas are even higher (more flexible or less distinctive) than for the Yazd courtyard houses.

These findings may pose challenges for two of the foundation assumptions in past research about the architectural planning of the Yazd courtyard house (Table 6). The first hypothesis should, in hindsight, not have needed detailed analysis, as a cursory review of archived plans and surveys for this architectural type clearly shows diversity of size and the presence of a number of courtyards in these houses. Certainly, some of the houses that are more commonly studied in the literature have either a single courtyard, or a major courtyard coupled with a minor or secondary one. This is not, however, the standard across the exemplar works, and often two or more courtyards are of similar size and have different orientations. While the orientation and scale of courtyards are not further considered in this topological approach, the results have implications for the classic environmental interpretation of the planning of these houses. The focus on a single courtyard is clearly a limitation of some research; however, the present study suggests that all courtyards might serve the type of environmental role postulated in past research.

Table 6. Summary of findings.

	Hypothesis	Result	Evidence
H1	*Yazd courtyard houses typically have a single courtyard and similar total numbers of spaces.*	Not supported	$\mu K_{G0} = 2.05$ $SD/\mu K \times 100 = 47\%$
H2	*The planning of Yazd courtyard houses is largely unchanged over time.*	Not supported	Longitudinal trendlines for TD, MD, i and RA demonstrate change
H3	*Yazd courtyard houses exhibit a clear pattern of planning relative to the courtyard.*	Partially supported	Clustering in $K_{G0}/K_{G1}/K_{G2\text{-}4}$ (Trendlines for RRA, i_{RRA} and CV (iii) $\mu H = 0.84$

The second hypothesis confirms that not only has this planning type changed over time, but it has evolved in interesting ways. For example, the Ilkhanid period (1256–1335 CE, 651–736 AH) has the highest RRA, RA, MD and CV results. The houses of this period were more hierarchical, segregated or isolated and controlled. In contrast, the Qajar period (1781–1925 CE, 1195–1344 AH) had the highest levels of spatial accessibility or permeability (reflected in i). Indeed, the data identify the Ilkhanid and Qajar periods as having quite divergent social (and possibly cultural or political) conditions. This result has implications for the classic environmental interpretation of these houses, as the depth of rooms around courtyards has decreased over time. This would appear to signal the need for further environmental modelling and consideration using longitudinal data. Conversely, the dominant socio-cultural interpretation of spaces in these houses suggests that they may be becoming less private, a position with some support in the paper.

The findings of the final hypothesis are perhaps the most significant. A traditional, hierarchical or arborescent spatial planning pattern around courtyards was not identified in the data. Instead, there is an almost cellular growth pattern, wherein as houses became larger, additional courtyards and perimeter rooms were included in a rich matrix of spaces. This limited the mean depth and isolation of rooms, and maintained a relatively consistent relationship between the number of courtyards, perimeter rooms and second gradient rooms. This would not, however, have reduced the spatial complexity, or the cognitive complexity of navigating around and using these houses [34]. There are also exceptions to this pattern. For example, it was mentioned previously in this paper that Vizier's House from the Safavid period could be considered both a courtyard and a four-platform house. Vizier's House is a statistical outlier in the data for hypothesis 3, suggesting that it might not be classified as a courtyard house at all. This case highlights the usefulness of this analysis approach, not only for testing hypotheses, but also for classifying different works. This capacity could be investigated in future research.

Author Contributions: Conceptualization, M.Z. and M.J.O.; methodology, M.Z.; formal analysis, M.Z.; investigation, M.Z. and M.J.O.; data preparation, M.Z.; writing—original draft preparation, M.Z.; writing—review and editing, M.Z. and M.J.O.; All authors have read and agreed to the published version of the manuscript.

Funding: This research received no external funding.

Institutional Review Board Statement: Not applicable.

Informed Consent Statement: Not applicable.

Conflicts of Interest: The authors declare no conflict of interest.

References

1. Merhrjerdi, Y.Z. *History of Yazd*; Lambert Academic Publishing: London, UK, 2012.
2. Amiriparyan, P.; Kiani, Z. Analyzing the Homogenous Nature of Central Courtyard Structure in Formation of Iranian Traditional Houses. *Procedia Soc. Behav. Sci.* **2016**, *216*, 905–915. [CrossRef]
3. Khajehzadeh, I.; Vale, B.; Yavari, F.; Fatemeh, Y. A comparison of the traditional use of court houses in two cities. *Int. J. Sustain. Built Environ.* **2016**, *5*, 470–483. [CrossRef]
4. Keshtkaran, P. Harmonization Between Climate and Architecture in Vernacular Heritage: A Case Study in Yazd, Iran. *Procedia Eng.* **2011**, *21*, 428–438. [CrossRef]
5. Hosseini, S.R.; Nik Ethegad, A.; Uson, E.; Armesto, A. Iranian courtyard housing: The role of social and cultural patterns to reach the spatial formation in the light of an accentuated privacy. *ACE Archit. City Environ.* **2015**, *10*, 11–30. [CrossRef]
6. Khākī Qaṣr, Ā. Barnāmih fażāyī khānih hā-yi sunatī Yazd dar nisbat bā sukūnat imrūzī. *Majalih Mi'mārī Iqlīm Garm va Khushk* **2018**, *8*, 23–46.
7. Ukhuvat, H. Barrisī-yi taṭbīqī-yi ab'ād-i ḥarīm dar maskan sunatī va mu'āṣir bā istifādih az mudil taḥlīlī BDRS. *Majalih Maskan va Muḥīṭ-i Rūstā* **2013**, *142*, 97–112.
8. Hillier, B. *Space Is the Machine*; Cambridge University Press: London, UK, 1984.
9. Ostwald, M.J.; Dawes, M. *The Mathematics of the Modernist Villa: Architectural Analysis Using Space Syntax and Isovists*; Birkhäuser: Basel, Switzerland, 2018.
10. Hanson, J. *Decoding Homes and Houses*; Cambridge University Press: Cambridge, UK, 1998.
11. Memarian, G.H.; Brown, F.E. Climate, culture and religion: Aspects of the traditional courtyard house in Iran. *J. Archit. Plan. Res.* **2003**, *3*, 181–198. Available online: https://www.jstor.org/stable/43030659 (accessed on 1 May 2021).
12. Ukhuvat, H.; Bimānīyān, M.R.; Ansārī, M. Bāzshināsī mafāhīm-i ma'navī-i (sukūnat) dar maskan-i sunatī iqlīm-i kavīrī. *Majalih Shahri-i Īranī Islamī* **2011**, *5*, 95–102.
13. Tavassoli, M. *Form, Space and Design: From the Persian to the European Experience*; Springer: Cham, Switzerland, 2019. [CrossRef]
14. Foruzanmehr, A.; Vellinga, M. Vernacular architecture: Questions of comfort and practicability. *Build. Res. Inf.* **2011**, *39*, 274–285. [CrossRef]
15. Irani, M.; Armstrong, P.; Rastegar, A. Evolution of Residential Building in Iran based on Organization of space. *Asian Cult. Hist.* **2017**, *9*, 46. [CrossRef]
16. Hejazi, M.; Hejazi, B.; Hejazi, S. Evolution of Persian traditional architecture through the history. *J. Arch. Urban.* **2015**, *39*, 188–207. [CrossRef]
17. Alkhansari, M.G. Analysis of the responsive aspects of the traditional Persian house. *J. Arch. Urban.* **2015**, *39*, 273–289. [CrossRef]

18. Sulṭān Zādih, Ḥ.; Yusifī, M.; Yusifī, M. Chigūnigī kārburd hindisih va tafkīk-i fażā hā dar mi ́mārī pīsh az tārīkh Īrān. *Majalih Andīshih Mi ́mārī* **2015**, *1*, 54–70.
19. Mazumdar, S.; Mazumdar, S.H. Religious traditions and domestic architecture: A comparative analysis of Zoroastrian and Islamic houses in Iran. *J. Archit. Plan. Res.* **1997**, *14*, 181–208. Available online: https://www.jstor.org/stable/43030208 (accessed on 1 May 2021).
20. Adeli, S.; Abbasi, M. Approaches to Nature in Iranian Traditional Houses in terms of Environmental Sustainability. In *Vernacular Architecture: Towards a Sustainable Future*; Mileto, C., Vegas, F., Garcia Soriano, L., Cristini, V., Eds.; Routledge: London, UK, 2015; pp. 33–38.
21. Soflaei, F.; Shokouhian, M.; Shemirani, S.M.M. Traditional Iranian courtyards as microclimate modifiers by considering orientation, dimensions, and proportions. *Front. Arch. Res.* **2016**, *5*, 225–238. [CrossRef]
22. Soflaei, F.; Shokouhian, M.; Soflaei, A. Traditional courtyard houses as a model for sustainable design: A case study on BWhs mesoclimate of Iran. *Front. Arch. Res.* **2017**, *6*, 329–345. [CrossRef]
23. Hillier, B.; Hanson, J. *The Social Logic of Space*; Cambridge University Press: Cambridge, UK, 1984.
24. Peponis, J.; Wineman, J. Spatial structure of environment and behaviour. In *Handbook of Environmental Psychology*; Bechtel, R., Churchman, A., Eds.; John Wiley & Sons: New York, NY, USA, 2002; pp. 271–291.
25. Memarian, G.H. Naḥv-i fażā-yi mi ́mārī. *Majalih Ṣuffih* **2002**, *35*, 74–83.
26. Ostwald, M.J. The Mathematics of Spatial Configuration: Revisiting, Revising and Critiquing Justified Plan Graph Theory. *Nexus Netw. J.* **2011**, *13*, 445–470. [CrossRef]
27. Bafna, S. Space Syntax a brief introduction to its logic and analytical techniques. *Environ. Behav.* **2003**, *35*, 17–29. [CrossRef]
28. Klarqvist, B. A space syntax glossary. *Nord. Arkit.* **1993**, *2*, 11–12. [CrossRef]
29. Lee, J.H.; Ostwald, M.J. *Grammatical and Syntactical Approaches in Architecture: Emerging Research and Opportunities*; IGI Global: Hershey, PA, USA, 2020. [CrossRef]
30. Alitajer, S.; Nojoumi, G.M. Privacy at home: Analysis of behavioral patterns in the spatial configuration of traditional and modern houses in the city of Hamedan based on the notion of space syntax. *Front. Arch. Res.* **2016**, *5*, 341–352. [CrossRef]
31. Mustafa, F.A.; Hassan, A.S.; Baper, S.Y. Using Space Syntax Analysis in Detecting Privacy: A Comparative Study of Traditional and Modern House Layouts in Erbil City, Iraq. *Asian Soc. Sci.* **2010**, *6*, 157–166. [CrossRef]
32. Manum, B.; Rusten, E.; Benze, P. AGRAPH, Software for Drawing and Calculating Space Syntax Graphs. Available online: https://www.ntnu.no/ad/spacesyntax (accessed on 1 June 2021).
33. Dawes, M.; Ostwald, M.J.; Lee, J.H. Examining Control, Centrality and Flexibility in Palladio's villa plans using Space Syntax measurements. *Front. Archit. Res.* 2021. [CrossRef]
34. Peponis, J. Building layouts as cognitive data: Purview and purview interface. *Cogn. Crit.* **2012**, *6*, 11–52.

 *buildings*

Article

Evolution Process of Scientific Space: Spatial Analysis of Three Groups of Laboratories in History (16th–20th Century)

Xinwei Zhang * and Tong Cui

Center of Architecture Research and Design, University of Chinese Academy of Sciences, Beijing 100190, China
* Correspondence: zhangxinwei17@mails.ucas.ac.cn

Abstract: Different disciplines need specialized spaces to ensure the smooth conduct of research, and the laboratory plays an important role as the physical carrier of knowledge production today. Reviewing history, it is found that the image of the laboratory and the organization of internal space have undergone great changes, which reflects people's cognition of scientific practice in different periods. This study uses space syntax tools to analyze relationships between scientific research activities and spatial forms over time, preliminarily discussing laboratories' spatial characteristics in different periods and the corresponding research modes. It is found that scientific research has undergone several phases, and the scientific paradigm deeply influences the spatial logic of scientific research buildings. The scientific research space in different periods presents unique syntactic results and topological structures, suggesting the trend of specialized research from closed to open and decentralization to centralization.

Keywords: spatial syntax; laboratory space; historical evolution; scientific research activities

Citation: Zhang, X.; Cui, T. Evolution Process of Scientific Space: Spatial Analysis of Three Groups of Laboratories in History (16th–20th Century). *Buildings* **2022**, *12*, 1909. https://doi.org/10.3390/buildings12111909

Academic Editors: Michael J Ostwald and Ju Hyun Lee

Received: 11 October 2022
Accepted: 5 November 2022
Published: 7 November 2022

Publisher's Note: MDPI stays neutral with regard to jurisdictional claims in published maps and institutional affiliations.

1. Introduction

The historian of science, Owen Hannaway, once mentioned that science was no longer simply a kind of knowledge; it increasingly became a form of activity. The laboratory is a place specially set aside for such activity [1]. From the secret underground experimental space of Tycho Brahe to the modern large-scale science facility, the internal space and external image of today's laboratories have undergone fundamental changes over the last few centuries with the development of science and innovation in construction technology. The definition of a laboratory has also been revised from the original "chemist's work-house" to "the building set apart for conducting practical investigations in natural science" [2]. From the sociological point of view, a laboratory is a place for knowledge production. The typical scientific activity in the laboratory is connecting the experimental materials to the instruments, generating records through a series of standardized operations, and then putting forward scientific propositions according to these records and modifying them repeatedly until they become facts [3].

Peter Morris believes that changes in function affect the form of the laboratory, and the development of a new field has often produced significant alterations in laboratory design [4]. The early scientific activities were mainly desktop experiments organized spontaneously by the elite based on their interests, usually in the study, kitchen, and other enclosed places. The increased complexity of scientific research and the interpenetration of disciplines have led to collaborative progress between scientific organizations, prompting the need for greater inclusivity and flexibility in the laboratory space.

In contrast to the past, the scope of scientific research activity is no longer confined to limited benches and rooms, and the emergence of continuous open areas encourages more creative activities and academic exchange. As a forward-looking institution, the evolution of the laboratory is closely related to the development of science and technology. In retrospect, the spatial configurations of laboratories in different historical periods make a statement about the status of academic teaching and research and researchers' attitude toward using space.

This study selects three groups of typical laboratories to provide new readings on scientific research space, revealing the topological type and genotype of the laboratory space in different periods, which is an abstract relational model that describes space configurations [5]. Given the complexity of laboratory layouts in various disciplines, this study is limited to the basic spatial configuration of experimental space and other functional spaces rather than a specific analysis of laboratory layouts by disciplinary classification. In this study, qualitative and quantitative types are determined by using justified graphs and their outcomes in the form of syntactic measures.

2. Literature Review

In the field of laboratory culture before the 20th century, few studies focused on the sites and space of experiments. It has been argued in the history and philosophy of science scholarship that the laboratory itself is merely a nodal point with no volume, materiality, or spatial identity [6]. Scholars before the 1960s tended to study etchings and paintings of early modern laboratories as credible sources to gather information about the equipment and working methods employed in historical laboratories but rarely included analyses of space or architectural structures [7–9]. Compared with laboratory design, the historical research on laboratory space from the architectural perspective accounts for a relatively small proportion, usually appearing as a brief historical review at the beginning of some design books.

From the perspective of disciplinary development, as early as 1975, scholars evaluated the global geographical distribution, personnel management, and knowledge production of physical research institutions at the beginning of the 20th century [10]. After the British scholar Maurice Crosland's study of early laboratory construction in the 16th–18th centuries [11], Morris systematically traced the evolution of laboratory facilities in chemistry from the Middle Ages to modern times [4]. Similar studies have been carried out in pharmacy and anatomy [12,13]. In other aspects, Ursula Klein traced the history of technoscience, mapping the relationship between today's cutting-edge disciplines and the development of the useful and technological sciences in Prussia from 1750 to 1850 [14]. Aileen Fyfe placed sciences in a broader cultural marketplace and discussed the evolution and presentation of "popular science" in Victorian Britain [15].

In addition, scholars discussed the spatial characteristics of scientific activities since the Renaissance from the perspective of engineering, sociology, and design [16]. The above research paid more attention to the material and cultural connotations reflected behind the experimental activities, ignoring the spatial characteristics of the laboratory. Furthermore, the historical development of laboratory buildings is not clear. Nancy B. Solomon once regarded laboratories established after World War II as modern laboratories different from the past based on flexibility. Peter Galison argued that with the improvement of industrialization at the end of the 19th century, flexibility had begun to take shape [17]. Andrew Cunningham believed that the 19th century was an important turning point in the history of science [18].

Depending on the historical shifts of scientific centers, the degree of institutionalization, and different participants, here we divide the evolution process of scientific space into four phases: Individual exploration, professional research, centralized research, and integrated interdisciplinary research (Figure 1). This study focuses on the first three phases. Several cases are selected for topological analysis to reveal the spatial logic of laboratories in different periods and the relationship between the public space and experimental units.

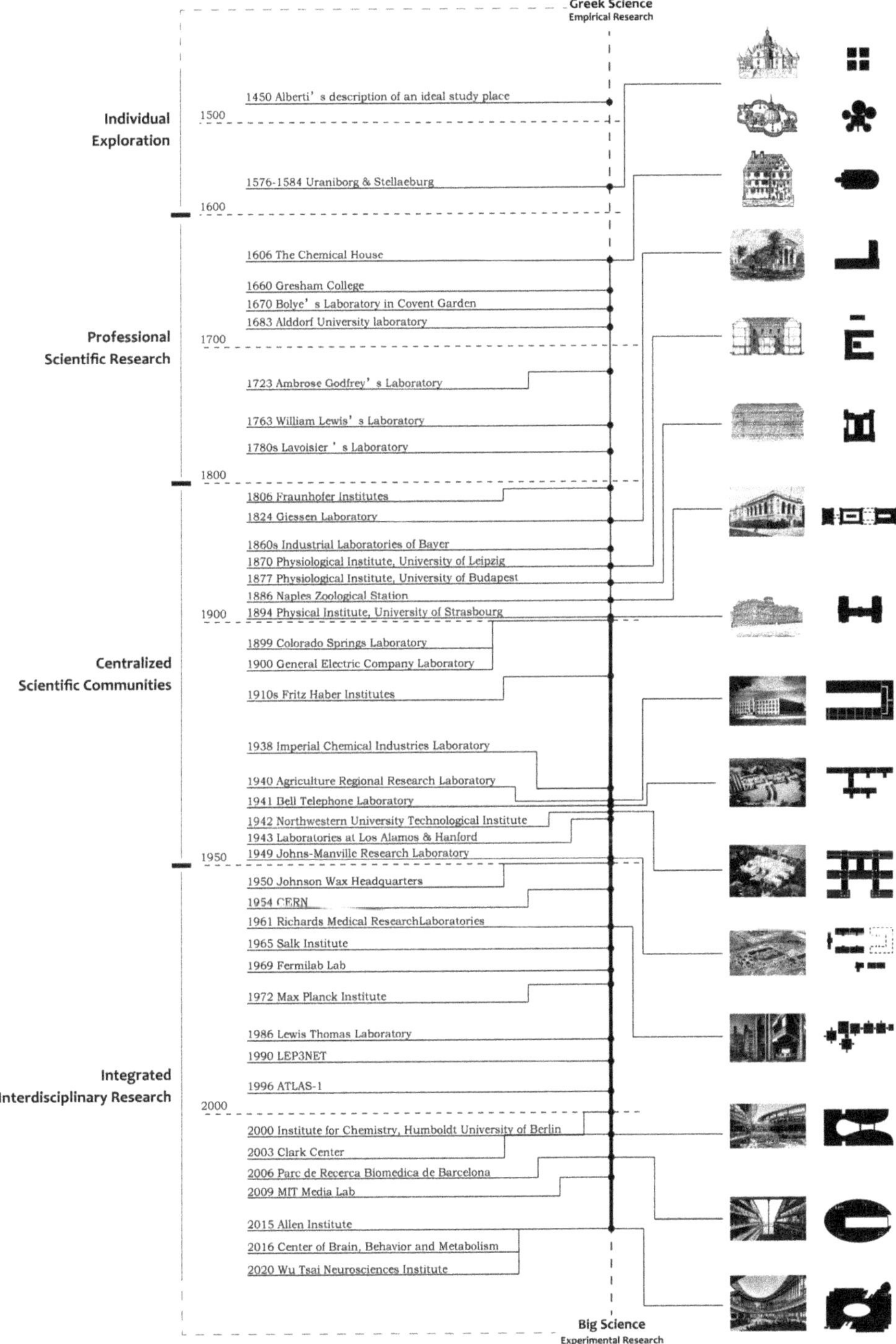

Figure 1. Evolution process of scientific space (source: The authors).

3. Method and Selection of Cases

According to space-syntax comprehension, space forms and organizations exist to coordinate various social activities, and the generation logic of space is consistent with human behavior. As a set of analytical measures for representing and quantifying spatial configuration, space syntax focuses on the potential relationship between form and function, space and users, by focusing on the specific connection between spaces.

This study establishes the topological relationship to obtain visual representation and values of each functional space using DepthmapX and JASS software. DepthmapX enables the depiction of buildings to generate a map of spatial elements, link them through relationships (visibility, intersection, or adjacency), and analyze the resultant network [19]. JASS is intended for convex space analysis to draw networks and convert graphs into justified graphs by transforming spaces and their connections into points and lines, which will be presented in the deep-branched tree form, shallow-ring bushy form, etc. [20–22].

Taking the Chemical House as an example, Figure 2 shows the transformation process from a plan to a justified graph. Firstly, a convex break-up map is created. The connections between spaces are extracted to form a set of topological networks, with each node corresponding to a room in reality. In turn, the spatial properties of all nodes are grouped according to functions such as public (P), experimental (L), courtyard (CRT), and service (Se). The convex map is then represented as a justified graph, and the exterior (public space) is taken as its root to connect from the bottom up. Here we obtain the spatial topology of one case and repeat the process to obtain the diagram of the remaining cases.

The following parameters are considered in this study:

Integration (HH) is the most widely used syntactic value, calculating how close the origin space is to all other spaces as the measure of relative asymmetry (RA) [23,24].

The base difference factor (BDF) is used to investigate the difference between various functional spaces. The value changes between 0 and 1. The closer to 0, the more differentiated and structured the spaces; the closer to 1, the more homogenized the spaces, and no configurational differences exist between them [20].

In addition, it is also important to derive certain in-depth syntactic measures related to the topological network, such as the distributedness index (ID) and space link ratio (SLR). The distributedness index (ID) is based on the four topological spaces (a, b, c and d), with 'a' and 'b' space types emphasizing tree-like configurational properties while 'c' and 'd' space types are conducive to ringiness [25]. Distributedness can be calculated by the formula (a + b)/(c + d) = distributedness, reflecting the available options for accessing all spaces in the system [20]. The space link ratio (SLR) is used to evaluate the distribution of the spatial layout, indicating the flexibility of movement in the system. The parameter expressions involved are as follows:

$$MD = TD/(k-1) \tag{1}$$

$$RA = 2\frac{MD-1}{k-2} \tag{2}$$

$$D_k = \frac{2\left(k\left(\log_2\left(\frac{k+2}{3}\right)-1\right)+1\right)}{(k-1)(k-2)} \tag{3}$$

$$RRA = \frac{RA}{D_k} \tag{4}$$

$$Integration(HH) = \frac{1}{RRA} \tag{5}$$

$$H = -\Sigma\left[\frac{a}{t}\ln\left(\frac{a}{t}\right)\right] + \left[\frac{b}{t}\ln\left(\frac{b}{t}\right)\right] + \left[\frac{c}{t}\ln\left(\frac{c}{t}\right)\right] \tag{6}$$

$$H^*(\text{BDF}) = \frac{H - \ln 2}{\ln 3 - \ln 2} \tag{7}$$

$$\text{SLR} = \frac{L + 1}{k} \tag{8}$$

The meanings of some parameters involved in the formula are as follows:
TD: Total Depth for actual node.
k: Number of nodes.
a: Max RA.
b: Mean RA.
c: Min RA.
t: a + b + c.
L: Number of links.

Using these parameters, this study aims to explore the evolution process of the scientific research space and understand the potential organization and cultural pattern of laboratory layouts in different periods. Through the case collection and study, we selected three representative groups of cases for analysis, designed in the late 16th century, the 19th century, and the mid-20th century. Information about the three groups of cases is shown below (Figure 2). The topological structure of each laboratory spatial configuration is represented by convex maps and justified graphs.

Given that syntactical results are directly affected by spatial delineations [26], the following decisions have been made in this study. Firstly, the external space is included in the topological structure as the original space (root) of the system to present the spatial organization, but not in the analysis of internal spatial differentiation. Secondly, the main floor is only considered because it contains all the essential functions and is available in the literature [27]. Finally, three types of space are abstracted from cases for differentiation analysis: Spaces for experiments, transition spaces (corridors and public spaces), and service spaces (vertical transportation, toilets, storage rooms, and equipment rooms).

The following paragraphs are intended to highlight the spatial features that exist within cases. The spatial pattern is considered according to the syntactic data and the differentiation degree between the integration values of functions (Tables 1 6). Table 1 shows the global integration and the base difference factor (BDF) for a horizontal comparison between cases. Table 2 shows the integration and the base difference factor (BDF) of the three types of spaces used for the differentiation analysis of each case, aiming to describe the difference in integration between functional spaces of each system and the degree of spatial differentiation within the same function. Table 3 shows the proportion of four topological spaces (a, b, c, and d), the resulting distributedness index (ID), and the SLR of each system for discussion on the topological type.

Furthermore, Tables 4–6 show the order of integration of the main functions and the red dotted line represents the average value. The information can be obtained, such as which rooms in the system have the highest degree of integration and what level (higher or lower than the average value) the integration of the experimental space may be in the system.

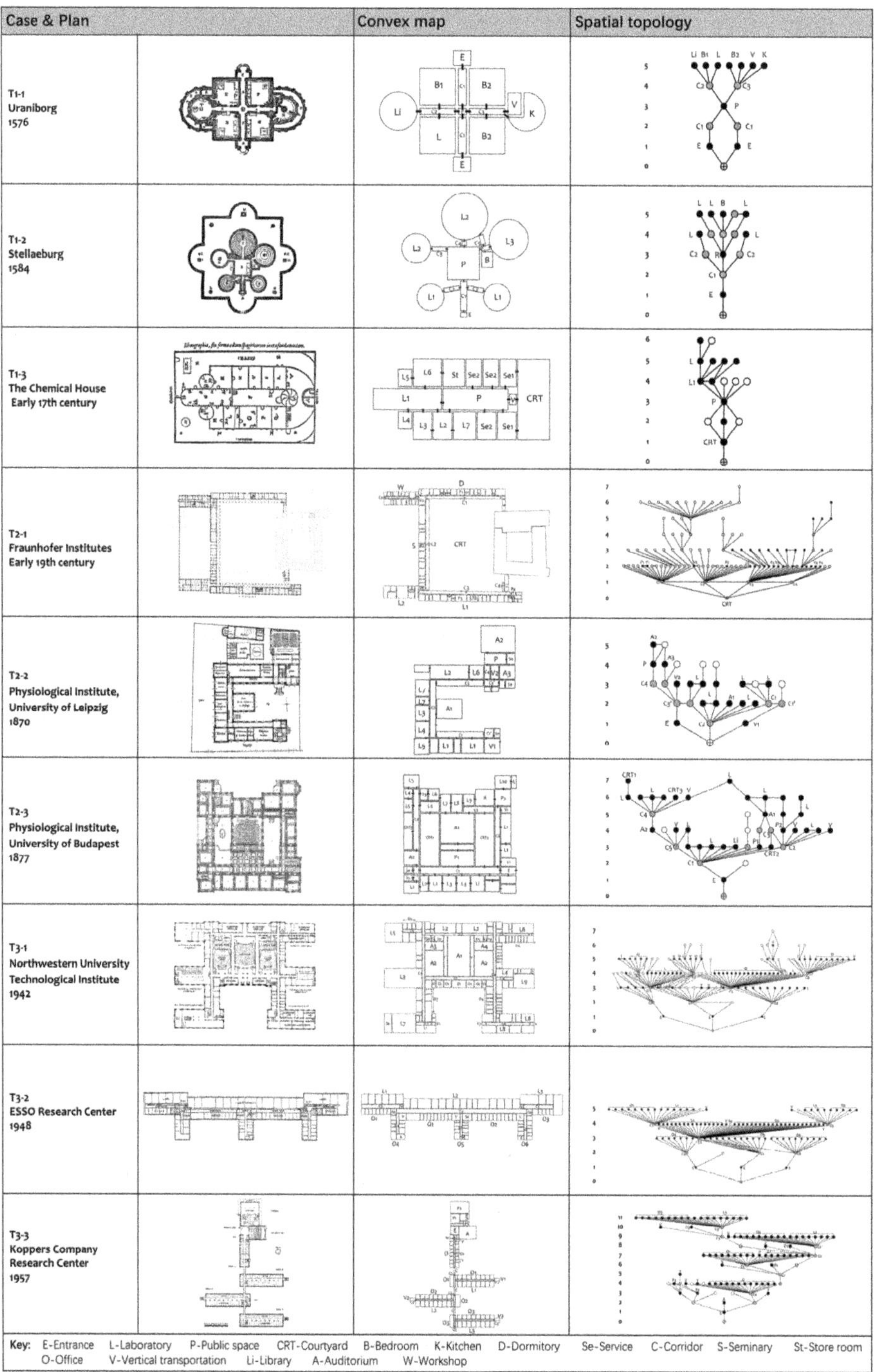

Figure 2. Background information of cases: Plans, convex maps, and spatial topologies (source: The authors).

Table 1. Syntactic Data of Cases (Integration (HH), RA, and BDF).

Sample	Integration (HH)			Relative Asymmetry (RA)			BDF (H*)
	Min	Mean	Max	Min	Mean	Max	
T1-1	0.671	1.034	2.312	0.115	0.291	0.397	0.744
T1-2	0.604	0.972	1.961	0.132	0.296	0.429	0.761
T1-3	0.753	1.323	2.929	0.086	0.217	0.333	0.696
T2-1	0.634	1.208	2.304	0.036	0.074	0.130	0.703
T2-2	0.601	0.945	1.700	0.108	0.210	0.307	0.805
T2-3	0.647	1.080	1.996	0.074	0.146	0.229	0.772
T3-1	0.897	1.903	3.638	0.020	0.044	0.081	0.657
T3-2	1.253	2.188	7.291	0.011	0.038	0.065	0.547
T3-3	0.466	0.876	1.454	0.058	0.101	0.182	0.751

Table 2. Differentiation Degree Between the Integration Values of Different Functions.

Sample	Mean Integration (HH)			BDF (H*)		
	Laboratory Spaces	Transition Spaces	Service Spaces	Laboratory Spaces	Transition Spaces	Service Spaces
T1-1	0.771	1.599	0.832	—	0.893	—
T1-2	0.681	1.119	—	0.986	0.924	—
T1-3	1.347	2.929	1.201	0.763	—	0.998
T2-1	1.101	1.571	1.516	0.864	0.866	0.973
T2-2	0.996	1.304	0.707	0.976	0.914	0.972
T2-3	0.997	1.421	1.036	0.965	0.922	0.993
T3-1	1.541	2.377	1.766	0.943	0.887	0.877
T3-2	2.359	3.833	2.257	0.966	0.852	0.967
T3-3	0.851	1.093	0.788	0.987	0.785	0.848

"—" Some difference factors are not calculated because there are only two types of interior space in the system or the absence of this type of space.

Table 3. Topological Types and Calculated Values of ID and SLR.

Sample	Node Count	Space a	Space b	Space c	Space d	Distributedness Index (ID)	SLR
T1-1	14	9	5	0	0	—	1.000
T1-2	15	6	4	5	0	2	1.070
T1-3	16	7	1	4	4	1	1.250
T2-1	103	58	20	17	8	3.12	1.160
T2-2	29	10	2	13	4	0.706	1.210
T2-3	42	9	5	15	13	0.5	1.310
T3-1	124	83	7	20	14	2.647	1.150
T3-2	105	98	7	0	0	—	1.000
T3-3	99	19	8	19	53	0.375	1.580

"—" Some values are not calculated because of the absence of this type of space.

Table 4. Order of Integration Values of Main Functions (T1).

T1
T1-1: Uraniborg— 11 function space
P: 2.23 > C3: 1.60 > C2: 1.39 > C1: 1.10 > **<u>1.034</u>** > B2: 0.83 = V: 0.83 = K: 0.83 > B1: 0.77 = L: 0.77 = Li: 0.77 > E: 0.67
T1-2: Stellaeburg—12 function space
P: 1.96 > C1: 1.57 > C5: 1.31 > C4: 1.07 > C3: 1.02 > **<u>0.972</u>** > C2: 0.91 > E: 0.84 > L2: 0.78 > L3: 0.76 = B: 0.76 > L2: 0.65 > L1: 0.60
T1-3: The Chemical House—13 function space
P: 2.93 > L1: 2.40 > L2: 1.88 > **<u>1.323</u>** > Se1: 1.26 = V: 1.26 = L6: 1.26 > L3: 1.15 = Se2: 1.15 > L4: 1.05 > L7: 0.94 > CTR: 0.85 > L5: 0.75 = St: 0.75
Key: E-Entrance; L-Laboratory; P-Public space; V-Vertical transportation; CRT-Courtyard; B-Bedroom; K-Kitchen; Li-Library; St-Store room; Se-Service; C-Corridor

The bold underlined value is the mean value of integration in the system. More information about the rooms (location and connection relationship) can be obtained from Figure 2.

Table 5. Order of Integration Values of Main Functions (T2).

T2
T2-1: Fraunhofer Institutes— 19 function space
C2: 2.30 > C3: 2.08 > C1: 1.99 > CRT: 1.96 > V2: 1.95 > C4: 1.54 > P2: 1.49 = S: 1.49 = V3: 1.49 > P4: 1.41 > P3: 1.39 = LI: 1.39 = V4: 1.39 = V5: 1.39 > P1: 1.36 = D: 1.36 > V1: 1.36 > **1.208** > L2: 0.9 > W: 0.81
T2-2: Physiological Institute, University of Leipzig—18 function space
C2: 1.70 > C3: 1.66 > C3′: 1.37 > C1: 1.24 > L4: 1.11 > L3: 1.07 > L2: 1.06 > L6: 1.04 > L7: 1.03 = A1: 1.03 > C4: 0.97 > **0.945** > E: 0.89 > C1′: 0.88 > L1: 0.85 > L5: 0.79 > A3: 0.76 > P: 0.75 > A2: 0.60
T2-3: Physiological Institute, University of Budapest—24 function space
C1: 2.00 > C2: 1.67 > C5: 1.62 > P1: 1.35 = CRT2: 1.35 > A2: 1.29 > L2: 1.24 = C3: 1.24 > L3: 1.23 = E: 1.23 > Li: 1.22 > P2: 1.17 > A1: 1.12 > C4: 1.11 > L1: 1.09 > **1.080** > L7: 0.97 > R: 0.92 > L6: 0.87 > L10: 0.86 > L9: 0.85 > L4: 0.82 > L5: 0.817 > CRT3: 0.81 > CRT1: 0.65
Key: E-Entrance; L-Laboratory; P-Public space; V-Vertical transportation; CRT-Courtyard; Li-Library; Se-Service; C-Corridor; A-Auditorium; R-Research space

The bold underlined value is the mean value of integration in the system. More information about the rooms (location and connection relationship) can be obtained from Figure 2.

Table 6. Order of Integration Values of Main Functions (T3).

T3
T3-1: Northwestern University Technological Institute—33 function space
C4: 3.63 > C2: 3.08 > C1: 3.01 > C3: 2.55 > C6: 2.25 > C8: 2.17 > C9: 2.15 > A4: 2.02 > O4: 2.00 > C: 1.97 > A1: 1.96 > **1.903** > L2: 1.85 > O2: 1.82 = V2: 1.82 > A2: 1.80 = O1: 1.80 = V1: 1.80 = E1: 1.80 > C5: 1.68 > C7: 1.67 > A3: 1.64 = L3: 1.64 > O3: 1.62 > L6: 1.51 > O6: 1.49 > L4: 1.46 = L8: 1.46 = E3: 1.46 > L5: 1.23 = L7: 1.23 > L1: 1.22 = O5: 1.22 = E2: 1.22
T3-2: ESSO Research Center—18 function space
C2: 7.29 > C3: 3.34 > C1: 3.15 > C5: 3.1 = C6: 3.1 > C4: 3.02 > L2: 2.68 = O2: 2.68 > **2.188** > L1: 1.87 = O1: 1.87 > O5: 1.81 = E2: 1.81 > L3: 1.79 = O3: 1.79 = O6: 1.79 = E3: 1.79 > E1: 1.78 > O4: 1.76
T3-3: Koppers Company Research Center—22 function space
C1: 1.45 > C2: 1.38 > C1′: 1.19 > C0: 1.141 > C2′: 1.138 > O1: 1.08 > O2: 1.03 > C3: 1.01 > L1: 0.93 = O1′: 0.93 > O: 0.9 > L2: 0.898 = O2′: 0.898 > **0.876** > C3′: 0.87 > O3: 0.81 > E: 0.8 > L3: 0.72 = O3′: 0.72 > A: 0.67 > P2: 0.6 > P1: 0.59 > P3: 0.52
Key: E-Entrance; L-Laboratory; P-Public space; V-Vertical transportation; O-Office; Se-Service; C-Corridor; A-Auditorium;

The bold underlined value is the mean value of integration in the system. More information about the rooms (location and connection relationship) can be obtained from Figure 2.

4. Historical Frameworks of Samples

4.1. Individual Exploration in Private Sites

The development of early scientific experimentation was closely related to alchemy. In the 6th century BC, there were discussions about the origin of things in Greece. Until the Middle Ages, famous scientists and nobles including Leibniz and Newton had studied alchemy [28]. Existing research in the history of science tends to regard the workplace of alchemists in the 16th century as the prototype of the early laboratory, which usually contained furnaces and distillation equipment. Scientific research in this period was predominantly individual exploration with decentralized approaches and available places to conduct experiments, even in the kitchen and living room [29]. Some experimental philosophers moved away from the secular space and placed knowledge in a restricted area at home [30].

The early laboratories with detailed drawings are Uranienborg and Stjerneborg (T1-1 and T1-2) on the island of Ven, described by the Danish astronomer Tycho Brahe in a book about astronomical instruments published in 1598. Influenced by popular astrology thoughts, the layout of the two astronomical observation facilities is symmetrical with the geometric center. In the case of Uranienborg, for example, the upper space was mainly used for living, with a few rooms used to store large instruments for astronomical observation.

The underground area includes the bedroom, kitchen, and laboratory space transformed from the dining room for alchemy activities [31].

Inspired by Tycho's laboratory, the German chemist Andreas Libavius described the Chemical House (T1-3) in detail, which is introduced as an ideal laboratory for researchers in his commentaries to his textbook of chemistry in 1606. In contrast to what he saw as the aristocratic seclusion of Uraniborg, Libavius viewed his Chemical House as being incorporated into the town and, thus, the public realm [32]. The plan unfolds around the central hall, and the privacy of the overall layout gradually increases from east to west. The main entrance of the building is on the east side, surrounded by a semicircular garden with galleries. The central part is composed of a central hall and experimental spaces, including the coagulatotorium and crystallization room, steam baths room, water baths room, and service spaces (toilets and storage rooms) [4]. In addition to experimental technicians, some professionals were allowed to visit. On the west side is a set of private laboratory spaces leading to the study and residential area on the second floor through the staircase [33,34].

Taking the entrance as the original space, a group of topological structures is obtained from the above cases, as shown in Figure 2 (T1). T1-1 and T1-2 are deep-branched tree forms in which 'a' and 'b' space types occupy a large proportion of the system, showing some local symmetry. In addition to experimental spaces, the two cases also mixed with large living areas. The distinction between functions is obscure, and the spatial logic is strongly centripetal, with the rooms arranged around the central space showing traces of residential accommodation. In the justified graph of T1-1, the laboratory room, bedrooms, and kitchen are in the deepest part of the building (topological depth is 5) side by side, and there is no connection between them. This structure is characterized by strong control of internal and external movement with limited movement options.

Moreover, due to the limited scale, the internal function configuration of T1-1 and T1-2 is relatively simple. The public space (P) at the core is the only control point in the system with the highest integration. The integration of the experimental space is close to the bedrooms and kitchen, with a difference of 0.1 or less, which is very weak in this case.

In contrast, T1-3 has several loops, providing occupants with more route choices. Although the 'a' type occupies a large proportion of space in the system, the ID values show that the case has a strong distribution. Several experimental rooms are arranged around L2 and multiple loops are formed according to the practical requirements. In addition to the central hall (P), L2 is another control point in the system, which has a high degree of integration and provides asymmetry. In addition, the BDF value of the experimental space in T1-3 is also the lowest of all the cases, which indicates a strong differentiation within the laboratory space. Through a series of rooms from east to west, the function gradually transits from the central hall near the entrance to the private space with low spatial integration, creating a privacy gradient in space configuration.

It is argued that there was a subdivision of scientific research in this period that the ordinary could not perceive: Trying, showing, and discoursing. Trying was an activity that occurred within relatively private spaces, whereas showing and discoursing were events in public spaces [35], distinguishing public knowledge from personal knowledge and general research from specialized research. The content of the research largely depended on the cognitive activities, intellectual interests, and personal preferences of the individual, in contrast to modern scientific research. The syntactic results of the above cases also confirm this to some extent. Functional differentiation within the experimental space shows apparent differences in integration and accessibility. Users consciously establish a privacy gradient to distinguish the research content that can be visited and displayed from private research activities in terms of spatial configuration, which meets the needs of the experimental process and reflects their will.

4.2. Centralized Organization Principles in Laboratories

In the early 19th century, laboratories increasingly became genuine research sites in chemistry, physics, and biology. The scientific practice gradually shifted from individual exploration for interests to cooperation research, closely integrated with industrial production. During this period, the laboratory as the production site of scientific knowledge was separated from the home. Private laboratories and government-led research institutions increasingly emerged, and scientific activities were organized and rapidly developed, especially in Germany.

In 1806, the entrepreneur Joseph Utzschneider gathered several experts to establish the Fraunhofer Institutes (T2-1) in the Benedictine Monastery, purchased from the government 5 km south of Munich. The cloister is the core of the building, which connects different spaces and forms a closed loop of daily life. Its spatial organization is reflected in a stable horizontal continuity, enclosing the central yard on three sides, with the dormitories on the north side (D), the research space on the south side (L), and the library and instrument storage space in the seminary on the west side (S).

The spatial organization of the three/four-sided yard enclosed by buildings frequently also appeared in the layout of the laboratories during this period, which corresponded to the collaboration for research and resulted in a batch of classic laboratories, such as the Leipzig Physiological Institute (T2-2) and the Physiological Institute in Budapest (T2-3). T2-2 is a three-sided layout connected by corridors, with an auditorium in the center toward the courtyard. A small lecture hall and auxiliary rooms are built on the north side as required for teaching. T2-3 is a four-sided layout. The central space is divided into three small courtyards by a lecture hall (A1) and a corridor (C4). The rooms for experiments are placed on the north and south sides. The entrance (E) of the building on the east side is directly connected to the corridor (C1).

The above cases reflect the centralized organization principle behind the enclosed layout with the following characteristics. Firstly, due to the growing demand for experimental spaces, the number of rooms increased and the space topology changed from the deep-branched tree form to the shallow-ring bushy form, which is reflected in the justified graph as several clusters of points that expand upwards at the bottom (near the entrance). Furthermore, the average integration of the experimental spaces is higher than in the past. Secondly, the cases in this period show a strong distribution feature with multiple control points in each case, especially in T2-2 and T2-3. The proportion of 'c' and 'd' types of spaces in the system increases significantly, which is reflected in the local formation of several small loops. The large number of 'a' type spaces in T2-1, unrelated to the experimental function, greatly impacts the ID values. Finally, compared with the past, the function of the public space is further divided into the lobby (distribution space), the lecture hall (exhibition space), and the conference room (discussion space). In some cases, the lecture hall occupies the courtyard space, visually becoming the core of the whole building. Its integration degree is slightly higher than the average, indicating that the efficiency of such space is not high in topology.

4.3. Modular Laboratory under Industrial Structure System

In the second half of the 19th century, industrial modes of production were widely spread in architectural culture, and efficiency became the focal point of architectural technology. Industrial buildings started to use massive standardized components, which were applied to the construction of large laboratories to meet the need for more extensive open areas and flexible space, in contrast to some prewar laboratories built with thick, immovable walls [36]. Moreover, science had gradually become a mainstream profession open to all sectors of society, and collaboration between scientists from different fields was required to carry out more complex research. Given the demand to effectively cope with the structural changes of space due to discipline reform and staff changes, the module was applied to the design of laboratories. Repetitive units were used to generate the overall plan of the building and adapt to different functional requirements and future expansion. Hundreds of laboratory buildings designed by module were built in the fields of communication, nuclear physics, aerospace, and industry around World War II.

Among the three selected cases, the Northwestern University Technical Institute (T3-1), built in the 1940s, is a typical multi-disciplinary research building, with different academic departments placed in six wings of the building. Each department is equipped with professional laboratories and independent research spaces connected to the public space in the center. The ESSO Research Center (T3-2) and the Koppers Company Research Center (T3-3) were designed by the same architects. T3-2 is composed of three T-shaped modules linked in sequence, with experimental spaces to the north and offices to the south. Each module has independent vertical transportation, service rooms, and an entrance. The layout of T3-3 is loosely arranged, with corridors linking five separate functional areas, including a public reception area, an office wing, and three wings of the laboratory.

The above cases show different characteristics in spatial organization compared with the past. Firstly, the collectivization of science led to the centralization of personnel and a strong relationship between researchers and the organization, which is implicit in the grid extended by standard modules and resulting in more efficient use of laboratory spaces. This spatial configuration tends to form its communities via corridors or other public spaces as the core of each department, maintaining a relatively independent state to focus on its scientific research work.

Secondly, the modular system brings laboratory space flexibility and weakens the differences between laboratories caused by the specific connection order between spaces. The BDF values of the experimental spaces in the three cases are above 0.9, indicating no apparent functional differences between the laboratory spaces in any of the cases. Thirdly, using the corridor as the only means of connection limits the potential organizational integration through spatial configuration. As is shown in Table 5, the integration of the laboratory space in T3-1 is lower than average (1.903), and the southwest (L7) and the northwest (L5) are the lowest (1.23). The cost of going to these two wings is higher than that of other wings due to the multiple transitions of corridors in topology. The same feature also exists in T3-2 and T3-3, in which there is a gap between the integration of the laboratory space and the transition space.

Finally, the transition space plays an important role in the building with high integration. Taking the corridor (C4) on the east side of T3-1 as an example, the total number of convex spaces is 124, with half (62) passing through C4 to reach the central public area. The high-frequency use of the transition space makes it one of the few places to create encounters except for the public space. Unless driven by a particular purpose or task, informal communication between researchers from different fields can only be triggered with a greater possibility at the junction of corridors and near the entrance.

5. Discussion

Space configuration that reflects potential social relations is the intermediary between the physical and cultural attributes of buildings, shaping the concept of community while inadvertently defining the identity and behavior of an individual [23]. This study quantitatively identifies the topology and spatial configurations of laboratory buildings in different periods, providing a new interpretation of the classical cases in history. The analysis method of space syntax used here is mainly based on the following points. Firstly, the analytical method of space syntax applied here allows for a deeper comprehension of historical design principles and the way in which the geometry of forms may be concealed in abstract experimental rules and physical characteristics, thus revealing the significance of spatial configurations for scientific activities. Secondly, space syntax can quantitatively describe the spatial structure of buildings, settlements, cities, etc. [37]. It has a broad scope of applications and can be used to reveal the genotypes inherent in laboratory space systems. Finally, the use of space syntax analysis methods here can make up for the lack of qualitative case studies to a certain extent. The combination of the two methodologies can promote the effective evaluation of laboratory typology.

Contemporary scientific research has become far more complex than in Galileo's era, and the laboratory has changed dramatically in many ways compared to the past. Research issues and practical problems often require more than one researcher or discipline to address them adequately. With the complexity of scientific research and the mutual penetration of disciplines, face-to-face interaction among interdisciplinary teams makes scientific research an increasingly social activity. Promoting communication and stimulating innovation through space has become a focus in laboratory design today. The space that can only realize the basic functions such as experiments and teaching can no longer meet the needs. We need to consider the initiative of space itself and the relationship between space, scientific activities, and scientific development. Therefore, under the influence of new experimental needs and changing scientific concepts, it is important to examine the evolution process of the laboratory building.

6. Conclusions

The following conclusions are drawn from the above analysis:

(1) The evolution of the laboratory space in history was a process from closed to open, decentralization to centralization. The above cases from different periods reflect the remapped relationship between scientific research activities and spatial forms.

In the early days, when the classification of disciplines was not clear, the laboratory was a space spontaneously organized by individuals. Layouts of the laboratories were usually symmetrical, with a deep-branched tree form. The public space at the core was the most integrated.

With the rise of scientific revolutions, the corresponding practical space gradually became specialized and standardized. The internal connection became complex, with a shallow-ring bushy form in topology. According to different discipline requirements, the laboratory functions were further differentiated.

The laboratory buildings after World War II were usually generated by modules to adapt to the cooperative research between scientists. The transformation space had an unshakable position in the system, and the difference between departments was decreased, which is reflected in the high BDF value of the experimental space.

This study evaluates the data obtained through a comparative analytical approach, revealing the relationship between the space configuration and functional efficiency of laboratory buildings in different periods. It provides valuable information for analyzing laboratories for traditional or experimental typology and references for future laboratory design.

(2) The above analysis shows that the laboratory space has undergone a process from closed to open, decentralization to centralization. It is manifested in the increasingly large laboratory rooms and centralized layout, as well as repetitive units of departments. All of these have led to an evolution in the form of laboratory buildings, with a correspondence between the form and function of laboratory buildings in different periods.

(3) The analysis results emphasize the importance of the transformation space. From the topological perspective, the corridor is also an effective means to increase space depth and create private areas. It is easy to see from the previous analysis of T3-1 that two experimental spaces in the same case are likely to have differences in integration and intelligibility due to the local intermediary space conversion.

(4) In the analysis of cases mentioned above, it is found that the high spatial connectivity does not necessarily indicate that the case is distributed, as is the case of T3-3. The experimental spaces of each department are connected, which leads to efficient connectivity. However, globally, there are few choices for movement from one point in the space to another, acting in a conservative mode. Therefore, it is necessary to make a comprehensive judgment in combination with the degree of integration and justified graph.

Author Contributions: Conceptualization, X.Z. and T.C.; methodology, X.Z.; software, X.Z.; validation, X.Z. and T.C.; formal analysis, X.Z.; investigation, X.Z.; resources, X.Z.; data curation, X.Z.; writing—original draft preparation, X.Z.; writing—review and editing, X.Z.; visualization, X.Z.; supervision, T.C.; project administration, T.C.; funding acquisition, T.C. All authors have read and agreed to the published version of the manuscript.

Funding: This research received no external funding.

Data Availability Statement: The data presented in this study are openly available in FigShare at [https://doi.org/10.6084/m9.figshare.21507813.v1], accessed on 4 November 2022, reference number [1].

Conflicts of Interest: The authors declare no conflict of interest.

References

1. Hannaway, O. Laboratory Design and the Aim of Science: Andreas Libavius versus Tycho Brahe. *Isis* **1986**, *77*, 585–610. [CrossRef]
2. Gooday, G. Placing or Replacing the Laboratory in the History of Science? *Isis* **2008**, *99*, 783–795. [CrossRef]
3. Latour, B.; Woolgar, S. *Laboratory Life: The Construction of Scientific Facts*; Princeton University Press: Princeton, NJ, USA, 2013; ISBN 1-4008-2041-3.
4. Morris, P.J. *The Matter Factory: A History of the Chemistry Laboratory*; Reaktion Books: London, UK, 2015; ISBN 1-78023-474-0.
5. Hillier, B.; Hanson, J.; Graham, H. Ideas are in things: An application of the space syntax method to discovering house genotypes. *Environ. Plan. B Plan. Des.* **1987**, *14*, 363–385. [CrossRef]
6. Aitchison, M. *The Architecture of Industry: Changing Paradigms in Industrial Building and Planning*; Routledge: New York, NY, USA, 2016; ISBN 1-317-04480-0.
7. Read, J. The Alchemist in life, literature and art. *Br. J. Philos. Sci.* **1952**, *3*, 88–89.
8. Holmyard, E.J.; Singer, C.; Hall, A.R.; Williams, T.I. *A History of Technology*; Clarendon Press: Oxford, UK, 1975.
9. Hill, C.R. The iconography of the laboratory. *Ambix* **1975**, *22*, 102–110. [CrossRef] [PubMed]
10. Forman, P.; Heilbron, J.L.; Weart, S. Physics circa 1900: Personnel, funding, and productivity of the academic establishments. *Hist. Stud. Phys. Sci.* **1975**, *5*, 1–185. [CrossRef]
11. Crosland, M. Early Laboratories c. 1600–c. 1800 and the Location of Experimental Science. *Ann. Sci.* **2005**, *62*, 233–253. [CrossRef]
12. Underwood, E.A. The early teaching of anatomy at Padua, with special reference to a model of the Padua anatomical theatre. *Ann. Sci.* **1963**, *19*, 1–26. [CrossRef]
13. Pickstone, J.V. *Ways of Knowing: A New History of Science, Technology, and Medicine*; University of Chicago Press: Chicago, IL, USA, 2001; ISBN 0-226-66795-2.
14. Klein, U. *Technoscience in History: Prussia, 1750–1850*; MIT Press: London, UK, 2020; ISBN 0-262-35948-0.
15. Fyfe, A.; Lightman, B. *Science in the Marketplace: Nineteenth-Century Sites and Experiences*; University of Chicago Press: Chicago, IL, USA, 2007; ISBN 0-226-15002-X.
16. Smith, C.; Agar, J. *Making Space for Science*; Palgrave Macmillan: London, UK, 1998; ISBN 978-1-349-26326-4.
17. Galison, P.; Thompson, E.A.; Edelman, S. *The Architecture of Science*; MIT Press: Cambridge, MA, USA, 1999; ISBN 0-262-07190-8.
18. Cunningham, A.; Williams, P. De-Centring the "Big Picture": 'The Origins of Modern Science' and the Modern Origins of Science. *Br. J. Hist. Sci.* **1993**, *26*, 407–432. [CrossRef]
19. Pinelo, J.; Turner, A. *Introduction to UCL Depthmap 10*; University College London: London, UK, 2010.
20. Hanson, J. *Decoding Homes and Houses*; Cambridge University Press: Cambridge, UK, 1998; ISBN 0-521-54351-7.
21. Mohamed, A.A. Space Syntax Approach for Articulating Space and Social Life. In *Handbook of Research on Digital Research Methods and Architectural Tools in Urban Planning and Design*; IGI Global: Hershey, PA, USA, 2019; pp. 223–249.
22. Kim, J.; Kwak, D. Extraction of spatial genetic characteristics and analysis of 1930s Korean Urban Hanok based on application of space syntax. *J. Asian Archit. Build. Eng.* **2022**, *21*, 197–210. [CrossRef]
23. Hillier, B.; Hanson, J. *The Social Logic of Space*; Cambridge University Press: Cambridge, UK, 1984; ISBN 1-139-93568-2.
24. Lee, J.H.; Ostwald, M.J. *Grammatical and Syntactical Approaches in Architecture: Emerging Research and Opportunities: Emerging Research and Opportunities*; IGI Global: Hershey, PA, USA, 2019; ISBN 1-799-81698-2.
25. Amorim, L. The sectors paradigm. In Proceedings of the First Space Syntax International Symposium; Bartlett School of Graduate Studies-UCL: London, UK, 1997; Volume 2, pp. 1–18.
26. Malhis, S. Narratives in Mamluk architecture: Spatial and perceptual analyses of the madrassas and their mausoleums. *Front. Archit. Res.* **2016**, *5*, 74–90. [CrossRef]
27. Malhis, S. The Spatial Logic of Mamluk Madrassas: Readings in the Geometric and Genotypical Compositions. *Nexus Netw. J.* **2017**, *19*, 45–72. [CrossRef]
28. Newman, W.R. *Newton the Alchemist: Science, Enigma, and the Quest for Nature's "Secret Fire"*; Princeton University Press: Princeton, NJ, USA, 2018; ISBN 0-691-17487-3.
29. Crosland, M. Difficult beginnings in experimental science at Oxford: The Gothic chemistry laboratory. *Ann. Sci.* **2003**, *60*, 399–421. [CrossRef]

30. Evelyn, J. *Diary and Correspondence*; H.Colburn: London, UK, 1857; Volume 1.
31. Schmidgen, H. Laboratory. *Encycl. Hist. Sci.* **2021**, *3*. Available online: https://lps.library.cmu.edu/ETHOS/article/id/450/ (accessed on 4 November 2022).
32. Morris, P.J.T. The history of chemical laboratories: A thematic approach. *ChemTexts* **2021**, *7*, 21. [CrossRef] [PubMed]
33. Grapí, P. *Inspiring Air: A History of Air-Related Science*; Vernon Press: Wilmington, DE, USA, 2019; ISBN 1-62273-614-1.
34. Kim, M.G. Chemical Laboratory and the Cosmic Space. In *Space: A History*; Janiak, A., Ed.; Oxford University Press: Oxford, UK, 2020; ISBN 978-0-19-991410-4.
35. Shapin, S. The house of experiment in seventeenth-century England. *Isis* **1988**, *79*, 373–404. [CrossRef]
36. Haines, C. Planning the scientific laboratory. *Archit. Rec.* **1950**, *108*, 107–127.
37. Bafna, S. Space syntax: A brief introduction to its logic and analytical techniques. *Environ. Behav.* **2003**, *35*, 17–29. [CrossRef]

 **buildings**

Article

Rethinking Art Museum Spaces and Investigating How Auxiliary Paths Work Differently

Jae Hong Lee [1,*] and Yong Seung Kim [2,*]

1   Division of Architecture, Gachon University, Seongnam 13120, Korea
2   Division of Architecture, Hanyang University, Ansan 15588, Korea
*   Correspondence: zenoar@gachon.ac.kr (J.H.L.); yskim@hanyang.ac.kr (Y.S.K.)

Abstract: It has been recognized that one of the key issues in designing museums is the interaction between the layout of space and the layout of objects, and spatial configurations are strongly related to didactic narratives, social implications, and curatorial intentions. However, it has not yet been examined thoroughly how museums work from a spatial perspective. Apart from the layout of objects, spatial configurations play an important role in creating various walking sequences, ranging from main routes to auxiliary paths. Art museums in particular can be characterized by such deviations generated by the auxiliary path, but they are hardly understood from this aspect. Therefore, this study aims to explore the auxiliary paths and examine how they work through in-depth theoretical analysis based on space syntax. By analyzing four art museums in terms of isovist attributes, syntactic measures, spatial sequences, and possible trails, it has been concluded that in the cases of the Uffizi Gallery and the Moderna Museet, spatial sequences work conservatively, so that auxiliary paths are channeled back to the gathering space. This is because the walking experience is strongly correlated with visual syntactic features such as connectivity, integration, and intelligibility. Conversely, walking sequences in the case of the Centre Pompidou and the Alte Pinakothek work generatively, and auxiliary paths are rarely related to the gathering space because the walking experience is strongly concerned with visual geometric properties such as isovist area/perimeter and occlusivity.

Keywords: art museums; auxiliary paths; syntactic measures; isovist attributes; spatial sequences

Citation: Lee, J.H.; Kim, Y.S. Rethinking Art Museum Spaces and Investigating How Auxiliary Paths Work Differently. *Buildings* 2022, 12, 248. https://doi.org/10.3390/buildings12020248

Academic Editors: Michael J. Ostwald and Ju Hyun Lee

Received: 29 December 2021
Accepted: 18 February 2022
Published: 21 February 2022

Publisher's Note: MDPI stays neutral with regard to jurisdictional claims in published maps and institutional affiliations.

1. Introduction

According to Macdonald, "collecting is fundamental to the idea of the museum," and "the idea of the museum has become fundamental to collecting practices" (p. 81, [1]). Collecting is a kind of social activity of re-contextualizing objects; therefore, museums play a role in regulating, forming, or creating some kind of meaningful "whole." On the basis of this conception, museum research has been conducted to identify types of collecting, analyze different motivations, examine collecting types in relation to social attributes, and explore the interaction between display layout and building morphology.

In terms of a museum's morphological features, according to Pevsner, there has been an important change: collecting was usually displayed in either "centrally planned rooms" or "long galleries" until the 17th century, but from the 18th century, there was a strong tendency to set out collections in a spatial layout (p. 113, [2]): that is, the trend to "separate the types of items." For example, L.C. Sturm suggested that an ideal museum is where "there are rooms for objects of natural history as well as one room on the top floor for small paintings, drawings and sculpture" (p. 114, [2]).

Regarding including and separating the types of items, one of the stereotyped museums was suggested through a series of museum competitions, such as the museums designed by Guy de Gisors and J.-F. Delannoy in 1778–79, E.-L. Boullée's museum in 1783, or J.-M.-L. Durand's design in 1802–09. These museums generally comprise a large square and set into the square is a Greek cross with four courtyards, with the four arms stretching out from a central "Pantheon rotunda" for works of art, natural history, a print room, and a library, etc. (pp. 118–123, [2]).

On the other hand, another stereotyped spatial morphology developed throughout museum history includes the Alte Pinakothek, designed by Leo von Klenze. The museum building is composed of a loggia, interfacing a series of many small rooms for displaying artworks on the one side, and leading outside to areas such as streets or gardens on the other side. Interestingly, the loggia provides access to all the small rooms, and it is somewhat limited for visitors to generate their own paths. This means that their movement is generated and controlled by the loggia. This may be termed as a lineally structured museum.

On the basis of this aspect, studies on museum architecture have focused on understanding museum space as a device for enhancing social interaction, shaping knowledge by classifying and arranging things, or creating a chronological sequence throughout the overall spatial organization [3,4]. Moreover, the space syntax community has focused on identifying the main dimensions of the variability of spatial layout and display strategies, proposing that both building and display layouts are intentionally designed not only to convey "a pre-given meaning" and reproduce "information" but also to create "possible meaning" and generate "a richer spatial structure" [5–8]. These theoretical aspects stress that there is a strong interaction between physical museum buildings and the process of collecting.

Thinking about the morphological features in terms of the walking experience, it has been suggested that a museum with traditional enfilades, such as Durand's Museum, offers several equal alternatives for continuing the visitor's route and, particularly, this spatial layout is often applied to museums that comprise different types of items or functions and deliver diverse themes. Thus, this type of museum does not need to have a long, narrow corridor. By contrast, a lineally structured museum, such as the Uffizi Gallery, the Alte Pinakothek, or the Moderna Museet, where spaces are rarely linked to one another but mostly adjoin with the loggia or corridor, creates a viewing order of items or themes, so that visitors might experience a single sequence along the loggia or corridor.

Brawne, however, has made an interesting point regarding the lineally structured museum: it provides auxiliary paths derived from the single main route [4]. This means that although the overall movement pattern is guided and even controlled by that single sequence at a global level, we can experience meaningful deviations in terms of walking sequences at a local level. Spatial experiences in this museum building could vary in terms of such auxiliary paths, and therefore such experiences could be a cumulative understanding of the form of the auxiliary paths.

However, it is not clear yet in what way those auxiliary paths form, how they are related to the main route, nor to what degree they are different from each other. Hence, this study particularly focuses on exploring and examining the auxiliary paths in lineally structured museums through in-depth theoretical analysis by using the space syntax technique; especially prominent art museums were analyzed to answer these questions.

2. Museum Studies

It has been recognized that one of the key issues in designing museums is the way the layout of space interacts with the layout of the objects [5] and, consequently, many researchers suggest that the spatial configuration provides "a structure of the exploitation of the collections and buildings by visitors" [9,10]. Others go even further to say that a visitor's experience in the museum is closely related to three elements: "didactic narratives, the ambient experience of the wealth of a collection including the richness of its architectural exposition, and the opportunities for social interaction afforded by the environment" [11].

Huang also suggested two key ideas in understanding museum buildings: "organized walking" (i.e., spatial sequences), which can lead to the process of constructing disciplinary knowledge throughout the plan and "a gathering space". By studying modern museums in Europe, he contends that there is "an underlying genotypical conflict between the need to congregate people and the need to organize their movement," and this genotypical conflict works as a bias in characterizing "the spatial types of the modern museum" [6].

Duncan and Wallach, conversely, highlighted the significance of totality in the museum experience, created by installations, the layout of exhibition rooms, and the sequence of collections. Particularly, they thought that the totality of art and architectural form makes visitors organize their museum experience as an "architectural script" [7].

Regarding the importance of spatial arrangement in museum architecture, Brawne stressed intelligibility, which means clarity and a sense of the order of the spatial layout [4]. In museums, spatial experiences tend to be determined by viewing images in sequence. This means that the museum experience is very similar to the experience elsewhere in a building or a town, and, in this perspective, Brawne transposed Lynch's five elements (paths, edges, districts, nodes, and landmarks) into a way of comprehending spatial experiences.

Psarra and Grajewski raised a fundamental question of how visitors understand museum buildings through movement, and defined museum buildings as systems of spaces that are seen sequentially through movement, and also as overall conceptual patterns that can be grasped at once. Based on these definitions, they suggested that experiencing buildings involve both levels of understanding: the architectural promenade and the conceptual structure [12]. In addition, Psarra, Wineman, Xu, and Kaynar argued that built space is understood as a relational pattern: a pattern of "distinctions, separations, interfaces, and connections", and a pattern that "integrates, segregates, or differentiates its parts in relation to each other" [13].

Salgamcioglu and Cabadak questioned whether the type of space, particularly a permanent or temporary one, affects a visitor's behavior in relation to objects and spatial configuration, and found a substantial distinction between permanent and temporary museum spaces: in permanent exhibitions, the visitor's behavior is influenced by the interfacing between interior and exterior space, whereas in temporary cases, artworks are a crucial parameter, affecting the behavior [14].

Hence, it can be said that museums, especially in the layout of space, provide somewhat different and distinct experiences by utilizing spatial arrangements, and it might be argued that, apart from the idea of the strong relationship between spatial layout and display layout, spatial arrangements themselves play a decisive role in structuring the museum experience. However, it has not yet been examined in what way and how distinctly the spatial arrangements work. Thus, this paper specifically aims to explore and discover substantial answers to these theoretical questions.

3. Research Methodologies and Cases

3.1. Research Methodologies

With these questions, the cases were analyzed by the following methodologies: visibility graph analysis (VGA), spatial sequence generated by c-sequence length and c-sequence depth, spatial choice by d-ring, and possible trails.

Visibility graph analysis, as one of the primary techniques in space syntax, has been derived from Benedikt's isovist and developed by the space syntax community. An isovist is defined as "the set of all points visible from a given vantage point in space" (p. 47, [15]), but it is not appropriate to assess the whole spatial layout of buildings because its geometric properties, such as the isovist area, perimeter, or shape, are purely local features of the viewpoint. Additionally, those properties miss the relationship between the viewpoint and the whole spatial environment. To overcome these problems, the space syntax community suggests VGA, which first sets out a grid (normally 0.6 or 1.0 m) throughout the spatial layout, and each one of the grid points is used as a viewpoint. After having completed this step, the VGA analyzes the relationship between a certain point and the others in both a geometric and a syntactic way using the depthmap X program (depthmapX-0.8.0_win64 version).

Geometrically, the isovist area is a measure of how much space can be seen from a given viewpoint and, conversely, how much the viewpoint can be seen from it; the perimeter is the perimeter of the isovist polygon; the compactness is a measure of the shape of the isovist (i.e., simple or complex, symmetrical or asymmetrical), so that the closer the isovist is to a circle, the more its compactness approaches a maximum value of 1; the occlusivity refers the occlusivity of the isovists, by measuring the length of the occluding radial boundary based on the perimeter of the isovists, and indicating "the depth to which environmental surfaces are partially covering each other as seen from the vantage point" (p. 53, [15]); and, lastly, the area/perimeter is obtained by the area to perimeter ratio, and it can be described as an adjective measure of "how 'spiky' or conversely how 'rounded' an isovist is" (p. 154, [16]).

In space syntax theory, spaces can be represented in two ways: depths and rings. Depth is the idea of how deep or shallow a given spatial structure is, and this conception is explained by connectivity, integration, and mean depth. The connectivity is meant to be the total number of direct connections to the other elements from a certain one at a local level; the integration, unlike the connectivity, is a measure of how many elements are related to the whole; and the visual mean depth is the average distance from each element to all elements within the spatial structure. Syntactically, therefore, VGA analyzes the visual connectivity as a local property, the integration as a global feature, and the mean depth. Additionally, we can identify how intelligible or understandable the spatial layout is by the correlation between visual connectivity and integration.

Contrary to the depths, the idea of rings is derived from two kinds of behavior: "occupation," which happens in a convex way, and "movement," which, conversely, occurs linearly. With these conceptions, spaces are defined as four different types (Figure 1a): an a-space is a space with a single link, and a b-space is one with more than one link, but it should be a part of "a connected sub-complex in which the number of links is one less than the number of spaces", so that it is "on the way to (and back from) at least one dead-end space"; a c-space is one with "more than one link which form a part of sub-complex" containing neither a-space nor b-space, and it should lie on a single ring. Lastly, a d-space is the space with "more than two links," which form a part of complexes containing neither a-space nor b-space, and it must lie on at least two rings (pp. 250–251, [17]).

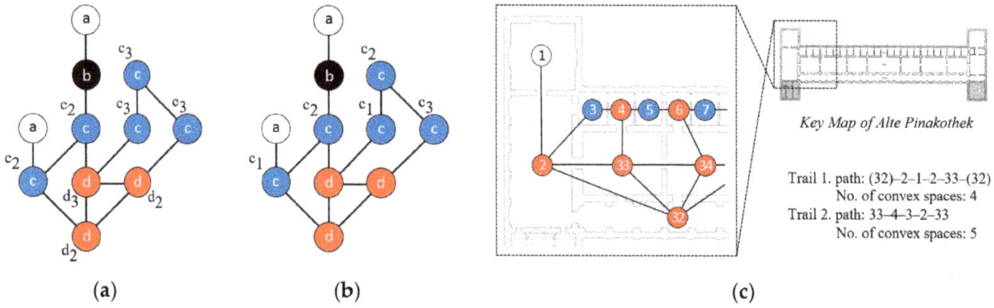

(a) (b) (c)

Figure 1. (**a**) example of c-sequence length and d-ring with space types of a-space (white), b-space (black), c-space (blue), and d-space (red); (**b**) example of c-sequence depth with space types; (**c**) example of possible trails in the Alte Pinakothek Museum.

Considering these four types of space, it is certain that a-space and b-space are strongly related to occupational activities, whereas c-space and d-space are involved in the overall movement patterns (pp. 296–298, [8]). Theoretically, it can be said that c-spaces are concerned with specifying sequences and producing structured movement patterns, so that this kind of spatial layout works conservatively. Hence, from the conception of the c-spaces, it can be possible to explore how strongly spaces are framed with "the degree of sequence which is constructed by the pattern of c-spaces" ([5,8,18]). The length of the sequence is numerically indexed and described by the c-sequence length (Total) and c-sequence depth (Total). First, the value of the c-sequence length (Total) tells us "how many spaces form the c-sequence", while the number of the c-sequence depth (Total) indicates "the depth into the c-sequence" ([5,18]). For example, there are five c-spaces in Figure 1a. Two of them are placed on one ring, constituting two c-spaces and two d-spaces. On this ring, people should pass through two c-spaces in order to finish their journey, so these c-spaces are labelled 'c2'; that is, a sequence of two spaces without choice on the path. Similarly, the other c-spaces on the other ring are labelled 'c3', meaning a sequence of three spaces without choice. When we add all c-spaces together, we can account for the c-ness in this layout (i.e., the c-sequence length (Total): 13; c-sequence length (Mean): 2.6 = 13/5). In a similar way, we can also calculate the c-sequence depth (Total). For example, the two c-spaces in the ring are labelled 'c1' and 'c2' (Figure 1b). On the other ring, the c-spaces are labelled 'c1', 'c2', and 'c3'. When we add the number of the c-spaces together, we can determine the depth of the c-sequence in this layout (i.e., the c-sequence depth (Total): 9; the c-sequence depth (Mean): 1.8 = 9/5).

On the other hand, d-spaces minimize those sequences, leading to the creation of many different paths, so that this layout acts in a generative way. With the idea of d-spaces, it can be examined "to what degree the spatial system allows us to take alternative routes", and, thus, the choices are measured by the pattern of d-spaces. Similar to the c-sequence, we can index d-spaces according to the number of rings they are on. For example, there are three d-spaces in Figure 1a: one of them has three rings, and it can be labelled 'd3, while the others have two rings, so that they are labelled 'd2'. When adding the numbers together, we can determine the amount of rings (i.e., the d-ring (Total): 7; the d-ring (Mean): 2.3 = 7/3).

With these conceptions, the spatial sequence of a layout can be described by the c-sequence length and the c-sequence depth, whereas the choice of the layout can be illustrated by the d-ring. It should be noted that the terms of spatial sequence and choice used here are not actual but theoretical.

The possible trails, according to the previous methodology, are created with the following rules. All trails should start from d-spaces, pass through a-/b-/c-spaces, and end at d-spaces where the trails begin. The trail should form a single ring. In Figure 1c, for example, two possible trails can be drawn from the case of the Alte Pinakothek. Trail 1 starts from 32, representing the loggia; passes through 2, 1, 2, and 33, and then ends at 32 so that the path of Trail 1 is 32-2-1-2-33-32. Trail 2, however, starts from 33, passes through 4, 3, and 2, and then returns to 33. After having constructed the trails, each trail is calculated by the average VGA values in a geometric (e.g., the isovist area, perimeter, compactness, occlusivity, and area/perimeter) and a syntactic way (e.g., connectivity, integration, and mean depth).

The reason for constructing these possible trails is to explore how many trails starting from the d-spaces can be constructed, to identify to what extent they are distinct in terms of walking sequence and walking choice, and finally in what way the galleries are spatially distinct.

3.2. Cases

Four galleries were chosen as case studies: the Uffizi Gallery, the Alte Pinakothek, the Museum of Modern Art at level 5 of the Centre Pompidou, and the Moderna Museet. (Figure 2). The Uffizi Gallery, designed by Giorgio Vasari and then developed by Alfonso Parigi and Bernado Buontalenti, was originally designed to be an administrative office, but it was gradually transformed into a museum to house the Medici family's art collections. One interesting point is that its edges against the adjacent buildings are very irregular. However, the loggia, which is the most dominant gathering space in this building, presents a powerful geometric shape, and it leads to gallery spaces in different shapes such as square, rectangle, octagon, oval, and polyhedron. Another interesting point is the arrangement of gallery spaces: for example, some galleries are arranged linearly so that visitors can cross the thresholds between gallery halls without going back to the loggia, but some others return visitors to the loggia.

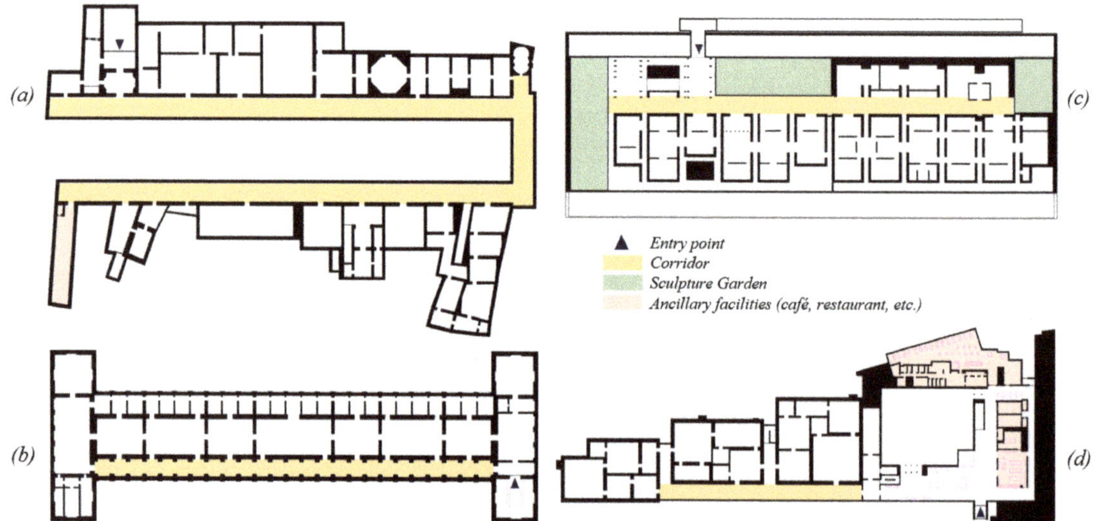

Figure 2. Cases: (**a**) second floor plan of the Uffizi Gallery in 2018, Italy; (**b**) second floor plan of the Alte Pinakothek in 1836, Germany; (**c**) fourth floor plan of the Centre Pompidou in 1977, France; (**d**) third floor plan of the Moderna Museet in 1998, Sweden.

The Alte Pinakothek, designed by Leo von Klenze and opened to the public in 1836, was built for displaying pictures only in Berlin. One of the remarkable features of this gallery, as we can see in the Uffizi Gallery, is the loggia, which plays an important role in providing and controlling access to the middle gallery halls for large paintings, and also to the twenty-five cabinets located in the northern part of this building to hang small paintings on the wall (pp. 129–130, [2]).

The Museum of Modern Art is located in the Centre Pompidou, designed by Richard Rogers and Renzo Piano. With the architectural conceptions of flexibility, transparency, and openness, any architectural elements or utilities are intentionally absent, and consequently, the interior space does not have any partitions or other vertical interruptions. However, the grand space had to be articulated by movable panels, placed in clusters, or dispersed in space. Particularly, the first transformation in 1974 was performed by the director of the museum, P. Hulton, to create "a spatial structure that resembles a city, with interlocking spaces, squares, paths and dead-ends" (p. 253, [5]). In 1985, the second transformation was introduced by G. Aulenti, with the idea of placing "the longitudinal and the transversal axes" across the floor plan and creating "room-like spaces that referred to the spatial conditions where the art of the first half of the twentieth century was conceived and to the domestic setting of private collections" (p. 256, [5]). One of the fascinating things from the transformations is that the long corridor leads us to such diverse urban-like elements. This space plays a role in providing access to gallery spaces, integrating spatial movements, and overlooks the outside of the building. More interestingly, the gallery spaces are grouped in several distinct ways and, as a result, substantial auxiliary paths are formed, although the movable panels are regularly placed.

Lastly, the Moderna Museet, designed by Rafael Moneo and opened in 1998, is located on the island of Skeppsholmen in Stockholm. The museum brings together different collections in the building, such as Swedish artworks, the traditional avant-garde pieces, and the outstanding works of the classical modern period and later on of the New York school. One of the interesting characteristics is found in its spatial layout: an over 100-m-long corridor forms "the spine of the whole complex" and provides access to "the three clusters of museum galleries" and a temporary exhibition room near the main entrance

(p. 69, [19]). This corridor plays a pivotal role in giving visitors a chance of choosing one of the clusters and in comprehending the spatial structure as a whole. The other particular characteristic is the clusters. There are three clusters, and each one is composed of six to seven gallery rooms that vary in size. Interestingly, all clusters have dead-end small gallery rooms similar to cabinets, so that it is possible for visitors to gain distinct experiences, not only through the serial gallery rooms, but also the clusters.

As we have discussed, these museums are understood to be representative of linearly structured museums: The Uffizi Gallery has a relatively long history as an art museum, and the outer edges in particular are irregular in form because of the adjacent buildings, but the loggia acts as an integral gathering space and provides access to the entire gallery space. We assumed that the spatial layout of the Alte Pinakothek might have been influenced by that of the Uffizi Gallery, as it has the same spatial structure, meaning that the loggia controls the overall accessibility to large gallery halls, but the halls, conversely, provide additional spatial sequences, leading to the cabinets located in the deepest part. The Museum of Modern Art in the Centre Pompidou, originally designed as an open multipurpose space, has been transformed into subdivided gallery spaces by placing partitions, but one of the interesting things is that the corridor, like a loggia, controls all access to the subdivided spaces and, unexpectedly, some extra sequences are discovered. The Moderna Museet also has a corridor giving access to three clusters, but the clusters do not lead to other spatial sequences.

4. Syntactic Analyses and Discussions

In this chapter, the cases have been explored quantitatively by space syntax techniques such as space types, spatial sequence and choice, isovist geometric features, and isovist syntactic properties. Then, possible trails derived from the cases have been theoretically examined to understand in what way the spatial configurations work distinctly and to what degree they are different.

4.1. Space Types, Walking Sequence, and Walking Choice

Table 1 shows the basic profiles of the four galleries. The Uffizi Gallery (UG) is composed of 53 convex spaces in total, and 41 are gallery halls (Figure 3). Regarding space type, the most prominent one is c-space (ratio: 0.55), and the others are a-space (0.19), d-space (0.17), and b-space (0.09) in sequence. In the Alte Pinakothek (AP), there are 39 convex spaces in total, and most of them (37 out of 39) are used as gallery halls. Concerning space type, the ratio of c-space (0.46) is very close to the ratio of d-space (0.44), and the others are a-space (0.08) and b-space (0.03). The Museum of Modern Art in the Centre Pompidou (CP) has a total of 67 convex spaces, which are defined, and 44 of them are designed for gallery space. Similar to the AP, both c-space and d-space are salient, 0.39 and 0.30, respectively, and the others are a-space (0.28) and b-space (0.03). Finally, in the Moderna Museet (MM), there are 36 convex spaces in total, and 20 spaces are used for exhibiting artworks. In this layout, the ratio of c-space is 0.44; a-space is 0.25; d-space is 0.17; and b-space is 0.14.

Let us look into where these spaces are placed across the plans. In the UG, except for the ancillary facilities, d-spaces are discovered in the loggia (e.g., Nos. 1 and 25) and five gallery spaces (e.g., Nos. 17, 18, 2, 4, and 5). The others are a-/b-/c-spaces. Interestingly, some a- spaces (e.g., Nos. 26, 30, 31, and 38) are directly connected to the loggia. In the AP, the spaces placed in two southern strips (i.e., the loggia and the large gallery halls) are d-spaces. By contrast, the northern strip has mostly c-spaces. Only the two wings on both sides marginally have a-/b-spaces. Unlike the case of the UG, a-spaces are placed in the most remote area from the loggia. In the CP, spaces are arranged similarly: the main corridor (i.e., No. 22) is a d-space; the gallery spaces located on the southern side are c-/d-spaces; similar to the AP, a-spaces are placed along the southern edge of the plan, whereas some a-spaces on the northern side have a direct connection to the main corridor. The MM seems to be similar in that the main corridor (i.e., No. 7), comprising three convex spaces (i.e., Nos. 7, 17, and 25), is defined as a d-space; the clusters are combined with mainly c-spaces and occasionally a-spaces and a d-space.

Table 1. Summary of basic profiles, space types, c-sequence length, c-sequence depth, and d-ring of the cases.

	Uffizi Gallery	Alte Pinakothek	Centre Pompidou	Moderna Museet
Total no. of galleries	44	37	46	19
Total no. of convex spaces	53	39	67	36
Total no. of a-space (ratio)	10 (0.19)	3 (0.08)	19 (0.28)	9 (0.25)
Total no. of b-space (ratio)	5 (0.09)	1 (0.03)	2 (0.03)	5 (0.14)
Total no. of c-space (ratio)	29 (0.55)	18 (0.46)	26 (0.39)	16 (0.44)
Total no. of d-space (ratio)	9 (0.17)	17 (0.44)	20 (0.30)	6 (0.17)
c-/d-space ratio	3.22	1.06	1.30	2.67
c-sequence length (T)	185	38	46	62
c-sequence length (M)	6.37	2.11	1.76	3.87
c-sequence depth (T)	107	28	36	39
c-sequence depth (M)	3.68	1.55	1.38	2.43
d-ring (T)	17	54	57	14
d-ring (M)	2.42	3.17	2.85	2.33

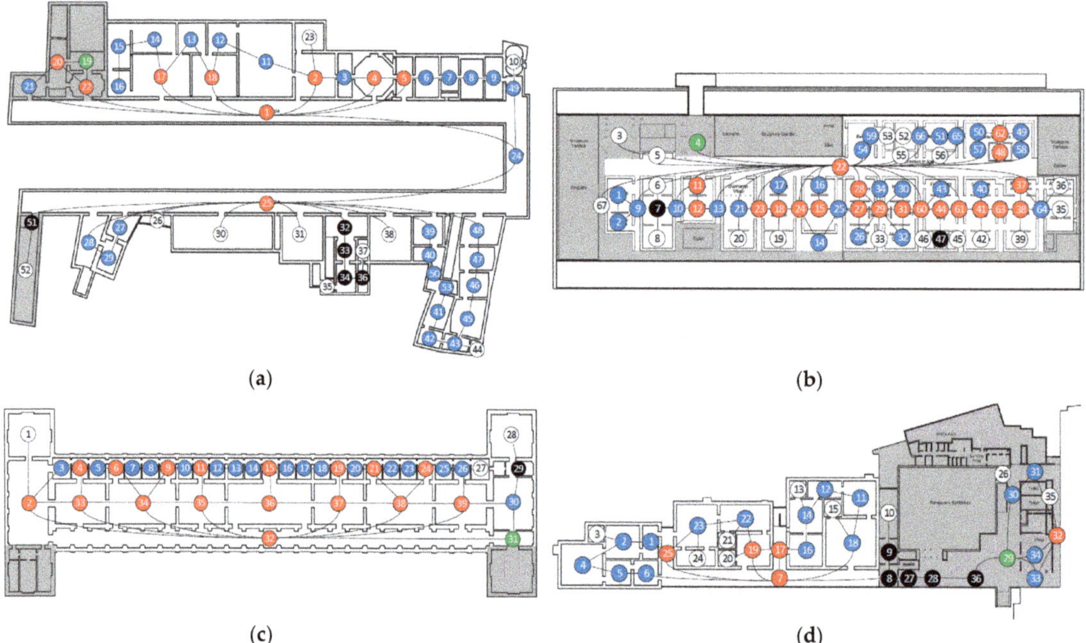

Figure 3. Space type analysis of the cases, and numbers indicate convex spaces: (**a**) Uffizi Gallery; (**b**) Alte Pinakothek; (**c**) Centre Pompidou; (**d**) Moderna Museet.

Considering these basic profiles and arrangements of space type across the plans, we go further and look into the ratios of c-space and d-space, which are simply obtained by dividing the ratio (or the total number) of c-space by that of d-space. In Table 1, the UG has the highest ratio: 3.22, whereas the AP has the lowest ratio: 1.06. Regarding walking sequence and choice, as expected, two extremes can be defined from the cases: the values of c-sequence length (M) and c-sequence depth (M) in the UG (6.37 and 3.68, respectively) and the MM (3.87 and 2.43) are higher than those in the AP (2.11 and 1.55, respectively) and CP (1.76 and 1.38). Particularly, the UG's length of the spatial sequence is two to three times longer than the others, and its depth is most three times deeper.

This tendency appears in a contrasting way on spatial choice: the value of the d-ring (M) of the AP (3.17) and the CP (2.85) are higher than those of the UG (2.42) and the MM (2.33). However, this result is not quite as great as that of the spatial sequence. This means that although the UG seems to have a strong linear spatial structure, it provides a small number of auxiliary paths starting from gallery spaces Nos. 17, 18, 2, 4, and 5.

From these results, it could be argued that gathering spaces such as loggias and corridors play an essential role in constituting the main sequences throughout spatial layouts. However, the two cases of the AP and the CP show distinctive features in that the halls, placed one step deeper from loggia and corridor, act as extra gathering spaces. This means that they work as key spaces for auxiliary paths at a local level within the main sequence. This characteristic has an impact on those two cases. However, the gathering spaces in both the UG and the MM work as the most powerful spaces constituting and controlling the whole sequences.

Unlike the cases of the AP, the CP, and the MM, the gathering spaces in the UG are connected directly to all different space types: for instance, No. 1 connects to c-spaces and d-spaces, and No. 25 connects to a-spaces, b-spaces, and c-spaces. Such connectivity to various space types leads us to unexpected walking sequences: sometimes the sequence ends in a single convex space (e.g., 25-30, 25-31, or 25-38), it lies on a relatively long sequence (e.g., 25-48-47-46-45-43-42-41-53-50-40-39-25), it provides an auxiliary path (e.g., 1-2-11-13-18-1 or 1-2-3-4-1), or it creates a backward sequence (e.g., 25-32-33-34-35-34-33-32-25). This can be discovered in the case of the CP, but it does not extend to the whole spatial layout. Thus, it can be argued that although the spatial layout of the UG seems to work conservatively due to the high value in c-sequence length (M), it provides diverse spatial experiences such as new paths or strongly ordered paths, dead ends, and enclosed spaces.

4.2. Visibility Graph Analysis

4.2.1. Visibility Graph Analysis in Terms of Geometric Measures

To what degree, then, are those spatial experiences regarding spatial sequence and choice differentiated from each other? To answer this question, we focus on looking into the geometric and syntactic results of VGAs (Table 2).

As one of its geometric properties, the UG has the highest value of isovist areas (mean value of 395.65), whereas the CP has the lowest (296.20). Looking at the result of the isovist perimeter, however, the UG has the lowest value (mean value of 160.63), whereas the AP has the highest (190.75). The reason that the UG has contrasting results concerning isovist area and perimeter can be understood by noting that the u-shaped loggia, compared with the gallery halls, is composed of three huge convex spaces and a great number of viewpoints can be constructed. This means that the viewpoints in such simple and large convex spaces are easily visible. However, when looking at the isovist perimeter, one of the significant points of difference between the UG and the AP is the fact that cross-visibility in the UG hardly takes place between the loggia and the gallery halls and between the gallery halls, but in the latter case, it occurs quite easily, not only across the three parallel strips but also between the gallery halls, especially when the longest perimeters of the isovist polygon are measured along with the straight doorways in the strips.

Next, concerning the compactness of the isovist, as expected, there is a strong pattern in terms of shape: the a-/b-space is simpler and more symmetrical because its isovist polygon forms in a compact way, whereas the c-/d-space is more complex and asymmetrical because it has at least three links or, at most, a huge number of links, and these links make the isovist shape more complex, similar to a star shape. Thus, the compactness value of the isovist in the case of the UG (mean value of 0.25), where the a-/b-space is adjacent to the gathering space is higher than the others. Conversely, the AP and the CP, in which it is placed far from the gathering space, have a low mean isovist compactness value (both mean values of 0.17).

Table 2. Syntactical result of VGAs of the cases.

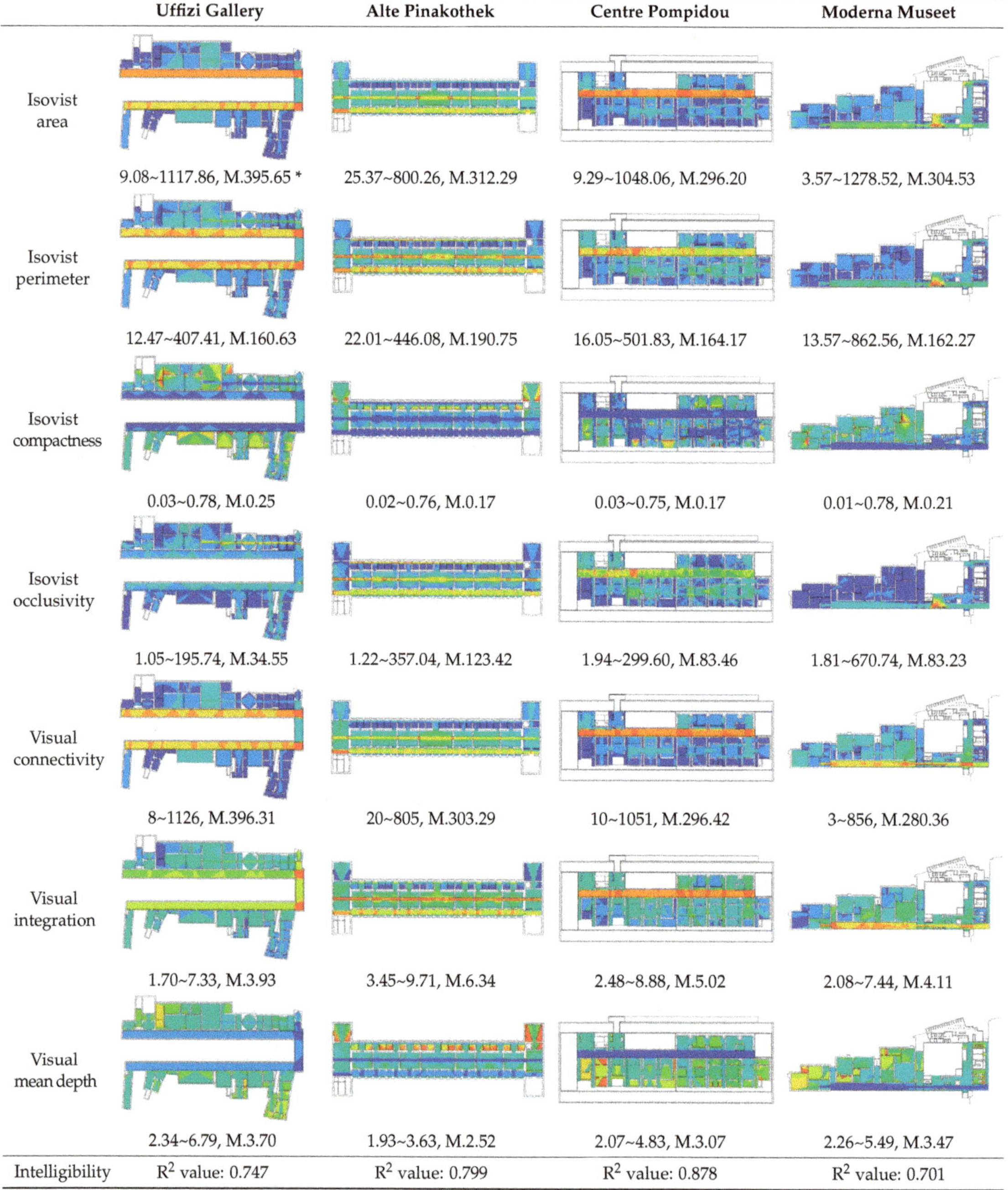

	Uffizi Gallery	Alte Pinakothek	Centre Pompidou	Moderna Museet
Isovist area	9.08~1117.86, M.395.65 *	25.37~800.26, M.312.29	9.29~1048.06, M.296.20	3.57~1278.52, M.304.53
Isovist perimeter	12.47~407.41, M.160.63	22.01~446.08, M.190.75	16.05~501.83, M.164.17	13.57~862.56, M.162.27
Isovist compactness	0.03~0.78, M.0.25	0.02~0.76, M.0.17	0.03~0.75, M.0.17	0.01~0.78, M.0.21
Isovist occlusivity	1.05~195.74, M.34.55	1.22~357.04, M.123.42	1.94~299.60, M.83.46	1.81~670.74, M.83.23
Visual connectivity	8~1126, M.396.31	20~805, M.303.29	10~1051, M.296.42	3~856, M.280.36
Visual integration	1.70~7.33, M.3.93	3.45~9.71, M.6.34	2.48~8.88, M.5.02	2.08~7.44, M.4.11
Visual mean depth	2.34~6.79, M.3.70	1.93~3.63, M.2.52	2.07~4.83, M.3.07	2.26~5.49, M.3.47
Intelligibility	R^2 value: 0.747	R^2 value: 0.799	R^2 value: 0.878	R^2 value: 0.701

* 9.08~1117.86, M.395.65: the range represents the minimum value (9.08) and the maximum (1117.86), and M stands for the mean value (395.65).

It might be said that the isovist occlusivity is one of the most fascinating measures in the geometric analysis because, as mentioned before, it tells us how previously unseen space, covered by any environmental elements, is discovered during movement ([18]). From the occlusivity analysis, it is quite obvious that the UG has the lowest value (mean value of 34.55), but the AP has the highest (123.42). The others, the CP (83.46) and the MM (83.23), are positioned in the middle of them. However, when we look closely at the result again, it is quite surprising that occlusive patterns are identified in the UG; for instance, the occlusive edges form in different ways: some of them at around Nos. 17 and 18 are configured diagonally; some galleries, Nos. 2, 3, 4, 5, 6, 7, 8, and 9, construct the occlusive edges linearly; or galleries such as Nos. 39, 40, 50, 53, 41, 42, 48, 47, 46, 45, and 43 have edges perpendicular to the loggia. Conversely, in the CP, the edges, particularly in the room-like spaces, are formed diagonally. However, the result of the MM is comparatively different from that of the CP, although the average value of the isovist occlusivity is quite similar to that of the CP. This is because the highest values are primarily found in the main corridor of the MM, but the occlusive edges do not extend to the gallery halls. This means that it is hard to expect dramatic visual changes in the clusters.

4.2.2. Visibility Graph Analysis in Terms of Syntactic Measures

Now we move on to the syntactic analysis of the VGA. In terms of connectivity as a local property, the UG shows, comparatively, the greatest mean value (396.31), whereas the MM has the least mean value (280.36). Regarding integration as a global feature, the UG has the lowest mean value (9.93), whereas the AP has the highest (6.34). From these two measures, it should be pointed out that a significant visibility shift is discovered in the case of the AP: the highly visible viewpoints at the local level are found in the loggia at the local level; however, the highly integrated viewpoints at the global level are in the large gallery halls in the middle strip. This means that the loggia in the UG works very well locally, but it does not make visitors comprehend the whole system globally. Conversely, the long lines of sight, which derive from the aligned doorways of the large gallery halls in the AP, make the whole spatial system more comprehensible and readable, although the loggia is intentionally designed to provide access to the gallery halls. In the other cases, concerning the CP and the MM, the main corridors work globally and locally as well.

Regarding mean depth, it is found that the UG has the highest mean depth value (a mean value of 3.70 and a maximum value of 6.79), but the AP has the lowest one (a mean value of 2.52 and a maximum value of 3.63).

4.2.3. Correlation between Space Types, Walking Sequence and Geometric Measures

Is there any possible correlation between the result of space types and the VGAs? If there is, what kinds of space types are strongly correlated with the geometric and syntactic properties? For this, a correlation analysis was performed. The data used in the correlation analysis are the basic profiles of the cases, such as the ratio of a-/b-/c-/d-space, the c-/d-space ratio, c-sequence length (M), c-sequence depth (M), and d-ring (M) defined in Table 1. Geometric measures of VGA analysis such as the mean value of the isovist area, perimeter, compactness, and occlusivity, and syntactic values of VGA, such as the mean value of the visual connectivity, integration, and mean depth are described in Table 2. It should be noted that the reason for carrying out the correlation analysis is to explore the relationships between space types, c-sequence, d-ring, geometric attributes, and syntactic features of the cases, and statistically explain how much the cases are different.

In Table 3, strong relationships can be seen between spatial sequences, choices, and the geometric measures of isovists. The isovist area is related to the values regarding c-space: for example, the r^2 value of the area to c-space ratio is 0.875, to a c-sequence length (M) of 0.811, and a c-sequence depth (M) of 0.811. The isovist compactness is even more strongly related to these values: the correlation of the compactness with a c-/d-space ratio, which has an r^2 value of 0.983, with a c-sequence length (M) of 0.999, and with a c-sequence depth (M) of 0.999.

Table 3. Correlation matrix between space types, c-sequence length, c-sequence depth, d-ring, and VGA.

	a-Space Ratio	b-Space Ratio	c-Space Ratio	d-Space Ratio	c-/d-Space Ratio	c-Sequence Length (M)	c-Sequence Depth (M)	d-Ring (M)
Mean Isovist area	0.000 *	0.071	0.875 **	−0.298	0.723	0.811 **	0.811 **	−0.172
Mean Isovist perimeter	−0.875 **	−0.383	0.008	0.813	−0.432	−0.288	−0.289	0.735
Mean Isovist compactness	0.051	0.454	0.757	−0.687	0.983 ****	0.999 ****	0.999 ****	−0.588
Mean Isovist occlusivity	−0.380	−0.248	−0.341	0.775	−0.812	−0.724	−0.726	0.580
Mean Visual connectivity	0.000	0.009	0.746	−0.208	0.581	0.658	0.659	−0.087
Mean Visual integration	−0.584	−0.661	−0.129	0.990 ****	−0.717	−0.573	−0.574	0.947 ***
Mean Visual mean depth	0.500	0.609	0.218	−0.991 ****	0.817	0.685	0.687	−0.910 ***

* This is r^2 value, and "−" means that there is a negative correlation between the two variables. ** $p < 0.1$, *** $p < 0.05$, **** $p < 0.01$.

However, the other two geometric measures are opposed: in terms of the perimeter, they are correlated with d-space ratio and d-ring (M), r^2 values of 0.757 and 0.735, respectively. Moreover, the isovist occlusivity is likely correlated with a d-space ratio with an r^2 value of 0.775, but there are negatively related with the occlusivity and the c-/d-space ratio, with an r^2 value of −0.812, a c-sequence length (M) of −0.724, and a c-sequence depth (M) of −0.726.

From those contrasting correlation results, the higher the ratio of c-space and c-/d-space, the greater the value of isovist area and compactness are. At the same time, the higher the ratio of d-space, the greater the value of isovist perimeter and occlusivity will be. Thus, a spatial layout where d-spaces are mainly occupied means that we are able to achieve not only open spatial relationships, but also rich cross-visibility through spaces that are particularly spiky, allowing "glimpses of other spaces past occluding surfaces" (p. 158, [19]).

These analyses once again strongly support the contention that both c-space and d-space play a decisive role in determining in what way spaces are related to each other, to what degree they are differentiated from one another, how strong or weak the spatial sequence is, and how distinctive spatial configurations are experienced globally or locally.

4.2.4. Correlation between Space Types, Spatial Sequence and Syntactic Measures

When we correlate syntactic values (e.g., visual connectivity, integration, and mean depth) with the types of space (Table 3), it is quite clear that the spatial layout of the UG, representing the highest ratio of c-space and the lowest of d-space, has a strong relationship between the visual connectivity and c-space ratio, c-sequence length (M), and c-sequence depth (M) with r^2 values of 0.746, 0.658, and 0.659, respectively. By contrast, in the case of the AP, described by the highest ratio of d-space and the lowest c-/d-space ratio, there are strong relations between the visual integration and the d-space ratio (r^2 value of 0.990) and the d-ring (0.947). There are meaningful correlations between the integration and the spatial sequence, but these are negatively related to c-sequence length (M), with an r^2 value of −0.573, and to c-sequence depth (M), with that of −0.574.

Another strong relationship exists between the visual mean depth and the values regarding d-space: c-/d-space ratio with an r^2 value of 0.817, c-sequence length (M) with 0.685, and c-sequence depth (M) of 0.687. Interestingly, the value of the c-space ratio is not likely to be related to the mean depth. To the values regarding d-space, the relationships are negatively related to the visual mean depth: −0.991 for the d-space ratio and −0.910 for the d-ring (M).

These syntactic measures, contrary to the geometric ones, are not properties of an individual isovist, but they are properties of the isovist's relation to all other isovists in the spatial system, so that the syntactic measures explain the overall structure of the building. From this conception, it can be suggested that c-space creates locally rich spatial sequences but leads to a less integrated spatial system; d-space, however, provides fewer spatial sequences but allows generating different movements, and consequently, it minimizes depths and makes the spatial system more integrated.

4.3. Examining Possible Paths

As mentioned before, possible trails based on the conception of walking sequence and choice were suggested to determine in what way the galleries are differentiated from each other and how distinctive spatial configurations are experienced. For this, possible trails were generated, starting from and ending at one of the gathering d-spaces allowing the possibility of different paths.

In the UG, a total of 13 trails are defined (Table 4) and they can be divided into two groups. The first group is the one that the trail starts from and ends at convex space No. 1: for example, trail 1 consists of a total of six convex spaces, excluding gathering convex space No. 1, with the string of (1)-18-12-11-2-23-2-(1); Trail 2 consists of three convex spaces, (1)-2-23-2-3-4-(1); trail 3 consists of two convex spaces, (1)-4-5-(1); trail 4 consists of five convex spaces, (1)-5-6-8-9-49-24-(1); trail 5 consists of three convex spaces, (1)-18-13-17-(1); and trail 6 consists of five convex spaces, (1)-17-14-15-16-17-(1). At the same time, the second group is the trails that set out from convex space No. 25, running parallel to convex space No. 1. Trail 7 is composed of 13 convex spaces with the string of (25)-48-47-46-45-43-44-43-42-41-53-50-40-39-(25); trail 8 consists of one convex space, (25)-38-(25); trail 9 consists of 11 convex spaces, (25)-32-33-34-36-37-36-34-35-34-33-32-(25); trail 10 is composed of one convex space, (25)-31-(25); trail 11 consists of one convex space, (25)-30-(25); trail 12 also consists of one convex space, (25)-26-(25); and trail 13 is composed of three convex space, (25)-27-29-28-(25). What is interesting in these defined possible trails in the case of the UG is the fact that convex space No. 1 leads to more or less moderate lengthy trails, but convex space No. 25 allows for extreme spatial sequences: simple experiences vs. lengthy and complex ones.

In the case of the AP, a total of 17 possible trails can be identified (Table 5). However, it should be noted that there is a significant difference between them. Having seen the trails, the first group of trails, Nos. 1–8, form one of the main sequences, meaning that the trail starts and ends in the gathering space so that these trails form the main sequences. By contrast, the other trails, such as Nos. 9–17, begin at one of the large gallery halls, such as convex spaces Nos. 33, 34, 35, 36, 37, and 38, instead of the gathering one, forming auxiliary paths. One of the distinctive characteristics of the main sequences compared with the auxiliary paths is the length of the spatial sequence.

Regarding the main sequences, for instance, the trails, except for Nos. 1 and 8, comprise two convex spaces. The others represent diverse stories: for example, trail 9 begins at one of the large gallery halls (i.e., convex space No. 33), passes through two cabinets (i.e., convex spaces Nos. 4 and 3), and then ends at the start; trails 10, 11, 12, 15, and 16 show identical sequences, starting from the large gallery halls, going through three cabinets and one large gallery hall, and ending at the start; trails 13 and 14, similarly, begin at one of the large halls, going through five cabinets, and returning to the starts; and trail 17 is somewhat different in that there is a retreat due to the dead-end a-space (i.e., convex space No. 27).

In the CP (Table 6), a total of 28 trails are produced. Similar to the case of the AP, the trails can be grouped according to whether the gathering space is included in spatial sequences or not. Hence, the first group is the one in which trails begin and end at the corridor, whereas the other is the one that begins at one of the d-spaces used for gallery halls. There is a total of 21 trails (i.e., trails 1–21) and all of them start from and end at the corridor through a series of gallery halls. In terms of the length of sequences, trails 1, 17, and 19 have the shortest length as they comprise a single convex space, except for the gathering space (i.e., convex space No. 22). By contrast, trails 3, 14, and 15 are explained comparatively as one of the longest trails. The others are sequenced in a series of two, three, four, or five convex spaces. Regarding auxiliary paths, a total of seven trails are defined. Interestingly, five of them are the ones that begin at the gallery spaces: for example, trail 22 starts from the convex space No. 41, continues through 42, 41, 40, and 63, and returns to the start; trails 24, 25, 26, and 27 begin and end at the convex space No. 29 via 26, 27, 30, 31, or 32. By contrast, the other two auxiliary paths trails (i.e., trails 23 and 28) start from and terminate in secondary corridors such as convex space Nos. 24 and 60, passing through galleries such as 14, 15, 25, 43, 44, 45, 46, or 47.

Table 4. Possible Trails and syntactic properties in the Uffizi Gallery.

No.	Possible Trails Total No. Convex Space and Trails	Mean Isovist Area	Mean Isovist Compactness	Mean Isovist Occlusivity	Mean Isovist Perimeter	Mean Isovist Mean Depth	Mean Isovist Area/Perimeter
1	6 {(1)-18-12-11-2-23-2-(1)} *	184.424	0.306	43.783	97.146	3.950	1.898
2	5 {(1)-2-23-2-3-4-(1)}	138.731	0.256	53.464	99.068	3.854	1.400
3	2 {(1)-4-5-(1)}	124.332	0.144	64.276	106.093	3.672	1.172
4	7 {(1)-5-6-7-8-9-49-(24)-(1)}	123.128	0.237	48.320	93.829	3.746	1.312
5	3 {(1)-18-13-17-(1)}	177.072	0.252	44.482	97.657	3.942	1.813
6	5 {(1)-17-14-15-16-17-(1)}	155.240	0.368	24.802	77.784	4.643	1.996
7	13 {(25)-48-47-46-45-43-44-43-42-41-53-50-40-39-(25)}	81.416	0.265	33.452	68.338	4.527	1.191
8	1 {(25)-38-(25)}	165.454	0.507	13.938	64.898	3.931	2.549
9	11 {(25)-32-33-34-36-37-36-34-35-34-33-32-(25)}	83.485	0.259	29.399	63.352	4.491	1.318
10	1 {(25)-31-(25)}	162.577	0.345	13.586	64.314	3.941	2.528
11	1 {(25)-30-(25)}	277.775	0.448	12.533	89.741	4.150	3.095
12	1 {(25)-26-(25)}	28.278	0.383	8.487	32.518	4.436	0.870
13	3 {(25)-27-29-28-(25)}	82.847	0.283	24.437	61.702	4.100	1.343

* The number in parentheses denotes the gathering space.

Table 5. Possible Trails and syntactic properties in the Alte Pinakothek.

No.	Possible Trails Total No. Convex Space & Trails	Mean Isovist Area	Mean Isovist Compactness	Mean Isovist Occlusivity	Mean Isovist Perimeter	Mean Isovist Mean Depth	Mean Isovist Area/Perimeter
1	4 {(32)-2-1-2-33-(32)} *	320.477	0.229	91.663	162.043	2.526	1.978
2	2 {(32)-33-34-(32)}	327.344	0.137	127.293	197.413	2.392	1.658
3	2 {(32)-34-35-(32)}	342.466	0.130	130.830	203.353	2.378	1.684
4	2 {(32)-35-36-(32)}	372.691	0.135	130.380	207.700	2.362	1.794
5	2 {(32)-36-37-(32)}	373.377	0.134	130.284	207.814	2.363	1.797
6	2 {(32)-37-38-(32)}	342.208	0.131	128.211	200.803	2.385	1.704
7	2 {(32)-38-39-(32)}	326.336	0.134	125.862	195.941	2.397	1.665
8	7 {(32)-39-30-29-28-29-30-31-(32)}	231.182	0.190	87.715	142.658	2.749	1.621
9	5 {33-4-3-2-33}	241.830	0.189	102.560	159.874	2.537	1.513
10	6 {33-4-5-6-34-33}	205.665	0.189	103.190	154.938	2.571	1.327
11	6 {34-6-7-8-9-34}	175.496	0.205	96.696	144.153	2.731	1.217
12	6 {34-9-10-11-35-34}	221.242	0.173	108.717	162.939	2.572	1.358
13	8 {35-11-12-13-14-15-36-35}	194.954	0.216	96.430	146.185	2.717	1.334
14	8 {36-15-16-17-18-19-37-36}	206.165	0.220	94.844	146.168	2.712	1.410
15	6 {37-19-20-21-38-37}	213.755	0.179	99.519	152.814	2.599	1.399
16	6 {38-21-22-23-24-38}	173.043	0.208	88.780	135.930	2.739	1.273
17	8 {38-24-25-26-27-26-39-38}	183.395	0.207	91.480	139.806	2.652	1.312

* The number in parentheses denotes the gathering space.

Table 6. Possible trails and syntactic properties in the Centre Pompidou.

No.	Possible Trails Total No. Convex Space & Trails	Mean Isovist Area	Mean Isovist Compactness	Mean Isovist Occlusivity	Mean Isovist Perimeter	Mean Isovist Mean Depth	Mean Isovist Area/Perimeter
1	3 {(22)-5-(22)} *	442.629	0.118	146.915	236.106	2.575	1.875
2	6 {(22)-9-1-2-9-(22)}	86.620	0.239	29.004	78.693	3.518	1.101
3	9 {(22)-10-7-6-7-8-7-10-(22)}	139.332	0.230	44.027	96.272	3.757	1.447
4	5 {(22)-10-12-11-(22)}	147.845	0.152	61.258	117.467	3.152	1.259
5	5 {(22)-11-12-13-(22)}	137.739	0.157	58.688	109.965	3.153	1.253
6	5 {(22)-21-20-21-(22)}	175.077	0.167	94.944	148.997	3.180	1.175
7	4 {(22)-21-23-(22)}	165.909	0.106	89.801	148.357	2.997	1.118
8	7 {(22)-23-18-19-18-17-(22)}	189.958	0.173	67.430	123.693	3.121	1.536
9	7 {(22)-17-18-19-18-24-(22)}	194.159	0.174	67.573	123.785	3.114	1.569
10	5 {(22)-24-15-16-(22)}	150.087	0.194	51.465	102.748	3.146	1.461
11	5 {(22)-28-29-34-(22)}	175.458	0.107	96.741	159.355	2.882	1.101
12	6 {(22)-34-29-31-60-(22)}	157.140	0.081	98.405	162.004	2.966	0.970

Table 6. *Cont.*

No.	Possible Trails Total No. Convex Space & Trails	Mean Isovist Area	Mean Isovist Compactness	Mean Isovist Occlusivity	Mean Isovist Perimeter	Mean Isovist Mean Depth	Mean Isovist Area/Perimeter
13	5 {(22)-60-44-61-(22)}	152.633	0.089	83.002	151.392	2.902	1.008
14	11 {(22)-61-41-42-41-63-38-39-38-37-(22)}	166.860	0.129	73.635	133.404	3.124	1.251
15	9 {(22)-37-64-36-64-35-64-37-(22)}	126.038	0.168	49.342	101.004	3.371	1.248
16	6 {(22)-54-59-53-59-(22)}	190.239	0.198	56.905	120.668	2.947	1.577
17	3 {(22)-55-(22)}	282.852	0.119	111.027	182.959	2.670	1.546
18	7 {(22)-66-52-66-51-65-(22)}	160.305	0.245	31.184	100.166	3.116	1.600
19	3 {(22)-56-(22)}	259.916	0.148	89.203	158.071	2.736	1.644
20	6 {(22)-57-50-62-48-(22)}	155.880	0.225	46.563	93.950	2.943	1.659
21	6 {(22)-48-62-49-58-(22)}	171.173	0.198	56.676	103.769	3.020	1.650
22	6 {41-42-41-40-63-41}	131.890	0.121	66.978	120.484	3.372	1.095
23	10 {60-43-44-47-46-47-45-47-44-60}	104.656	0.188	54.090	98.602	3.368	1.061
24	4 {29-30-31-29}	162.821	0.114	97.823	158.945	3.189	1.024
25	4 {29-32-31-29}	164.968	0.108	100.209	161.647	3.125	1.021
26	4 {29-26-27-29}	144.819	0.131	87.483	144.907	3.211	0.999
27	4 {29-27-28-29}	167.253	0.114	94.463	155.451	3.060	1.076
28	5 {24-14-25-15-24}	122.173	0.209	40.588	92.401	3.483	1.322

* The number in parentheses denotes the gathering space.

Lastly, in the case of the MM, three possible trails are defined, and all of them are rooted in the gathering space (Table 7). Specifically, trail 1 is made up of seven gallery spaces, excluding the spaces used for the gathering purpose, with the string of (7)-6-5-4-2-3-2-1-(25)-(7). The others, trails 2 and 3, are composed of nine galleries with the strings of (7)-(25)-23-24-23-22-21-22-19-20-19-(7), and (7)-18-15-18-11-12-14-13-14-16-(17)-(1).

Table 7. Possible Trails and syntactic properties in the Moderna Museet.

No.	Possible Trails Total No. Convex Space & Trails	Mean Isovist Area	Mean Isovist Compactness	Mean Isovist Occlusivity	Mean Isovist Perimeter	Mean Isovist Mean Depth	Mean Isovist Area/Perimeter
1	10 {(7)-6-5-4-2-3-2-1-(25)-(7)} *	170.995	0.277	40.737	93.082	4.252	1.837
2	12 {(7)-(25)-23-24-23-22-21-22-19-20-19-(7)}	182.092	0.262	44.475	97.152	3.773	1.874
3	12 {(7)-18-15-18-11-12-14-13-14-16-(17)-(1)}	198.500	0.287	42.913	96.397	3.948	2.059

* The number in parentheses denotes the gathering space.

5. Discussion

From these results, it is unclear if the trails are somehow related to particular spatial sequences or choices. For instance, although a total of 13 trails in the UG are conceived, it has a low ratio in d-space (0.17) and the highest ratio of both c-space and c-/d-space (0.55 and 3.22, respectively). The AP has the highest ratio of d-space (0.44) with the lowest c-/d-space ratio (1.06), and a total of 17 trails are defined. The CP, however, has a somewhat lower d-space ratio (0.30) with a moderately lower ratio of c-/d-space (1.30), but at 28 trails, the greatest numbers among the cases are identified. Lastly, the MM has a low d-space ratio (0.17) with a comparatively higher ratio of c-/d-space (2.67), but it only has three trails.

Nevertheless, there is one significant finding from the possible trails. That is, the trails delineated in both the UG and the MM are strongly related to one of the gathering spaces (i.e., the loggia in the UG and the main corridor in the MM), so that all trails are main sequences, meaning that spatial sequences are closely associated with the gathering spaces. Hence, it can be said that the gathering space plays an essential role in providing variant spatial sequences: sometimes the sequence is very short and simple such as in trails 8, 10, 11, and 12 in the UG, and at other times, it is very long and complex such as in trails 7 and 9 in the UG and the trails in the MM.

In both the AP and the CP, however, trails are categorized into two groups, namely, main sequences and auxiliary paths, indicating that trails begin not at the gathering space but at one of the d-space gallery halls. For instance, more than half of the trails in the AP (i.e., nine out of 17) belong to auxiliary paths, therefore the large gallery halls running along the middle "strip" play an additional role in providing unexpected paths leading to cabinets, which are situated in an area of the gallery that is furthest away from the gathering space. Similarly, seven auxiliary paths in the CP allow extra journeys, and convex space No. 29, in particular, acts as a nodal point for providing diagonal experiences.

The conception of the main sequence and auxiliary paths is strengthened by the result of visual mean depth. In Table 2, the average value of visual mean depth in both the UG and the MM is relatively higher than that of the AP and the CP. This means that the d-space gallery halls, which are placed a step away from the gathering space, provide unforeseen journeys and also decrease the depth.

In what way, then, do they differ? To answer that question, a box and whisker plot was constructed to investigate the trail's differences in terms of the isovist area, isovist compactness, isovist occlusivity, isovist perimeter, isovist shape (area/perimeter), and visual mean depth (Figure 4). The findings are summarized as follows:

- Regarding the isovist area, it is certain that the values for the trails of the AP are significantly higher than the others, and they are also widely distributed: mean = 261.860, SD = 71.579, $p < 0.001$. By contrast, the values for the UG are lower than the others: mean = 137.289, SD = 59.827.
- In terms of the isovist compactness, the values for the trails of the UG (mean = 0.314, SD = 0.089, $p < 0.001$) are significantly higher than the others. By contrast, the values for the AP and the CP are lower. Hence, it can be explained that the shapes of isovists in the UG are comparatively simple and symmetrical, but in the cases of the AP and the CP, they are complex and asymmetrical.
- With the isovist occlusivity, the values for the trails of the AP (mean = 107.909, SD = 16.268, $p < 0.001$) are significantly higher than the others. However, when we look more closely into the values for the CP, it is quite interesting that the values of the trails are evenly distributed beyond the lower and upper quartiles. This means that the trails are strongly associated with the occlusive lines. Therefore, it can be said that, although the mean value of the CP is neither higher nor lower, the occlusivity is considered one of the key aspects in describing its configurational features.
- Considering the perimeter, it is definite that the values for the trails of the AP (mean = 168.267, SD = 26.175, $p < 0.001$) are significantly higher than the others. By contrast, the values for the UG are lower than the others.
- In terms of the isovist shape, defined with the division of the isovist area by the isovist perimeter, the values for the trails of the MM are significantly higher (mean = 1.924, SD = 0.097, $p < 0.001$), so it can be said that the isovist shape is likely rounded. By contrast, the values for the CP are lower (mean = 1.309, SD = 0.255, so the shape is more likely spiky.
- On the subject of the visual mean depth, the values for the trails of the UG are significantly higher (mean = 4.106, SD = 0.306, $p < 0.001$), but those for the AP are lower (mean = 2.552, SD = 0.144).

Taken together, it can be argued that from both geometric and syntactic perspectives, the trails of the UG are characterized as being symmetrical, compact, non-occlusive, rounded, non-cross-visible, and deep; the trails of the AP as asymmetrical, loose, occlusive, spiky, cross-visible, and shallow; the trails of the CP as asymmetrical, (partly) loose, (partly) occlusive, spiky, (partly) cross-visible, and shallow; and the trails of MM as symmetrical, compact, non-occlusive, rounded, non-cross-visible, and deep. It should be noted here that the characteristics of the CP's trails are somewhat vague because of a great range of the isovist area and the isovist occlusivity. Despite the relatively lower mean values of these two measures, the values are running from the lowest to the highest.

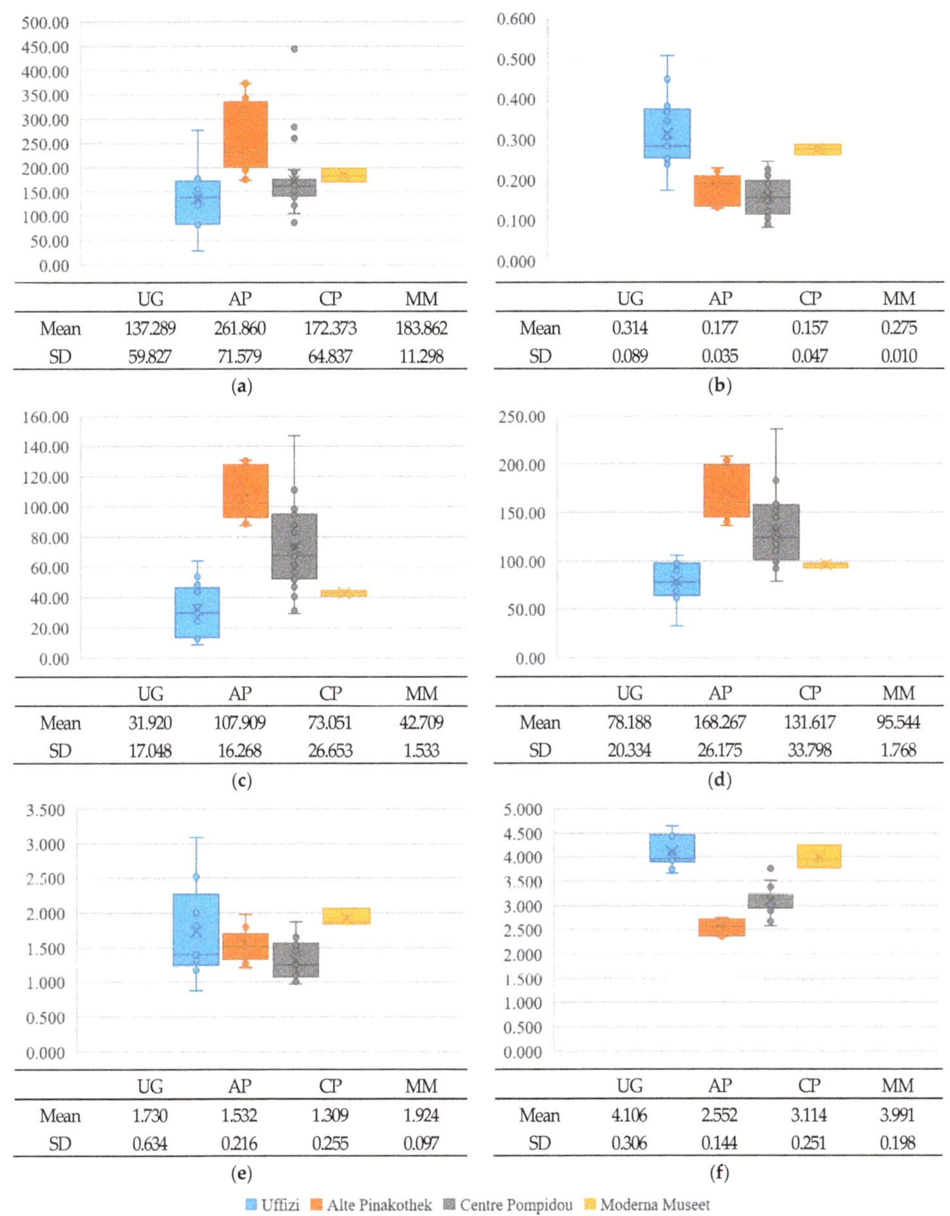

Figure 4. Box and whisker plot for the possible trails: (**a**) isovist area; (**b**) isovist compactness; (**c**) isovist occlusivity; (**d**) isovist perimeter; (**e**) isovist area/perimeter; (**f**) isovist mean depth.

6. Conclusions

This study aims to explore in what way linearly structured museums play a role in differentiating spatial configurations and spatial experiences and then to identify important underlying principles throughout with in-depth geometric and syntactic case studies, using space syntax techniques such as space type, spatial sequence and choice, isovist, and VGA.

Particularly, possible trails, derived from the conception of space type and spatial sequence, are used to find out in what way spatial configurations work.

Through an in-depth investigation of four well-known art museums, namely, the Uffizi Gallery, the Alte Pinakothek, the Centre Pompidou, and the Moderna Museet, we have reached the following conclusions.

First, spatial sequence, leading to structured movement patterns, is strongly correlated with the isovist area and compactness but hardly corresponded to the isovist perimeter and occlusivity. Due to compressed and shortened features, the collection of visible surfaces (i.e., isovist) are simple, rounded, symmetric, and hard to discern, therefore working conservatively. This was discovered primarily in the cases of the UG and, partly, the MM. By contrast, spatial choice, enabling us to create our paths throughout the layout of space, is significantly related to isovist perimeter and occlusivity but not to the isovist area and compactness. Due to the lengthy and omnipresent properties, the collection of visible surfaces were asymmetric, complex, spiky, and easily seen, therefore working in a generative way. This was seen in the cases of the AP and partly in the CP.

Second, the spatial sequence is specifically defined as two types of sequences: the main sequence is derived from gathering space, yet with the auxiliary paths created from one of the gallery halls. The former is strongly associated with the gathering space, and it is mainly developed in the case of the UG and the MM. Here, we can argue that gathering space plays two roles: it provides syntactic information regarding gathering properties such as connectivity, integration, and intelligibility, and also leads us to return regularly to the same place (i.e., gathering space) after completing several distinct trails (e.g., a long journey composed of a series of c-spaces, a very short one made of only a single a-space, or the ones between them). Therefore, this spatial sequence is strongly correlated with visual syntactic features.

At the same time, the auxiliary paths are seldom related to gathering space. This means that it is independent of the main axis. For instance, the large gallery halls placed along the second strip in the AP work in articulating naïve trails via cabinets, and the galleries, located at the second row from the main corridor in the CP, organize a small walk. Particularly, this journey is strongly concerned with visual geometric properties such as occlusivity and the area/perimeter. Particularly, one should note that occlusivity is one significant attributes making the auxiliary paths distinctive. The AP is characterized by high values of occlusivity. This means that the occluding radial surfaces are longer than the other cases, so that the paths reflect perceptual uncertainty, although its spatial structure is powerfully linear. Additionally, the CP has relatively high values of occlusivity, and the auxiliary paths are quite unpredictable, which is also due to the occluding radial surfaces. On the other hand, the UG features low values of occlusivity, meaning that the occluding radial surfaces are shorter than the others, and, therefore, it can be said that the paths are quite predictable, in spite of a number of auxiliary paths.

Finally, we should note that space itself works in distinct ways: without a doubt, the gathering space provides access to all spaces across the floor plan and plays a role in generating different walking sequences that are mainly composed of c-spaces. These sequences are easily accessible and recognizable. However, when we look into the cases where d-spaces are the prominent space type, certain gallery halls play the same role of generating spatial sequences (i.e., auxiliary paths). Therefore, they perform two distinct functions: occupation and local movement. More importantly, local movement is facilitated by occlusive edges.

From these findings, it can be said that art museums work with distinctly different spatial sequences, and particularly, the auxiliary paths generated from the gathering spaces generate spatially different experiences. Thus, it might be argued that it seems that art museums work in the same way, while gallery spaces are configured in distinct ways. This is the reason that spatial experiences are substantially different from one another.

Author Contributions: Conceptualization, J.H.L. and Y.S.K.; methodology, J.H.L.; validation, J.H.L.; formal analysis, J.H.L. and Y.S.K.; investigation, J.H.L.; data preparation, J.H.L.; writing—original draft preparation, J.H.L.; writing—review and editing, J.H.L. and Y.S.K.; visualization, J.H.L. All authors have read and agreed to the published version of the manuscript.

Funding: This research received no external funding.

Data Availability Statement: The data presented in this study are available on request from the corresponding author.

Conflicts of Interest: The authors declare no conflict of interest.

References

1. Macdonald, S. Collecting Practices. In *A Companion to Museum Studies*; Macdonald, S., Ed.; Wiley-Blackwell: West Sussex, UK, 2011; pp. 81–97.
2. Pevsner, N. *A History of Building Types*; Princeton University Press: Princeton, NJ, USA, 1976.
3. Brawne, M. *The New Museum: Architecture and Display*; Frederick A. Praeger: New York, NY, USA, 1965.
4. Brawne, M. *The Museum Interior: Temporary and Permanent Display Techniques*; Thames and Hudson: London, UK, 1982.
5. Tzortzi, K. The Interaction between Building Layout and Display Layout in Museums. Ph.D. Thesis, University College London, London, UK, 2007.
6. Huang, H. The Spatialization of Knowledge and Social Relationships: A study on the Spatial Types of the Modern Museum. In Proceedings of the 3rd International Space Syntax Symposium, Atlanta, GA, USA, 7–11 May 2001.
7. Duncan, C.; Wallach, A. The Museum of Modern Art as Late Capitalist Ritual: An Iconographic Analysis. *Marx. Perspect.* **1978**, *1*, 28–51.
8. Hillier, B.; Tzortzi, K. Space Syntax: The Language of Museum Space. In *A Companion to Museum Studies*; Macdonald, S., Ed.; Wiley-Blackwell: West Sussex, UK, 2011; pp. 282–301.
9. Choi, Y.K. The Morphology of Exploration and Encounter in Museum Layouts. *Environ. Plan. B Plan. Des.* **1999**, *26*, 241–250. [CrossRef]
10. Peponis, J.; Dalton, R.C.; Wineman, J.; Dalton, N. Measuring the Effects of Layout upon Visitor's Spatial Behaviours in Open Plan Exhibition Settings. *Environ. Plan. B Plan. Des.* **2004**, *31*, 454–473. [CrossRef]
11. Penn, A.; Martinez, M.; Lemlij, M. Structure, Agency, and Space in the Emergence of Organisational Culture. In Proceedings of the 6th International Space Syntax Symposium, Istanbul, Turkey, 12–15 June 2007.
12. Psarra, S.; Grajewski, T. Architecture, Narrative and Promenade in Bensen and Forsyth's Museum of Scotland. *Archit. Res. Q.* **2000**, *4*, 122–136.
13. Psarra, S.; Wineman, J.; Xu, Y.; Kaynar, I. Tracing the Modern: Space, Narrative and Exploration in the Museum of Modern Art, New York. In Proceedings of the 6th International Space Syntax Symposium, Istanbul, Turkey, 12–15 June 2007.
14. Salgamcioglu, M.E.; Cabadak, D. Permanent and Temporary Museum Spaces: A Study on Human Behavior and Spatial Organization Relationship in Refunctioned Warehouse Spaces of Karakoy, Istanbul. In Proceedings of the 11th International Space Syntax Symposium, Lisbon, Portugal, 3–7 July 2017.
15. Benedikt, M. To take hold of space: Isovists and isovist fields. *Environ. Plan. B Plan. Des.* **1979**, *6*, 47–65. [CrossRef]
16. Conroy, R. Spatial Navigation in Immersive Virtual Environment. Ph.D. Thesis, University College London, London, UK, 2001.
17. Hillier, B. *Space is the Machine*; Cambridge University Press: Cambridge, UK, 1996.
18. Lee, J.H. The Impact of Maps on Spatial Experience in Museum Architecture. Ph.D. Thesis, University College London, London, UK, 2014.
19. Mack, G. *Art Museums into the 21st Century*; Birkhäuser: Basel, Swizerland, 1999.

Article

A Study on Mega-Shelter Layout Planning Based on User Behavior

Young Ook Kim [1,*], Joo Young Kim [1], Ha Yoon Yum [1] and Jin Kyoung Lee [2]

[1] Department of Architecture, Sejong University, Seoul 05006, Korea
[2] Department of Architecture, Catholic Kwandong University, Gangneung 25601, Korea
* Correspondence: yokim@sejong.ac.kr

Abstract: We explore the spatial layouts of mega-shelters and suggest better spatial planning strategies. A mega-shelter for refugees contains multiple functions, such as dormitory, dining, medical, kitchen, storage, and community areas. Post-disaster refugees often suffer from PTSD that affects their mental health and spatial cognitive ability. The spatial configuration of a mega-shelter can accelerate their recovery by providing an environment that not only satisfies the basic needs, but one that can improve their spatial cognitive ability and promote a sense of community in this new, albeit temporary, small society. Four mega-shelters in the U.S., Australia, and Japan were analyzed using space syntax methods, specifically axial line analysis and visibility graph analysis (VGA), as well as justified graph analysis. The comparative analysis shows that while specific spatial layouts are different, all shelters were designed from a manager's perspective. The movements of the refugees were sometimes unnecessarily exposed to supervision and control, and community areas were often found in locations with low accessibility. By incorporating strategies such as siting community space in areas with high global integration values and adopting transition areas, mega-shelters can create an environment that can enhance the refugees' will to recover and rebuild by promoting communications with neighbors and various community activities.

Keywords: disaster; mega-shelter; spatial behavior; planning guideline; space syntax

Citation: Kim, Y.O.; Kim, J.Y.; Yum, H.Y.; Lee, J.K. A Study on Mega-Shelter Layout Planning Based on User Behavior. *Buildings* **2022**, *12*, 1630. https://doi.org/10.3390/buildings12101630

Academic Editors: Michael J Ostwald and Ju Hyun Lee

Received: 15 August 2022
Accepted: 3 October 2022
Published: 8 October 2022

Publisher's Note: MDPI stays neutral with regard to jurisdictional claims in published maps and institutional affiliations.

1. Introduction

Natural disasters such as typhoons, earthquakes, and floods and social disasters such as terrorism, wars, and fires occur frequently. In such cases, mega-shelters are built to protect the refugees in a safe environment. The term 'mega-shelter' was first used in the Mega-shelter Planning Guide [1] prepared by the International Association of Venue Managers (IAVM) and the American Red Cross (ARS). A mega-shelter is a type of congregate shelter, generally capable of accommodating 2000 to 25,000 people. Such a facility, usually set up in a very short time by the government, provides a safe and sanitary environment to temporarily accommodate refugees [1] (p. 10). Mega-shelters are usually set up during disasters and other emergency situations in large public facilities, such as sports stadiums, convention centers, schools, and churches [1] (p. 10). The spatial planning of mega-shelters needs to take the following two perspectives into account.

First, while a mega-shelter usually operates for less than a year, it contains all functions for everyday life. In addition to basic residential space for sleeping, dining, and sanitary (laundry room, restroom, and shower) uses, there are also spaces designated for community activities, education, child care, medical treatment, and management. Therefore, a mega-shelter is a complex space with a diverse range of functions, and the spatial planning of a mega-shelter is, in fact, creating a new social environment where refugees temporarily live. Given the relationship between spatial configuration and social behavior, which has been the subject of extensive research following the study by Hillier and Hanson [2], how to create a better social environment for refugees through spatial planning needs to be considered. Second, multiple studies have reported that individuals who experienced disasters suffer from PTSD (post-traumatic stress disorder) [3–7]. Many refugees become psychologically unstable during this time of uncertainly and often have a declined level of spatial cognitive ability because of stress [8]. PTSD influences human spatial behavior, and a high level of stress can reduce the level of spatial cognitive ability. In a shelter, refugees face an unfamiliar environment that confuses them further, and their spatial-use behavior is different from in normal times. Hence, the spatial planning of mega-shelters needs to take into account the fact that many refugees have reduced spatial cognitive ability.

Sanderson and Burnell [9] pointed out the two perspectives listed above and argued against focusing on only the supply of mega-shelters without reflecting the demands of the refugees. They stated that temporary residential facilities that do not consider shelter functions, supporting refugee communities, and other social, environmental perspectives can reduce the refugees' will to rebuild their lives and reduce such opportunities [9]. They emphasized that since temporary residential facilities influence the refugees' will to rebuild, the planning guideline for such a facility that incorporates key considerations was extremely important [9]. A review of recent research indicates that while spatial plans of shelters emphasize the importance of community facilities or study shelter layouts, an objective standard for allocating space has not been proposed [10,11]. In the same context, Zhang and Dong [12] stressed that based on Maslow's hierarchy of needs theory, a shelter needs to consider a wider range of perspectives in addition to the functional design focused on the demand for a sanitary and safe environment. Instead of a minimal, functional shelter that provides a sanitary and safe environment, they emphasized a more advanced design, one that is consumer-focused and human-centered [12].

A review of the literature on shelter planning uncovered that there have been a number of research studies on layouts of residential facilities, designs of residential units space, and planning guidelines [11,13–15]. For example, Hirata et al. [16] presented a checklist for the design and operation of shelters. These studies mostly focused on residential functions, and there are almost no studies that examined the spatial planning of an entire shelter—a multi-functional space with sanitary, medical, and other uses in addition to residential use. One study classified the layout of shelters with multiple functions [17] but did not consider the appropriateness of a layout based on human behavior. Most are post-occupancy evaluation studies [17,18] and almost none examined the characteristics of spatial behavior connected to the shelters' spatial configurations, or their relationship to the creation of a social environment. In addition, the shelter guidelines prepared by the U.S. Federal Emergency Management Agency (FEMA) [19], the Japanese Ministry of Education, Culture, Sports, Science, and Technology (MEXT) [20], and the Australian government [21] all focus on space size and functions from a management and operations perspective. These design guidelines also do not consider the spatial behavior of refugees, who are the shelter users.

In this context, the goal of this study is to examine the effect of the spatial configuration of a mega-shelter on users' spatial and social behaviors, and based on the results, provide an academic basis for preparing a mega-shelter planning guideline. Accordingly, first, the effect of spatial configuration on a shelter user's spatial behavior and social environment is studied, and second, spatial configurations of selected mega-shelters are analyzed. Finally, strategies to improve existing mega-shelter planning guidelines are proposed by considering the impact of spatial configuration on human behavior.

2. Literature Review

2.1. Behavior and Psychology after Experience of Disaster

Frequently, individuals who experienced disasters suffer from PTSD. Two points are especially relevant to this study.

First, the symptoms of post-disaster PTSD include chronic tiredness, loss of interest in everyday activities, anxiety, depression, and anger [3–5]. Hence, for disaster victims to recover from the traumatic event and restore their lives, substantial assistance and support is needed from the surrounding environment—for example, from neighbors and the local community. Many studies have reported that community support is an important force that helps refugees return to their normal lives [6,7,22,23]. In particular, Littleton et al. [22] showed that individuals who receive less support after experiencing disaster are more vulnerable to losses in community solidarity and may potentially suffer persistent PTSD symptoms. Second, the stress suffered by individuals who experienced disaster reduces their spatial cognitive abilities [8]. As a result, refugees travel on familiar routes within a shelter rather than using shortcuts. Tang et al. [24] argued that chronic tiredness and loss of interest in everyday activities can lead to physical inaction and increase the risk of suicide. These studies show that given the close relationship between spatial configuration and spatial behavior presented by Kim and Penn [25], a careful spatial planning of shelter facilities that incorporate the spatial behavior of refugees is very important.

2.2. Spatial Planning of Shelters

Several studies have been conducted on the spatial planning of temporary shelters. Biswas [26] showed that the planning and design method of post-disaster housing program affects the mid- to long-term recovery of residents. Nappi and Souza [10] focused on the potential formation of a community based on a user–environment relationship when planning a local shelter. They emphasized the need for a clear axis that can promote the bond between neighbors by expanding the bond of families—the smallest unit of a community. They also argue that various functions in a shelter need to be spatially aligned, and that a shelter should be designed so that users are able to perceive the entire structure, and that the visual information in regard to spatial configuration enables users to confirm his/her location and current direction. Similar research examined functional planning considerations when utilizing a school as a shelter, including operation, residential area, and minimizing movement circulation for medical treatment activities [17].

In addition to academic studies, mega-shelter planning guidelines prepared by public institutions in various countries also need to be reviewed. Standards for shelter space in the U.S. [19], Japan [20], and Australia [21] were examined. Based on these guidelines, mega-shelters contain 10 key function categories. They include entrance, registration, management, welfare, medical and mental health service, dormitory, sanitary, dining, community, and logistics. Table 1 shows guideline elements related to spatial planning. The guidelines list criteria for connectivity and access control of the space for each function.

Table 1. Mega-shelter planning guideline elements that influence spatial configuration.

Function	Use	Guideline
Entrance	Entrance	- There should be only one main entrance for better control of resident movements - Entrance should be located where the number of users is the largest, to control the flow of people
Registration	Registration Space	- Should be easy to find, close to the main entrance
	Information Center	- Information center should be close to control room so that updated information can be rapidly distributed to the residents
Management	General	- Staff-only corridor is needed for efficient movement
	Control Room	- Control room should be located at the center of communication, between inside and outside of the facility
	Staff Resting Area	- Staff resting area should be in a quiet location - Away from major traffic areas - Should be located close to the control center

Table 1. *Cont.*

Function	Use		Guideline
Welfare	Space for Vulnerable Groups	-	Should be located close to the entrance for good accessibility
Medical and Mental Health Services	General	- - - -	Should have easy access to outside (parking lot) for ambulances and medical supplies Medical service area should be close to the registration space Registration space manager should perform screening test with medical space manager and create a list of residents who need assistance Medical service area should be close to restrooms
	Counseling Room	-	The locations of medical and mental health counseling space should be planned to be away from residential space to protect privacy of the refugees
	Quarantine Facility	-	Quarantine space should be located away from residential/common areas to prevent spread of contagious diseases and to protect privacy of the patients
Dormitory	General	- - - - -	Of all uses in the shelter, requires the highest level of privacy Should be close to sanitary facilities in a low-traffic area Should be close to other amenities Should not be located to areas with high level of noise Should have internal corridor access
	Prayer/Reading Rooms, Workspace, etc.	-	Individual activity room should be located in low-accessibility locations to protect privacy
Sanitary	Public Restrooms	- -	Restrooms should be close to residential space and waiting space of the registration space Sanitary facilities close to residential space should be located in low-traffic areas
	Temporary Restrooms	- -	Outside space for installation should be reserved in advance Should be accessible by vehicles
Dining	Dining	-	For efficient distribution of meals, movement plan needs to be prepared for dining area
	Kitchen	-	Kitchen should be close to kitchen storage space
	Kitchen Storage	-	Kitchen storage should be located close to outside to enable convenient disposal of trash
Community	General	-	Should be located in an area that does not disturb shelter residents
	Community Recreation Room	- -	Should be located in a closed space for activities that require minimum level of exposure and access control Recreation activities should be monitored by managers
Logistics	Supply Storage	- -	Should be located close to outside for convenient delivery of materials and goods Controlled access and management are needed
	Logistics Center	-	Logistics center should be easily accessible by the residents and at the same time be in a secure space for distribution and storage

Review of both academic studies and shelter planning guidelines prepared by public institutions indicate that a human-centered design approach for users is required for mega-shelter planning. Rather than simply helping people survive by satisfying minimal sanitary and safety desires, a mega-shelter needs to enable users to overcome their trauma and restore psychological stability through the community that can form and develop in a shelter. There have been multiple studies on layouts of residential facilities in shelters. The United Nations High Commissioner for Refugees (UNHCR) [13], Park [14] and Kim et al. [27] proposed a form of modular planning for residential units when setting up a mass shelter. Jung et al. [15] lists key strategies for spatial allocation when using an existing public community facility as a temporary shelter, with the residential function occupying the central area connected to other functions horizontally and vertically. Lee et al. [11] proposed construction and spatial design guidelines for essential non-residential functions in long-term, large-scale temporary shelters. In a non-quantitative guideline, they suggested that a street network first be established along residential units, followed by allocation of space for other functions depending on street types. However, there have been almost no study that examined spatial layout of an entire shelter containing a diverse range of functions such as community facility, medical facility, and education/childcare space in a comprehensive manner. One study classified the layout of shelters with multiple functions [17] but did not provide an objective basis for the proposed layout. In addition, most government guidelines were prepared from a facility manager's perspective, focusing on controlling the access and activities of the refugees, as well as logistics. There is little guidance for spatial planning of the facility. Finally, there have been almost no studies that quantitatively examine the effects of a spatial configuration on users' spatial behavior and community formation using spatial syntax methodology.

3. Research Methodologies and Cases

3.1. Spatial Configuration Analyses of Shelters

To assess the validity of various uses in a mega-shelter, the characteristics of each space and suitability of spatial arrangement—which influence a user's behavior—need to be understood. To this end, spatial configurations of shelters were analyzed to examine the characteristics of each space, and based on the results, the suitability of the spatial arrangement was evaluated. Specifically, the perceptibility and accessibility for users of each space in a shelter were analyzed. This procedure reveals the overall layout of a shelter, and depending on the layout, the level of visibility, accessibility, and probability of use of each space from a user's perspective. Specifically, axial line analysis, visibility graph analysis (VGA), and justified graph were created using the space syntax method. For the four selected case studies, spatial topology and global integration values were assessed, as well as space syntactic properties. The results provide the quantitative basis for assessing the accessibility and probability of the use of each space within the mega-shelters.

3.1.1. Calculation of Integration Value Using Axial Graph Analysis and VGA Analysis

The space syntax method is used to study the relationship between spatial configuration and social phenomena based on space visibility. Spatial configuration affects movement patterns and encounters between residents, and such encounters in turn have a positive impact on the development of social relationships [28]. Integration, one of the space syntax properties, is closely correlated to encounters and interactions between people [29,30].

Global integration is an important indicator in space syntax theory that shows ease of accessibility from each space to all other spaces and is calculated in depth value. A high level of integration indicates a small depth value, whereas a low level of integration indicates a high depth value. A high global integration value of a specific space shows that accessibility is high from this space to all other spaces, and this may be used to analyze accessibility and assess suitable uses for each space in a shelter.

Global integration is calculated using 'Axial Line Analysis' and 'VGA' by depthmapX software (Version 0.8.0, Turner, A., 2020) based on space syntax theory. An axial map represents space based on visibility and is used to analyze the relationships between spaces. VGA was additionally conducted to further examine the spatial configuration characteristics of the shelters. It is based on the isovist theory proposed by Tandy [31]. To overcome the limitations of Tandy's isovist theory, Turner [32] proposed VGA methodology in which the visibility graph was replaced by the concept of visual depth in space syntax theory. Using VGA, the accessibility, visibility, and spatial configuration characteristics of all spaces in the study area can be examined.

3.1.2. Analysis of Spatial Depth and Hierarchy Using Justified Graph

Justified graph analysis is a method to study the relationship between a given space and other spaces by examining how they are connected. A justified graph is an expression of relationships of spaces from a specific point topologically, and spatial hierarchy and relatively depth from the given point can be assessed.

As seen in Figure 1, the justified graphs of the diagrams, having the same geometric form but including different spatial transitions, are indicated. Although the geometries are the same in the diagrams, it is clearly seen that the configuration differentiates with the change of spatial relations and depth changes in the justified graph. In the images, Figure 1a is a deep tree (6 depth), Figure 1b a shallow tree (3 depth), and Figure 1c a shallow ring (4 depth) justified graph model. By 'tree' we mean that there is one link less than the number of cells linked, and that there are therefore no rings of circulation in the graph. All trees, even two as different as in the two in the figures, share the characteristic that there is only one route from each space to each other space—a property that is highly relevant to how building layouts function. However, where 'rings' are found, the justified graph makes them as clear as the 'depth' properties, showing them in a very simple and clear way as what they are, that is, alternative route choices from one part of the pattern to another [33].

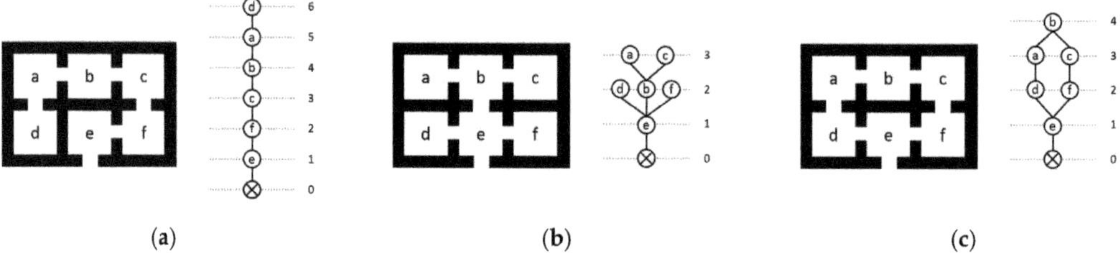

(a)   (b)   (c)

Figure 1. Examples of different justified graph: (**a**) deep tree formed graph model; (**b**) shallow tree formed graph model; (**c**) shallow ringy graph model.

The results can show which spaces are connected closely, the accessibility of each space from the entrance or other key spaces, and which spaces have similar levels of accessibility.

3.2. Selected Case Studies

Mega-shelters can be set up either by rapidly constructing temporary structures or utilizing existing buildings. This study will examine both cases. First, case studies from the U.S. [19] and Australia [34] are presented where shelters were newly constructed. In the mega-shelter guidelines, both countries provide a standard floorplan for setting up a shelter either in an indoor gym or on school grounds. Second, case studies from Japan [20] are presented where existing buildings are used as shelters. Japan utilizes school facilities across the country as local shelters, and the guideline prepared by the Japanese government provides how to allocate various functions of a shelter in existing school buildings.

In Figure 2, Cases 1 [19] and 2 [34] show a standard floorplan for new construction, while Cases 3 and 4 [20] show plans for the utilization of existing school buildings.

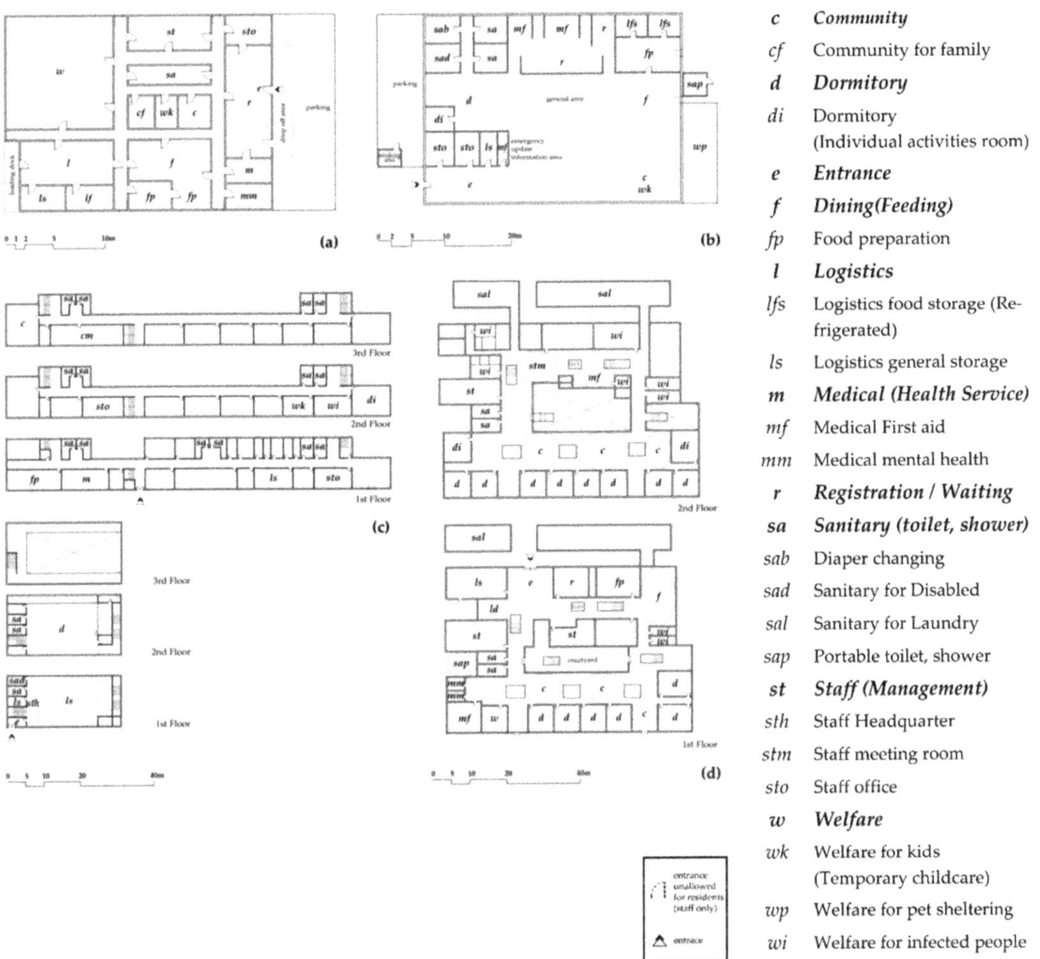

Figure 2. Mega-shelter floor plan: (**a**) case 1 (U.S.); (**b**) case 2 (AU); (**c**) case 3 (JP1); (**d**) case 4 (JP2).

In Case 1, as shown in Figure 2a, all functions are clearly defined and are connected by a corridor. The corridor along the *y*-axis divides the shelter into three zones: management zone, common area zone, and residential zone. In Case 2, as shown in Figure 2b, the general area, dining area, and recreation area are located in the central open space. Other functions are separated, including registration, medical, sanitary, staff, and kitchen. Case 3, as shown in Figure 2c, shows an existing school facility used as a temporary shelter, and the teachers' quarter is used for management and other common uses while the annex buildings are used as residential areas. Case 4, as shown in Figure 2d, shows a courtyard and community facilities on the ground floor of the central area. It also has a welfare space for patients with colds and other mild diseases, in case the duration of facility use is extended.

4. Results

4.1. Depth of Space and Hierarchical Analysis by Justified Graph

To understand spatial structure, it is necessary to comprehend how the space is experienced from user's perspective. Justified graph analysis was conducted to examine the depth and hierarchy of the shelter space, from the entrance to its deepest interior space. The results are presented in Figure 3.

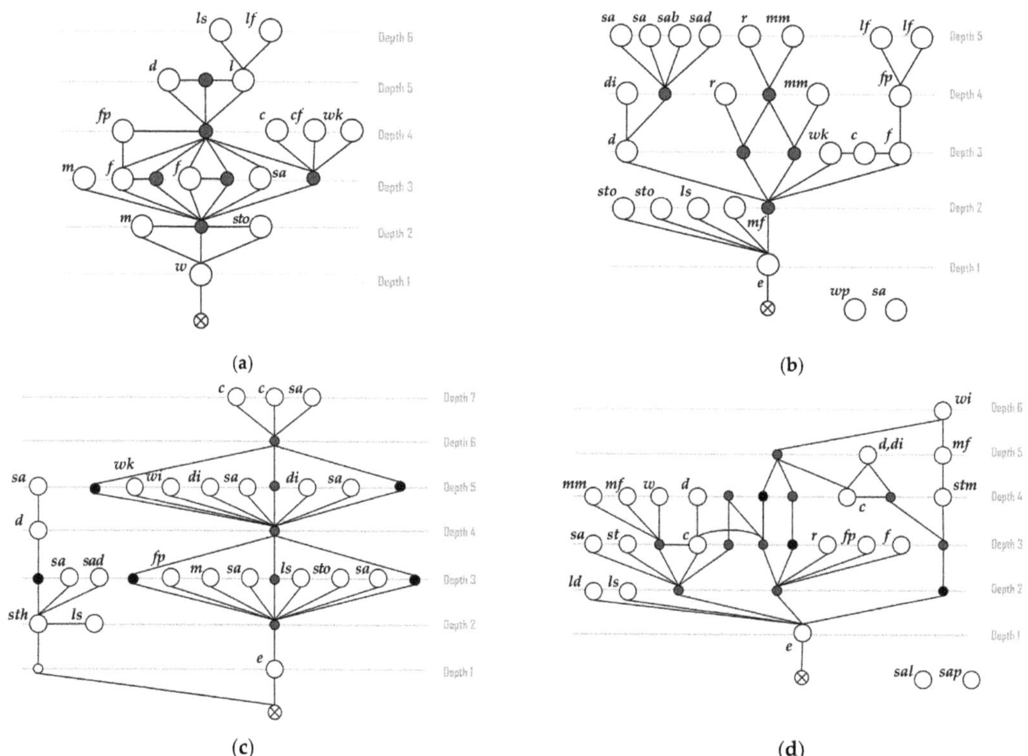

Figure 3. Justified graph analysis (see Figure 2 for key to room functions; black nodes indicate verticals, while grey nodes indicate corridors): (**a**) case 1 (U.S.); (**b**) case 2 (AU); (**c**) case 3 (JP1); (**d**) case 4 (JP2).

In Case 1 (U.S. shelter), as shown in Figure 3a, the registration space near the entrance is the shallowest space, most easily accessible from outside. It is followed by staff office and medical space. Three functions (registration, staff, and medical) are connected in a ring structure in the justified graph. The sanitary function, with depth value of 3.0, connects programs in inner and outer spaces. In the same hierarchy exist feeding and other staff facilities, followed by community and welfare (childcare) spaces for residents. At the end of the corridor there are dormitory and logistics facilities. The analysis shows that spatial layout of the U.S. shelter is efficient from a manager's perspective since it allows access to all spaces with only few depths. However, from a resident's perspective, even visiting a sanitary facility requires moving through a corridor shared with the manager. Since space where the residents can move freely and privately is limited to the dormitory, some may feel that their movements are monitored and controlled.

In Case 2 (Australia shelter), as shown in Figure 3b, staff, medical (first aid station), and logistics storage spaces near the entrance are the shallowest space, with average depth of 1.67. With the general area in the center, spaces are connected in a tree-shaped structure. Main functions include dormitory, registration, and community space. The dormitory's individual activity room and sanitary space are connected to the dormitory. Registration space is located relatively inside, at depth of 4.5. It is adjacent to the medical (mental health) space, where refugees who need special supervision are screened. The feeding space has the same depth as welfare (childcare) and community spaces, and is adjacent to food preparation and logistics (food storage) spaces. The analysis shows that with the general area in the center, the Australian shelter separated facilities where the privacy of the residents must be protected and those where it does not. Hence, even when a space may show the same depth value in the justified graph, the characteristics of the Australian facility may be very different with the general area acting as a buffer zone. The functions with similar goals are connected on the same branch in the tree structure starting from the general area—dorm–sanitary spaces, registration–mental health spaces, and dining–welfare (childcare), and community. All three function clusters are heavily used by the refugees or are required to promote refugee communities.

In Case 3 (Japan shelter #1), as shown in Figure 3c, the main building and the annex building are separated, containing primarily management functions and resident service functions, respectively. Functions needed for the management of the shelter are concentrated in the main building, and all functions on each floor have equal depth values. In the main building, the facilities with the shallowest depth are food preparation, medical, logistics storage, staff office, and sanitary use at a depth of 3.0. On the other hand, the community space at a depth of 7.0 is the deepest interior space. In the annex building, staff headquarters and logistics are located in the shallowest space. In the main building, the lower floors house service functions related to sustenance of life (dining, medical, and logistics) and the upper floors house service functions related to social desires of a higher level (welfare, dorm (individual activity room), and community spaces). While the dormitory facility in the annex building may have privacy since it is located in the innermost space at a depth of 4.0, residents need to move through the headquarters to go outside and some may feel that they are being controlled.

In Case 4 (Japan shelter #2), as shown in Figure 3d, logistics storage, logistics distribution center, and registration with uses are located closest to the entrance. Management functions such as food preparation, staff office, and dining facilities at a depth of 3.0 share a corridor with registration. Staff office and sanitary use are located along the corridor to the courtyard, and welfare–medical space and community–dormitory space are located on different branches. The corridor leading to a community space with a depth value of 3.0 is connected to dining use. The dormitory and medical and welfare are the deepest space with a depth of 4.0 on the first floor. On the second floor, the courtyard divides the management zone and refugee zone. Staff headquarters, medical (first aid station), welfare for infected people, and sanitary spaces are contained in the management zone, while the dormitory and community are the primary use in the refugee zone. In this case, a total of four vertical circulation routes are creating various circulation routes via corridors. The vertical circulation route in front of the entrance leads directly to the management facility on the second floor, and essential functions for shelter management are located along the corridor on the first floor. Resident facilities—dormitory and community spaces—are located inside the courtyard, thus completing the corridor–community–dorm loop. This design is applied to the second floor in same manner, thereby promoting communication and strengthening bonds among refugees.

From the analyses of mega-shelters, the common characteristics are as follows. First, the spatial layout of the mega-shelter closely follows the process by which the refugees are admitted, beginning with registration and ending with the residential space. The spatial order and depth increase as a new resident moves through the entrance, then through the spaces of the management (registration (or staff headquarters), medical (first aid station), and staff space), and then finally to the dormitory, which is a private space. However, the depths of welfare, medical (mental health), and community spaces varied by shelter, indicating that considerations for facilities required for long-term shelter operation were not fully integrated into the planning of the shelters. Second, the most essential functions to sustain the lives of the refugees, such as dining and logistics, were located in shallow spaces. However, sanitary facilities showed varying depths in all cases, since they were built as single facilities for both refugees and the managers or were built as multiple facilities—depending on the size and design of the shelter. Third, the average depths of the management facilities (dining, logistics, medical, and staff) were shallower than refugee-centric facilities (dorm and community). This indicates that management facilities are located near the entrance to enable strict control of movement of the refugees to and from the outside. Fourth, the dormitory is located in either a high-depth space or away from core functions to protect the privacy of the residents. Finally, complementary functions can be found on the same branches of the justified graphs. It indicates that those programs have been sited in proximity for efficient operation of shelters.

Similarities found between the shelters are as follows. In Case 1 (U.S. shelter) and Case 3 (JP1 shelter), various functions are located without separation of the circulation routes of the management and the users. In Case 2 (AU shelter) and Case 4 (JP2 shelter), programmed and private areas are separated using the general area and the courtyard as transition areas, respectively. Here, areas are separated into areas for refugees and areas for management, or areas where privacy needs to be protected and areas where it does not. From the management's perspective, some space needs to be separated from the circulation routes of the refugees—for example, for security. From a refugee's perspective, separation from the management's space can minimize unreasonable surveillance. A layer of semi-public, semi-private space contributes to residents' comfort and satisfaction [35]. The refugees' psychological level of comfort can be enhanced by providing space that can be used freely. As the privacy of the individuals is respected and the social distance between individuals is guaranteed, psychological and social benefits are maximized [10].

4.2. Analysis of Shelter Layout Characteristics by Axial Line Analysis and VGA

Spatial layout has a direct impact on users' spatial cognition and movement, as well as on space-use patterns. For spatial configuration analysis of a shelter's interior space, axial line analysis and VGA were conducted. The results are presented in Table 2, Figures 4 and 5. Since the objective is to compare levels of global integration, corridors (which connect functions) have been excluded from the analyses.

Axial line analysis of Case 1 (U.S.), as shown in Figure 4a and the first row of Table 2, indicates that logistics (1.31), dining and snack area (1.29), and staff area (1.21) belong to the integration core group with the highest integration values. In the next group are management-related facilities such as registration/waiting (1.14), medical health services (1.07), shelter manager's office (1.04), and food preparation (1.04), as well as dormitory (1.13) and sanitary (1.18) facilities. The segregation core group, with the lowest level of integration, were counseling (1.00), community space (0.92), welfare space (0.92), and logistics storage (0.75), indicating these spaces have low accessibility and usability. VGA, as shown in Figure 5a, showed that the dormitory space (excluding corridor) has relatively high visibility. The analysis also revealed that community and welfare spaces, while located in the center of the shelter, unexpectedly have rather low visibility and were relatively inactive spaces.

In sum, axial line analysis and VGA of Case 1 (U.S.) showed that the global integration values of all spaces except the dormitory were within a similar range. Logistics and dining are functions in the integration core group, while counseling, community, welfare, and logistics storage spaces are segregated from other functions. Overall, spaces where staff worked had higher potential for activities, and facilities for the refugees were in more isolated areas. As an essential function for the operation of a shelter, the dining area has a high level of integration with a large number of users; however, the community space has a low level of integration even though the space is capable of accommodating a large number of people.

Table 2. Order of global integration of functions from axial line analysis.

Case	Order of Global Integration of Functions																				
Case 1 US (functions)	*l* >	*f* >	*st* >	*sa* >	*r* >	*d* >	*m* >	*sto* >	*fp* >	*mm* >	*wk* =	*c* =	*cf* >	*lfs* =	*ls*						
Case 1 US (values)	1.31	1.29	1.21	1.18	1.14	1.13	1.07	1.04	1.04	1.00	0.92	0.92	0.92	0.75	0.75						
Case 2 AU (functions)	*e* =	*c* =	*wk* >	*f* >	- >	*d* >	*m* >	*fp* >	*r* >	*sto* =	*sto* =	*ls* =	*di* >	*sap* >	*wp* >	*sab* =	*sa* =	*sad* >	*lfs* =	*lfs* >	-
Case 2 AU (values)	1.49	1.49	1.49	1.48	1.43	1.28	1.16	0.85	0.80	0.78	0.78	0.78	0.75	0.64	0.63	0.63	0.63	0.63	0.56	0.56	0.52
Case 3 JP1 (functions / floors)	*sth* (*1) >	*fp* (1) =	*m* (1) =	*sa* (1) =	*ls* (1) =	*sto* (1) >	*ls* (*1) >	*sto* (2) =	*sa* (2) =	*wk* (2) =	*wi* (2) >	*di* (2) >	*sad* (*1) =	*cg* (3) =	*c* (3) =	*sa* (3) >	*d* (*2) >	*sa* (*2) **			
Case 3 JP1 (values)	0.73	0.65	0.65	0.65	0.65	0.65	0.63	0.54	0.54	0.54	0.54	0.54	0.47	0.47	0.47	0.47	0.43	0.35			
Case 4 JP2 (functions / floors)	*ld* (1) =	*cd* (1) >	*f* (1) >	*sal* (o) >	*ls* (1) >	*sa* (1) >	*ca* (2) >	*sto* (1) >	*stm* (2) >	*sal* (2) >	*wi* (1) >	*r* (1) >	*w* (1) >	*sto* (2) >	*sa* (2) >	*mm* (1) >	*m* (2) >	*wi* (2) >	*d* (2) >	*di* (2) >	*d* (1)
Case 4 JP2 (values)	1.10	1.10	1.00	0.99	0.97	0.97	0.95	0.94	0.93	0.92	0.87	0.86	0.86	0.85	0.85	0.82	0.80	0.77	0.76	0.75	0.75

■: value of integration—top 0~20% (integration core) ■: value of integration—top 21~60% ■: value of integration—top 61~100% (segregation core) * For key to room functions, see Figure 2. ** In Case 3 and 4, number below function indicates the number of floors. An asterisk in front the number indicates an annex building.

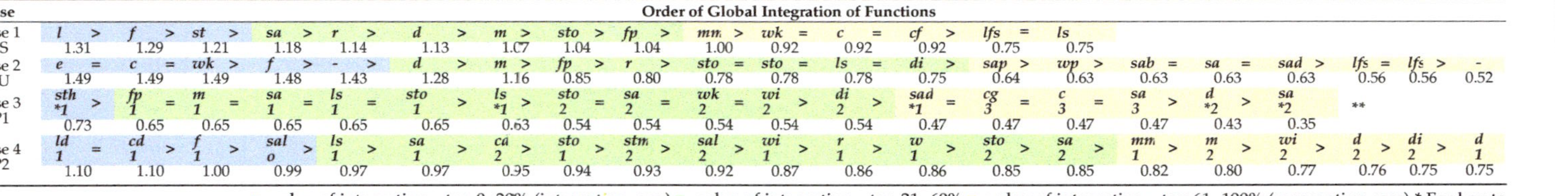

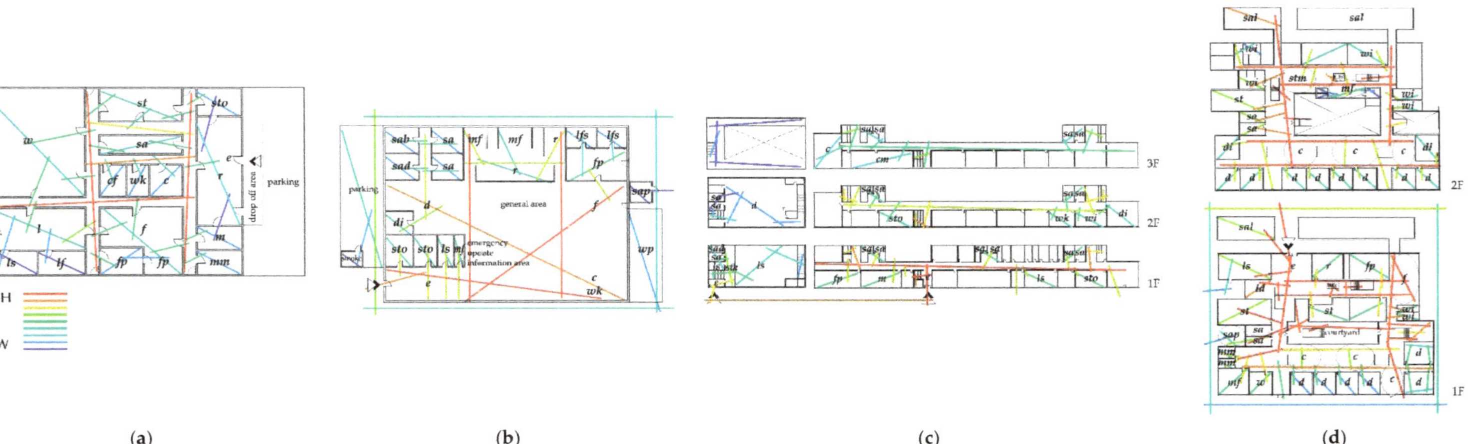

Figure 4. The axial line analysis results on cases (see Figure 2 for key to room functions): (**a**) case 1 (U.S.); (**b**) case 2 (AU); (**c**) case 3 (JP1); (**d**) case 4 (JP2).

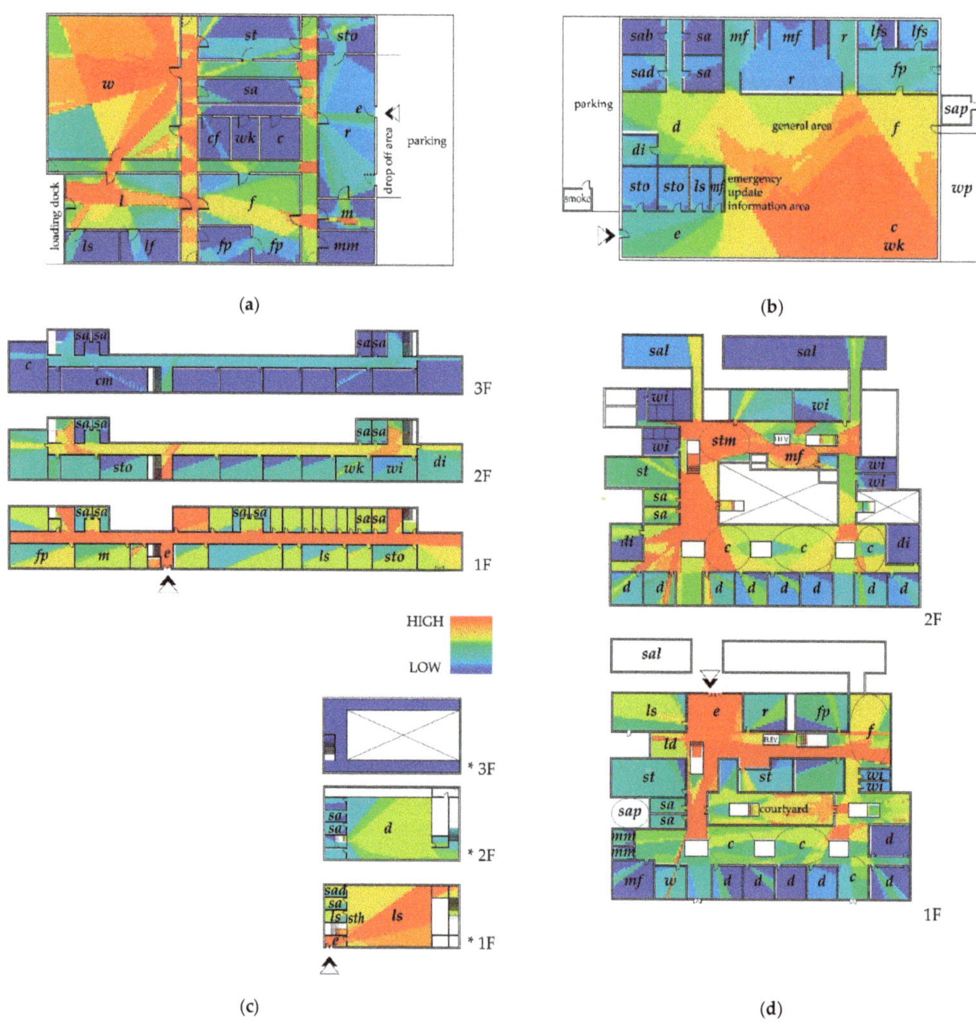

Figure 5. Visibility graph analysis of the cases (see Figure 2 for key to room functions; an asterisk in front the number indicates an annex building); (**a**) case 1 (U.S.); (**b**) case 2 (AU); (**c**) case 3 (JP1); (**d**) case 4 (JP2).

Axial line analysis of Case 2 (Australia), as shown in Figure 4b and the second row of Table 2, indicates that the entrance/meeting area, recreation area, childcare area (1.59), and dining area (1.48) have the highest integration values. Functions related to the integration core function, such as dormitory (1.28), medical (mental health) (1.16), food preparation (0.85), registration/waiting (0.80), staff office, logistics storage, and first aid station (0.78) have mid-level integration values. Finally, sanitary (0.63), kitchen storage (0.56), portable sanitary (0.64), pet sheltering (0.63), and smoking area (0.52) have the lowest accessibility. In VGA, as shown in Figure 5b, the recreation area and childcare area were shown to be the most open spaces to users, both visually and by accessibility. Functions with low accessibility were staff area, sanitary, registration, mental health services, food preparation, and storage—spaces that are visually isolated from open space.

Both axial line analysis and VGA of Case 2 (Australia) showed that community, welfare (temporary childcare), and dining spaces in the central open space were the core integration functions. These functions are connected to the center of the shelter through the general area, a visually open area. The general area is a transition space without a specific designated function for the shelter. This transition space also plays the role of providing privacy for the dormitory space. The privacy of users needs to be protected—for manager, office and food storage, and for refugees, the sanitary space. The general area, acting as a transition space, functions as a buffer space between facilities that require high accessibility and uses that require the privacy of users.

Axial line analysis of Case 3 (Japan #1), as shown in Figure 4c and the third row of Table 2, shows that the level of global integration of the main building is higher than the annex building. The staff headquarters (0.73) space shows the highest level of integration. The mid-level integration facilities are food preparation, medical, logistics, staff, and sanitary (0.65) facilities on the first floor of the main building, and welfare (temporary childcare and infected people spaces), staff office, individual, sanitary (0.54) facilities, and finally, logistics storage space (0.63) on the first floor of the annex building. The facilities with the lowest level of integration were community and sanitary (0.46) spaces on the third floor of the main building, as well as sanitary for disabled (0.47) on the first floor of the annex, and dormitory (0.43) and sanitary (0.35) on the second floor of the annex. VGA, as shown in Figure 5c, revealed that food preparation, medical center, sanitary, and logistics storage on the first floor of the main building were the most active spaces. This was followed by welfare (temporary childcare and infected people) on the second floor and individual activities spaces for residents, such as reading rooms. The most isolated functions were community and sanitary spaces on the third floor.

Both axial line analysis and VGA of Case 3 (Japan #1) showed that the level of global integration was highest for the facilities on the first floor, which includes food preparation, medical center, sanitary, and logistic storage in the main building and logistics and staff headquarter in the annex building. The lowest level of integration was found in the community space on the third floor of the main building. The operating food preparation space and medical services on the first floor of the main building are convenient for the refugees since they can meet their demands for meals and health services easily. This is also convenient for staff and logistics to receive supplies and manage them. However, having a community space on the third floor with low accessibility can reduce opportunities for communications among users and lower the probability of psychological recovery through community activities. Protecting the privacy of the dormitory on the second floor of the annex building is desirable, but it has low access to individual activity rooms on the second and third floors of the main building, which are spaces for refugees. It also reduces opportunities for refugees to initiate communications.

Axial line analysis of Case 4 (Japan #2), as shown in Figure 4d and the fourth row of Table 2, shows that the functions with the highest level of global integration were logistics (distribution center) close to the entrance, the community space (1.10) in front of first floor dormitory, dining (1.00), and sanitary (laundry) (0.99) outside. The facilities with the lowest level of integration were medical/mental health (0.82), dormitory and individual activity rooms (0.75) on the first floor, dormitory (0.76) and medical care space for infected people (0.80), and welfare dormitory for infected people (0.72) on the second floor, mostly related to the refugees. VGA, as shown in Figure 5d, predicts that the facilities along the open corridor on the first floor dining and staff meeting room, medical center, and community space in front of the dormitory on the second floor will have large numbers of users. Finally, logistics (distribution center), courtyard, and the community space in front of the dormitory on the first floor have a high level of integration.

Both axial line analysis and VGA of Case 4 (Japan #2) showed that the most segregated spaces were the dormitories on each floor and the welfare for infected people space on the second floor. Both analyses also found that dining and community spaces were most active with a high level of connectivity. Most functions in both high global integration areas and the segregation core are used by the refugees. Programs were distributed according to the characteristics of the rooms of the existing structure, as well as the degree of activity required for each function. The central courtyard contributes to creating an efficient circulation route in the shelter. On a smaller scale, existing school amenities such as student shoe racks help to distinguish the classroom space from the space outside the classroom near the door. The manager space, while it may be difficult to access the refugees physically, has visual access to the living area of the refugees through the courtyard. The dormitory of the refugees is located in the shelter's most internal space, protecting their privacy. The community space in front of the dormitory can strengthen its privacy and help refugees start their own community activities.

The analyses of the four shelters revealed the following (Table 3). In all cases, logistics- and dining-related functions have excellent accessibility and are classified as active space. Logistics and dining are both essential functions directly related to maintaining a minimum quality of life, and they require substantial space area, while a dormitory is a function for which the protection of privacy is paramount but that has relatively good level of integration since it also has good access to essential functions such as sanitary and dining. While different shelters have different segregation core functions, logistics storage, welfare, and community spaces fall into this category in general.

Table 3. Global integration of shelter rooms by functions.

Value (Global Integration)	Case 1 (U.S.)	Case 2 (AU)	Case 3 (JP1)	Case 4 (JP2)
High-level rank (top 0–20%)	- management - dining - logistics	- welfare (child care) - community - dining	- management (headquarters) - dining (preparation) - medical - logistics	- dining - logistics
Mid-level rank (top 21–60%)	- registration - medical	- dormitory	- welfare	- logistics - management - medical - registration - community
Low-level rank (top 61–100%)	- medical (counselling) - community - welfare - logistics	- sanitary - management - logistics (storage)	- community - dormitory	- medical (counselling) - dormitory - welfare (infected people)

5. Discussion

Based on the literature review and case studies of mega-shelters, several issues can be raised, as follows.

First, the planning guides used for the planning and management of shelters provide instructions for spatial layout and allocation of functions only for efficient operation and assistance. For example, the movement of refugees should be controlled at the entrance, and the control center needs to be located in the center of communications between the outside and inside of the shelter. Some guidelines even stipulate that the recreation space requires supervision by managers. Given such guidelines, the spatial plans for mega-shelters have been prepared from a manager's perspective.

Second, in spaces where the activities of the two user groups are segregated, spatial configuration and the level of integration of the spatial layout show clear hierarchy. In general, logistics- and dining-related functions have been analyzed as spaces with good accessibility and high levels of activity. In contrast, spaces exclusively used by the residents were located in high-depth areas compared to spaces only used by the managers. However, in this context, welfare and community spaces used by the residents showed a low level of integration, contradicting the purpose of these spaces. It shows that the locations of uses within shelters are determined from a manager's perspective, with the goal of stable construction and operation of the shelter. In addition, the dormitory is located in the most internal space from the entrance, often segregated from adjacent spaces to emphasize protection of privacy.

Overall, only facilities required to maintain life, such as dining and logistics, are included as core integration functions in the guidelines' spatial plans. The dormitory space is mostly classified as a segregation function, for which the protection of privacy is the foremost standard. However, there remains shortcomings such as lack of clear distinction between non-dormitory refugee uses and management uses and a lack of consideration to create a community that looks beyond individual refugees and takes into account the psychology of the refugees and their spatial behaviors and plans for locations of medical (mental health) and other needed uses.

Third, when building a new shelter or utilizing existing buildings, it is important to precisely understand the spatial structure of the mega-shelter before determining how to allocate various uses. In other words, deciding on spatial hierarchy and uses in a newly constructed shelter and determining use allocation in an existing school facility based on the appropriate understanding of spatial hierarchy are in fact from the same perspective. Therefore, it is important to clearly understand circulation routes of managers and users regardless of the shape of the building space and then decide how to allocate core functions and relevant uses for the two groups of users. A discussion on the spatial hierarchy of a shelter layout, relative accessibility of various uses of a shelter, and spatial connectivity between relevant functions is needed with an objective perspective.

6. Conclusions

Analysis of the previous literature and guidelines on the management of mega-shelters have all revealed that the spatial layouts of the shelters emphasize only efficiency for the managers. However, such a spatial configuration does not play a positive role for the recovery of refugees from their traumas, and in fact, as Sanderson and Burnell [9] argued, it has a negative impact on the development of the refugees' sense of community that is essential to strengthen the refugees' will for restoration and recovery. Specifically, the following points should be noted.

First, in both academic research and the shelter planning guidelines prepared by public institutions, the spaces used by refugees are recommended to be located in a segregated space to protect the privacy of the refugees. While there is recent research that emphasizes the importance of community facilities in addition to simply accommodating residents [10,11], almost no consideration is given as to how to allocate space in an objective way for these and residential uses in a shelter [11,14]. However, such a spatial layout stems from a shelter manager's perspective, placing control and supervision of the refugees foremost. In this context, Markus [36] pointed out that the spatial layout in which patients are segregated in the periphery while the care team is located in the integrated core is related to the building types, such as almshouses, that express and enforce power. Spatial configuration has significant impact on spatial cognition and spatial behaviors [25]. When isolated in a segregated space, the user cannot easily comprehend the spatial structure of the entire shelter. As the analyses of shelters show, areas with the highest integration values were mostly areas for the management of the shelter, followed by high-traffic corridors and public spaces. If a shelter's major corridors and public spaces are located in spatially central areas with high integration values, they can facilitate communications among the residents in the entire shelter and also assist wayfinding. It can also contribute to the safety of the residents as well. In other words, a high level of integration of key public spaces can affect an individual's spatial ability, including wayfinding, and in the end has a negative impact on his/her sense of spatial control. Lynch [37] also supported this argument and articulated that illegible space is not a desirable space for users. Hillier [33] and Kim [38] described that unintelligible space hinders spatial cognition formation, which in turn acts as a constraining factor for the full manifestation of space-use patterns. In this context, the community space is very important from a space configuration perspective. Therefore, while dormitory space may be located in a segregated space, the common space of the refugees—community space in particular—needs to be located in a space where the spatial layout of the entire shelter can be grasped easily and intuitively. As Nappi and Souza [10] argued, the community space needs to be located in the center of the shelter with a high level of integration with visibility into the deepest spaces of the shelter. This will ensure that the users will perceive the entire spatial structure easily. Such a layout will increase opportunities among residents to meet and communicate, and the users can experience psychological stability and a sense of spatial control.

Second, a mega-shelter is a small society where individual refugees and their families live with other neighbors. The shelter must become a place where people with post-disaster stress and trauma can together restore their health. However, not only public institutions' guidelines [19–21] but also studies on guidelines for the planning and management of shelters [11,13–16] indicated that most studies focus on either accommodation of refugees or efficient construction and management of shelters, rather than on refugees' recovery of health through provision of user-centered community spaces. As the four case studies in this paper showed, even shelters set up with the same goals and functions can have different effects on refugees, depending on spatial hierarchy and the emphasis of spatial configurations. The results of this study indicate that the spaces used by the managers and the refugees need to be appropriately differentiated and located. Rather than allocating a manager's space to a higher level in the spatial hierarchy to enable unilateral supervision of users, both need to be at a similar level while appropriately segregated. In addition, a transition area or general area needs to exist between the two spaces so that, while managers can efficiently take care of the refugees, the refugees do not feel that they are being monitored. A transition space with no specific function can separate circulation routes of both groups and control the level of activities in a space. In other words, the managers can take care of the residents visually through the transition space; the residents can enjoy more free movement. Perhaps it is in such a spatial configuration that human-centered, natural spatial patterns can emerge, as Zhang and Dong [12] argued.

Albeit temporarily, a shelter is a space where people who suffered disaster constitute a new group, a new society, and live together. In particular, when planning for a large-scale facility such as a mega-shelter, a careful approach is required. The plan for such a facility should not only focus on accommodating and managing refugees in the short term, but also recognize that it is creating a social environment where people act in accordance with their needs and wants. The spatial planning of a shelter must meet the basic needs of the refugees, such as physiological and safety needs. However, this is not enough and as some may be suffering from PTSD, a shelter's spatial plan is important for their recovery. To this end, the spatial configuration of a shelter must be planned given the psychological state and spatial behavior of the refugees in a post-disaster, stressful situation. This study is meaningful in that it complements the shortcomings of previous studies that emphasized user-centric spatial planning but did not provide specific strategies for the design of a better spatial configuration. This study is meaningful in that it complements the shortcomings of previous studies, which focused on the efficient management of a shelter or, even when they did emphasize user-centric spatial planning, did not provide qualitative basis for spatial behaviors of the users. The academic evidence presented in this study can have practical application for future mega-shelter guideline preparation, since the criteria for the spatial planning of shelters are provided.

Author Contributions: Conceptualization, Y.O.K. and J.K.L.; methodology, Y.O.K. and J.Y.K.; validation, Y.O.K.; formal analysis, H.Y.Y.; investigation, H.Y.Y.; resources, H.Y.Y.; writing—original draft preparation, H.Y.Y. and J.K.L.; writing—review and editing, Y.O.K. and J.Y.K.; visualization, H.Y.Y.;

supervision, Y.O.K. and J.Y.K.; project administration, Y.O.K.; funding acquisition, Y.O.K. All authors have read and agreed to the published version of the manuscript.

Funding: This research was supported by the Korea for Infrastructure Technology Advancement (KAIA) grant funded by the Ministry of Land, Infrastructure and Transport (Grant 22TSRD-C151228-04).

Data Availability Statement: Data derived from the current study can be provided to readers upon request.

Conflicts of Interest: The authors declare no conflict of interest.

References

1. American Red Cross. *Mega-Shelter Planning Guide: A Resource and Best-Practices Reference Guide*; International Association of Venue Managers, Inc.: Coppell, TX, USA, 2010.
2. Hillier, B.; Hanson, J. *The Social Logic of Space*; Cambridge University Press: Cambridge, UK, 1989.
3. Ruzek, J.; Walser, R.D.; Naugle, A.E.; Litz, B.; Mennin, D.S.; Polusny, M.A.; Ronell, D.M.; Ruggiero, K.J.; Yehuda, R.; Scotti, J.R. Cognitive-behavioral Psychology: Implications for Disaster and Terrorism Response. *Prehosp. Disaster. Med.* **2008**, *23*, 397–410. [CrossRef]
4. Guterman, P.S. Psychological Preparedness for Disaster. *Retrieved March* **2005**, *22*, 2016.
5. Valent, P. Disaster Syndrome. *Encycl. Stress* **2007**, 706–709. [CrossRef]
6. Frueh, B.C.; Ebrary, I. *Assessment and Treatment Planning for PTSD*; John Wiley & Sons: Hoboken, NJ, USA, 2012.
7. Green, B.L.; Lindy, J.D. Post-traumatic Stress Disorder in Victims of Disasters. *Psychiatr. Clin. N. Am.* **1994**, *17*, 301–309. [CrossRef]
8. Brown, T.I.; Gagnon, S.A.; Wagner, A.D. Stress Disrupts Human Hippocampal-Prefrontal Function during Prospective Spatial Navigation and Hinders Flexible Behavior. *Curr. Biol.* **2020**, *30*, 1821–1833.e1828. [CrossRef]
9. Sanderson, D.; Burnell, J. *Beyond Shelter after Disaster: Practice, Process and Possibilities*; Routledge: New York, NY, USA, 2013.
10. Nappi, M.M.L.; Souza, J.C. Temporary Shelters: An Architectural Look at User-Environment Relationships. *Arquitetura Rev.* **2017**, *13*, 112–120. [CrossRef]
11. Lee, B.Y.; Han, D.K.; Jang, S.B.; Choi, H.K.; Lee, S.W. Facility Standards and Layout Typology of Temporary Shelters in Response to Disasters—Focused on Size Standard and Block Layout of Public Facilities. *Crisisonomy* **2019**, *15*, 1–14. [CrossRef]
12. Zhang, T.; Dong, H. Human-centred Design: An Emergent Conceptual Model. In Proceedings of the Include2009, London, UK, 8–10 April 2009.
13. United Nations High Commisioner for Refugees. *Handbook for Emergencies*, 3rd ed.; United Nations High Commissioner for Refugees: Geneve, Switzerland, 2007.
14. Park, S.H. *Technology Development of Integrated Shelter Management and Evacuee Support in Shelter*; National Disaster Management Research Institute: Ulsan, Korea, 2016.
15. Jung, J.Y.; Yeom, T.J.; Park, M.J. A Case Study on the Improvement of Residential Environment on the Shelter in Case of Disaster. *J. Archit. Inst. Korea* **2018**, *38*, 157–158.
16. Hirata, K.; Ishikawa, T.; Furukawa, Y.; Murata, A.; Notake, H.; Makizumi, T.; Shigematsu, H.; Watanabe, T.; Hamano, M.; Ikutomi, N. Creation and Practice oF an Unprecendented Large Shelter Management Model -Establishing a Shelter Management Program for 10,000 People Assuming Practical Use. *J. Hous. Res. Found. JUSOKEN* **2020**, *46*, 263–272. [CrossRef]
17. Ohno, A.; Kakino, Y. A Study on Management as Shelters oF Elementary Schools from the Viewpoint of Function and Equipment: A Case of Elementary School in Toyohashi City. *Archit. Plan.* **2013**, 371–372. Available online: https://www.city.toyohashi.lg.jp/secure/18786/houkokusyo.pdf (accessed on 8 April 2022).
18. Kim, M.; Kim, K.; Kim, E. Problems and Implications of Shelter Planning Focusing on Habitability: A Case Study of a Temporary Disaster Shelter after the Pohang Earthquake in South Korea. *Int. J. Environ. Res. Public Health* **2021**, *18*, 2868. [CrossRef]
19. Federal Emergency Management Agency. *Shelter Field Guide*; Department of Homeland Security, FEMA: Washington, DC, USA, 2015.
20. The Ministry of Education, Culture, Sports, Science and Technology. Compilation of "Regarding Disaster-Resistant School Facilities—Strengthening Disaster Prevention Functions as Tsunami Countermeasures and Evacuation Centers". Available online: https://www.mext.go.jp/b_menu/shingi/chousa/shisetu/013/toushin/1344800.htm (accessed on 23 April 2022).
21. Department of Human Services. *Emergency Relief Handbook: A Planning Guide 2011–2012*; DHS: Victoria, Australia, 2011.
22. Littleton, H.; Haney, L.; Schoemann, A.; Allen, A.; Benight, C. Received support in the aftermath of Hurricane Florence: Reciprocal relations among perceived support, community solidarity, and PTSD. *Anxiety Stress Coping* **2022**, *35*, 270–283. [CrossRef]
23. Kaniasty, K. Social support, interpersonal, and community dynamics following disasters caused by natural hazards. *Curr. Opin. Psychol.* **2020**, *32*, 105–109. [CrossRef] [PubMed]
24. Tang, T.C.; Yen, C.F.; Cheng, C.P.; Yang, P.; Chen, C.S.; Yang, R.C.; Huang, M.S.; Jong, Y.J.; Yu, H.S. Suicide Risk and its Correlate in Adolescents who Experienced Typhoon-induced Mudslides: A Structural Equation Model. *Depress Anxiety* **2010**, *27*, 1143–1148. [CrossRef]
25. Kim, Y.O.; Penn, A. Linking the Spatial Syntax of Cognitive Maps to the Spatial Syntax of the Environment. *Environ. Behav.* **2004**, *36*, 483–504. [CrossRef]

26. Biswas, A. Exploring Indian Post-disaster Temporary Housing Strategy through a Comparative Review. *Int. J. Disaster Resil. Built Environ.* **2020**, *10*, 14–35. [CrossRef]
27. Kim, K.; Moon, H.; Cho, Y.; Kim, M.; Kim, S.; Lee, H. *Set Minimum Standards for Temporary Housing in Case of Disaster*; Korean Institute of Architects: Seoul, Korea, 2011.
28. Hillier, B.; Penn, A. Visible colleges: Structure and randomness in the place of discovery. *Sci. Context* **1991**, *4*, 23–50. [CrossRef]
29. Penn, A.; Desyllas, J.; Vaughan, L. The space of innovation: Interaction and communication in the work environment. *Environ. Plan. B Plan. Des.* **1999**, *26*, 193–218. [CrossRef]
30. Hillier, B.; Penn, A.; Hanson, J.; Grajewski, T.; Xu, J. Natural movement: Or, configuration and attraction in urban pedestrian movement. *Environ. Plan. B Plan. Des.* **1993**, *20*, 29–66. [CrossRef]
31. Tandy, C.R.V. The isovist method of landscape survey. *Methods Landsc. Anal.* **1967**, *10*, 9–10.
32. Turner, A.; Doxa, M.; O'sullivan, D.; Penn, A. From isovists to visibility graphs: A methodology for the analysis of architectural space. *Environ. Plan. B Plan. Des.* **2001**, *28*, 103–121. [CrossRef]
33. Hillier, B. *Space Is the Machine: A Configurational Theory of Architecture*; Cambridge University Press: Cambridge, UK, 2007.
34. Lake, E.F. Emergency Relief Handbook: A Planning Guide 2013. Available online: https://prezi.com/6ztdbllhp5hv/emergency-relief-handbook-a-planning-guide-2013/ (accessed on 1 August 2022).
35. Obeidat, B.; Abed, A.; Gharaibeh, I. Privacy as a motivating factor for spatial layout transformation in Jordanian public housing. *City Territ. Archit.* **2022**, *9*, 1–17. [CrossRef]
36. Markus, T.A. *Buildings and Power: Freedom and Control in the Origin of Modern Building Types*; Routledge: London, UK, 1993.
37. Lynch, K. *The Image of the City*; MIT Press: Cambridge, MA, USA, 1960; Volume 11.
38. Kim, Y.O. The role of spatial configuration in spatial cognition. In Proceedings of the Third International Space Syntax Symposium, Atlanta, GA, USA, 7–11 May 2001.

 *buildings*

Article

Basic Analysis of the Correlation between the Accessibility and Utilization Activation of Public Libraries in Seoul: Focusing on Location and Subway Factors

Xiaolong Zhao [1] and Kwanseon Hong [2,*]

[1] BK21 FOUR Service Design Driven Social Innovation Educational Research Team, Dongseo University, Busan 47011, Republic of Korea
[2] College of Design, Dongseo University, Busan 47011, Republic of Korea
* Correspondence: kshong@gdsu.dongseo.ac.kr; Tel.: +82-051-320-1865

Abstract: In the past, the utilization rate of public libraries in Seoul could be estimated based on their accessibility. However, several issues emerge if we apply this correlation to the present day. Therefore, we re-examined the causal relationship between accessibility and the utilization rate of public libraries to provide directions for improving the use of public libraries in densely populated cities with growing cultural demand. After investigating the utilization rate of public libraries in Seoul from 2015 to 2019, the degree of utilization activation (DUA) was set as the dependent variable, and the integration of public libraries (derived by the quantification of urban space with space syntax) was set as the independent variable. A hypothesis was established to examine the causal relationship using statistical techniques. According to the results, the derived index values had independence and normality, but the accessibility index of public libraries did not exhibit a causal relationship with DUA. It was verified that the causal relationship recognized in the past (where accessibility was the sole predictor of utilization rate) cannot be applied to public libraries in the present day. Modern factors affecting DUA may involve either user motivation or the recent developments in public libraries compared to the past.

Keywords: urban hierarchy; accessibility control; utilization rate; public library; space syntax; Seoul

Citation: Zhao, X.; Hong, K. Basic Analysis of the Correlation between the Accessibility and Utilization Activation of Public Libraries in Seoul: Focusing on Location and Subway Factors. *Buildings* **2023**, *13*, 600. https://doi.org/10.3390/buildings13030600

Academic Editors: Michael J Ostwald and Ju Hyun Lee

Received: 1 February 2023
Revised: 22 February 2023
Accepted: 22 February 2023
Published: 24 February 2023

1. Introduction

1.1. Background

In July 2021, the United Nations Conference on Trade and Development (UNCTAD) upgraded South Korea's status to "developed country" [1]. Since its founding, the UNCTAD has never upgraded a member state's status from "developing country" to "developed" [2]. Economic factors are the standard criteria for assessing a country's development level [3,4]. However, in major cities in Europe, the US, and Japan, the promotion of cultural infrastructure [5,6] with the purpose of improving the quality of life of their residents [7,8] is also significant [9,10]. On the one hand, museums and art galleries (among other cultural infrastructures) not only promote domestic culture but also convey a nation's knowledge to the outside world [11,12] through exhibits [13,14]. On the other hand, libraries focus on providing users with cultural knowledge, social education, and free access to information [15–17]. Typically, exhibits are presented in foreign languages to promote museum and art gallery exhibitions [13,18,19]. In contrast, most public libraries in countries with specific language cultures provide literature information services in their native languages, even if it is necessary or more appropriate to use foreign languages [17,20–22]. This phenomenon explains why museums and art galleries focus on drawing foreign visitors, which is, of course, in accordance with their intended use. It, likewise, infers that public libraries are usually oriented toward their native populations. In addition, people visit museums or art galleries less frequently, and for shorter periods of time, than public libraries [20,23].

This, in turn, demonstrates the potential for public libraries to boost their usage, since they provide a wider variety of information sources compared to exhibitions, which usually have specific themes.

In the late 1990s, public libraries in Korea shifted toward user-centered systems as the demand for cultural activities increased [24–26]. Policies and practices to establish and reorganize public libraries have already been initiated in Seoul [27]. In accordance with the increased trend toward improving and developing new cultural infrastructures, the Seoul Institute's City Urban Research Information Center revealed, in 2016, that the rate of opening new public libraries in Seoul had increased by 104.5% from 2007 to 2015, representing the highest introduction rate for new cultural infrastructure in Seoul [28]. This trend indirectly shows that people's demand for public libraries has continued to increase, even after the 2000s. In particular, the government's comprehensive plan to initiate information services for libraries since the 2000s has promoted the development of user-centered public libraries [29]. Since then, communication spaces, such as open-complex cultural spaces, have maintained their previous functions as informational, educational, and cultural resources while providing new cultural experiences to users. Moreover, the number of community centers serving as community hubs and makerspaces has increased [29–31]. Therefore, while still serving their original purposes, public libraries in Korea are consistently evolving into places where individuals may gather in comfort. However, despite the fact that the number of public libraries in Seoul has rapidly increased since 2015 to meet citizen demands [28,29], most of the newly established libraries are located in geographical locations that are difficult for residents to access, an issue that negatively affects their respective usage [32–34]. Some scholars even thought that the marginalization of public libraries in urban spaces might be the cause of their reduced daily usage [27,34,35]. Based on these claims, many previous scholars in Korea regarded the urban accessibility of public libraries in Seoul as a factor affecting their degree of utilization activation (DUA). In other words, DUA can be considered to correspond to the accessibility of the location of public libraries. In addition, physical accessibility can be affected by various aspects, such as travel distance, transportation means, and lead time [36]. Therefore, the use of efficient transportation means can be an important step toward reducing the time required to reach libraries located at greater distances [37]. Therefore, accessibility-related factors must be considered when investigating the accessibility of public libraries. However, some researchers have stated that changes in user motivation induced by functional changes in public libraries may also affect users' DUA [38–40], as the trend toward research in Korean public libraries has recently shifted toward integrating space and designing functions and services to provide information to users [41–44]. Consequently, although a causal relationship between accessibility and utilization of public libraries has been previously established, it is now necessary to re-examine this relationship due to changes in DUA caused by functional changes in public libraries and social factors.

1.2. Research Gap, Purpose, and Novelty

Since 2020, research on public libraries has focused on their operation, the provision of non-face-to-face services, and the role of digital technologies in the context of social emergencies [45–47]. This proved to be particularly important during the recent COVID-19 outbreak because it enabled the optimal operation of public libraries through non-face-to-face services and other alternatives. Prior to the COVID-19 pandemic, most studies focused on specific social issues. Since then, international research trends have begun to emphasize public libraries' provision and popularization of information services as well as their user satisfaction, spatial storytelling, complexation, and placement [48–50]. Similarly, Korean-based research studies have been focusing on public libraries from around 2015 [51–53]. This trend shows that most of the research conducted by Korean and foreign scholars has focused on the development of public libraries compared to the previous years. In contrast, most studies on public libraries related to urban spaces in Korea have excluded remote library locations, as opposed to administrative districts, central regions, and easy-to-access areas [54–56], and, hence, the importance of transportation factors and the entire urban space is usually not included.

However, a recent study by Zhao evaluated the absolute change in the utilization rate of public libraries and found that this rate was higher in the outskirts of Seoul compared to the average utilization rate of public libraries in Seoul [25]. Although Zhao's study did not explicitly mention the causal relationship between the utilization rate of public libraries and location factors, many preceding studies have interpreted the decline in accessibility in urban outskirts as a hierarchical relationship of urban structure [57–61]. Although the causal relationship between the accessibility of public libraries and their DUA was already established in the past [32–35], Zhao's findings suggest that this causal relationship may not be valid in the present context. As this situation obscures the relationship between the geographic location of public libraries and their usage, it is necessary to review whether accessibility can still be used to estimate utilization rates and improve the usage of public libraries. Further, various studies have been conducted on the urban spatial structure of Seoul around 2010 [62–64]; however, the introduction of new urban plans dictates that we need to update the urban hierarchy of Seoul accordingly.

Therefore, this study investigated the urban hierarchy of Seoul, reviewed the development, distribution, and usage of public libraries, and re-examined the causal relationship between accessibility and utilization of public libraries with respect to the present day. Consequently, our findings aimed to present novel directions to improve the usage of public libraries in densely populated cities characterized by rapidly growing cultural demand, and provide baseline data for establishing and opening new public libraries in the future.

1.3. Range and Hypothesis

According to Article 2-1 of the Libraries Act of Korea [65], public libraries are classified into governmental public libraries founded by the state or a local government and nongovernmental public libraries established by private organizations, and both fall under the category of cultural infrastructure intended to be used freely by citizens. In addition, they are classified into general public libraries, small libraries, libraries for the disabled, hospital libraries, barrack libraries, prison libraries, and children's libraries, according to their characteristics. However, in this study, the libraries for the disabled, hospital libraries, barrack libraries, prison libraries, and children's libraries are excluded because they have specific purposes and user groups. Generally, this study only focused on general public libraries and small libraries, which are likely accessible to the general public, and analyzed their location accessibility, accessibility considering control factors, and DUA. The DUA of public libraries is based on the number of visitors for five consecutive years (from 2015 to 2019 in this study). Consequently, libraries opened after 2016 were also excluded from the scope of this study because the DUA can only be calculated when the same public library operates continuously year after year and because the number of visitors to public libraries has significantly decreased since 2020 as a result of the COVID-19 pandemic [66].

The causal relationship investigated in this study is a concept that has been widely used and mentioned in the field of philosophy [67,68]. More specifically, causal relationships are generally divided into temporal heterochronous (heterochronous causality) and spatial synchronic (synchronic causality) causal relationships [69–71]. Among them, the synchronic causality presented in philosophy has stimulated numerous debates and issues in the academic community due to the development of the quantum entanglement theory in physics [72,73]. However, this study raises hypotheses and validates them by applying heterochronous causality instead of synchronic causality. As mentioned in Section 1.1, the first reason is that this study considers academic arguments on the inevitable causal relationship in which the accessibility of public libraries in Seoul (cause) affects the DUA (result) [27,34,35] and the fact that the causal relationship of events exists in a temporal order. Second, the interpretation of causal relationships in David Hume's ontology is viewed as a process of inferring other beings from an individual being. In addition, human movements are not determined by reason but, instead, by customs and habits [67,70,71].

Consequently, these reasons can be applied to the present study to confirm this causal relationship by referring to the relationship between DUA and the accessibility of public libraries in Seoul in the past. However, according to Hume [71], the changes in usage caused by social factors and functional changes in public libraries correspond to the changes in human customs and habits. Hence, it is necessary to review whether a causal relationship is established due to past causes in anticipation of the changed results.

Therefore, this study sets the accessibility factor of public libraries as an independent variable, corresponding to the "cause," to verify the causal relationship between accessibility and usage. In contrast, the DUA of public libraries in Seoul is set as a dependent variable corresponding to the "result." Since this study attempts to verify this causal relationship in the present, the null and alternative hypotheses below were put forward by assuming that past causal relationships could also be applied in the present day. The logical process for verifying causality is shown in Figure 1.

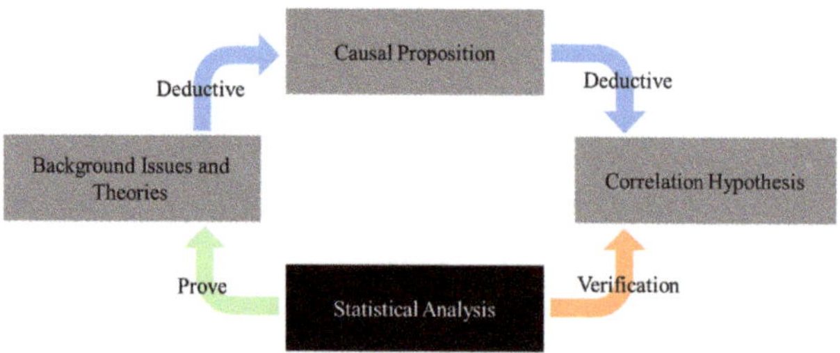

Figure 1. The process of validating causality.

The null hypothesis: The accessibility of public libraries in Seoul affects the DUA and establishes a causal relationship between accessibility and usage.

Alternative hypothesis: The accessibility of public libraries in Seoul does not affect the DUA and does not establish a causal relationship between accessibility and usage.

In addition to the location accessibility of public libraries mentioned above, the controlling factors of accessibility are also essential and must be considered. Thus, in this study, accessibility was divided into the urban accessibility of public libraries and the accessibility of public libraries considering controlling factors, and both cases were examined.

2. Space Syntax and Research Method

2.1. Space Syntax

Hillier and Hanson collaboratively created a set of methods for investigating spatial arrangements (known as space syntax) [74]. According to Figure 2 [75], a space is usually analyzed using convex spaces and axial maps in space syntax. Correlation studies were conducted to evaluate spatial usage patterns and spatial accessibility by calculating their integration with spatial depths as fundamental units [76–78], which have been proven to have significant relevance to social, cultural, economic, and political phenomena in cities [79–84]. Among them, a convex space is a closed polygon in which all interior angles are less than 180°, and an axial map reproduces a space with lines connecting the visual maximum points of the connected spaces [85]. However, when the interior angle of a convex space exceeds 180° or when a refracted space must be expressed with an axial map, the space is typically divided into two or more corresponding spaces [75]. In addition, spatial depth should not be perceived as the actual distance but instead as the proportionate number of spaces necessary to go from one location to another. Based on these rules, Equation (1) shows how to calculate the integration between a convex space and an axial map [74], and Table 1 summarizes the meaning of each symbol [25].

$$I_{(i)} = \frac{D_n}{\frac{2}{n-2}\left(\frac{\sum_{k=1}^{n} d(i,k)}{n-1} - 1\right)} \tag{1}$$

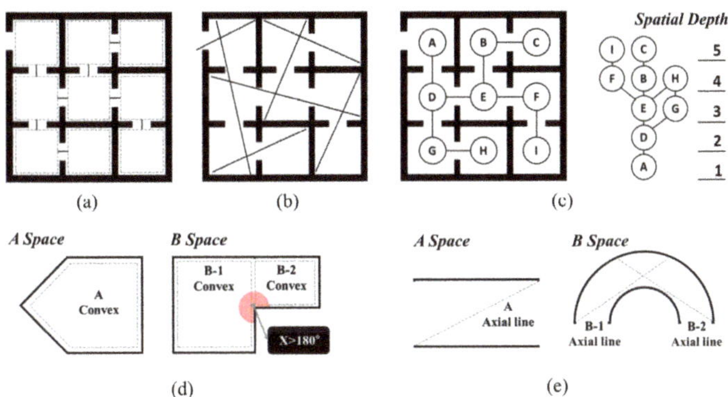

Figure 2. An example of a spatial analysis form: (**a**) convex space; (**b**) axial map; (**c**) J-Graph; (**d**) creating a convex space; (**e**) creating an axial map.

Table 1. The interpretation of space–syntax expressions.

Type	Interpretation
MD	Average spatial depth
TSDi	Total spatial depth in space i
S	The number of steps taken through space i
m	The number of steps from space i to the deepest space
K	The total number of spaces
Ks	The number of spaces in Step S
Dn	Correction factor
d(i, k)	The depth from space i to space k
n	The total number of nodes

Space syntax is used to create an urban hierarchy integration index based on axis maps and, thus, investigate the location accessibility of public libraries in Seoul. In terms of the accessibility of public libraries, which is the controlling factor, the present study assumes subway stations as a single convex space, subway stations were connected according to the subway lines, and the distance within 2000 m of each station was set as one spatial depth. The integration was calculated using a topological structure formed by connecting a subway station to the public libraries within a 2000 m radius from the origin. This depth was selected due to the following reasons: First, public libraries have a relatively low probability of being located adjacent to subway stations, as most of them are distant from residential areas [32–34]. Second, according to a 2012 press release by the Ministry of Land, Transport, and Maritime Affairs [86], a relatively low rate of people use transportation to travel within a distance of 2000 m in the city of Seoul, which corresponds to a walking distance for most Koreans. Accordingly, Figure 3 demonstrates an example of an axis map and a convex space of Seoul.

2.2. Research Method

The assessment of the country's cultural infrastructure released by the Ministry of Culture, Sports, and Tourism [87] in 2021, served as a foundation for compiling fundamental data on public libraries in Seoul. The status of operating general public libraries and small libraries from 2015 to 2019 was investigated by analyzing data from the national library statistics system [88]. Among them, the public libraries that opened after 2015 but were then closed, or did not provide a record of the number of visitors, were excluded from the study. Since this study calculated DUA based on the number of visitors for five consecutive years, there are deviations with regards to the quantification of the data obtained. In addition, this study started with the premise of reflecting both the location accessibility of public libraries and the accessibility considering the control factors. Therefore, public libraries outside a 2000 m radius from each subway station and those distributed in blind spots were excluded from this study.

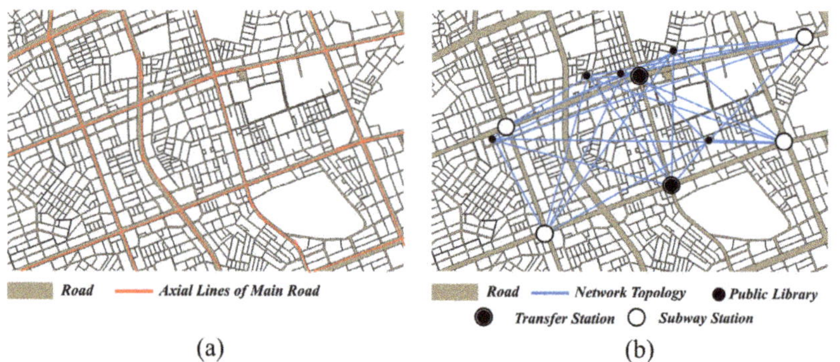

(a)   (b)

Figure 3. Example of creating an axial map and convex space in an area of Gangnam-gu, Seoul: (**a**) the axial lines represent major roads; (**b**) a convex space topological structure of public libraries around a subway station within a 2000 m radius.

According to the above process, there was a total of 1099 public libraries (195 general public libraries and 904 small libraries) in Seoul as of 2021. Among them, 783 were relevant to the scope of this study. Those excluded comprised 296 libraries that closed, stopped operating, or did not have any records on the number of their visitors for five consecutive years. Twenty libraries located in blind spots outside the areas covered by subway lines were also excluded. Information on the subway stations comprising convex spaces was collected on the basis of public data from the Seoul Subway Operation Status Statistics [89]. Based on the Seoul Subway Lines 1 to 9, namely the Gyeongui–Jungang, Gyeongchun, Airport Railroad Express, Bundang, and Ui-Sinseol Lines, 218 general stations and 67 transfer stations were connected to 783 public libraries to create a topological structure consisting of 1068 convex spaces. The National Spatial Information Portal's GIS public data on the status of ground roads, issued in October 2022, were used as basic data for the Seoul city axial map to examine the accessibility of public library locations [90]. The current state of the roads in Seoul was derived as a DXF file using QGIS and then adjusted to a scale of 1:1 based on a metric system in Auto CAD to arrange a total of 167,292 axial lines. Figure 4 shows the final axial map and the topological structure connecting the convex spaces. Based on the above, the integration of public libraries derived using a space syntax was an independent variable in this study.

In cases where the period for measuring the DUA of public libraries was short, an absolute method was used by expressing the rate of change in the number of visitors to public libraries with the previous year. However, as this study measured the DUA based on the number of visitors for five consecutive years, a relative method was instead used by assuming the starting year as 100% and reviewing the number of visitors for the next four years compared with the starting year to calculate the average value [25]. Therefore, as shown in Equation (2), the final DUA was calculated on the basis of the number of visitors N. Assuming that the value of 2015 was 1, the sum of each ratio, namely N_{2016}/N_{2015}, N_{2017}/N_{2015}, N_{2018}/N_{2015}, and N_{2019}/N_{2015}, was determined by dividing with the total measurement period of five years to obtain DUA, and the closer this value was to 1, the more stable the retention rate was. A value less than 1 indicated the degree of inactivation, and a value greater than 1 indicated the degree of activation. Using the above process, DUA was set as a dependent variable corresponding to the result in this study.

$$DUA = \frac{1 + \sum (i = N_{2016}, N_{2019}) \frac{N_i}{N_{2015}}}{n} \tag{2}$$

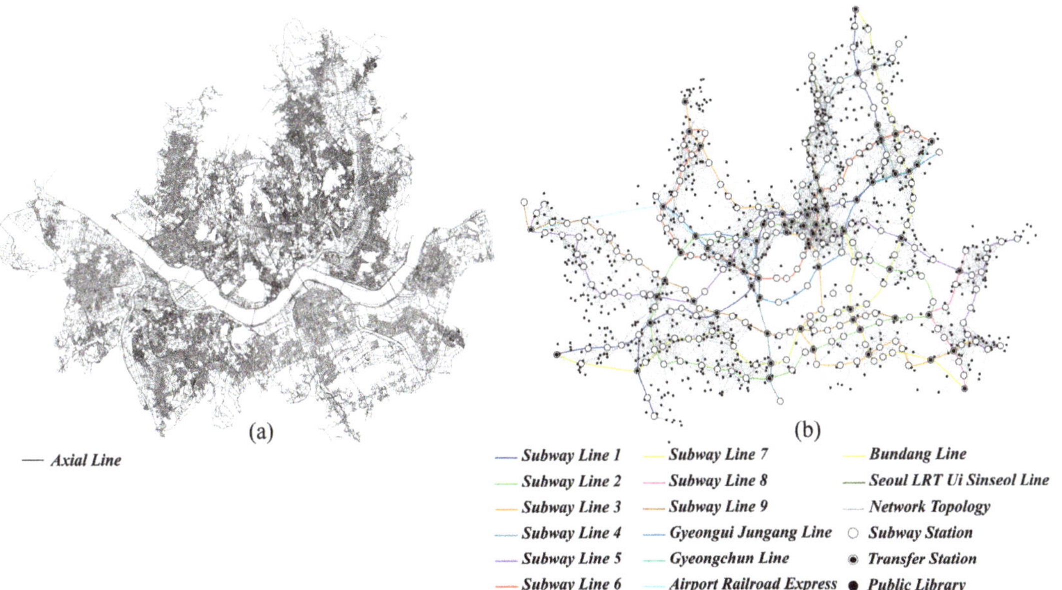

Figure 4. Spatial analysis for calculating public library accessibility indicators: (**a**) an axial map of Seoul; (**b**) a convex space topological connection structure of public libraries along the subway lines.

The location accessibility of public libraries, accessibility considering the control factors, and DUA were calculated using the above-mentioned procedure. This study used statistical techniques to verify the normality and independence of data, confirmed the correlation between the independent and dependent variables, and applied regression analysis to determine the influence of the respective causal relationships. However, a linear or nonlinear correlation between the independent and dependent variables could not be obtained from the correlation analysis; the null hypothesis was rejected, and the alternative hypothesis was adopted. The reason was that the premise of conducting regression analysis to statistically confirm causality must be preceded by a correlation. In addition, to compare the index values derived in this study on the same scale, the index values derived using Equation (3) [55] were organized in descending order and converted into relative values within a range of 1.000–0.000. Table 2 demonstrates how to interpret each symbol in the equation, and Figure 5 shows the overall research procedure. In addition, Depthmap, which was developed by UCL, was used to derive the integration indicators using space syntax, and a multiple regression analysis was conducted using IBM SPSS Statistics 25.

$$A_{(r)} = 1 - \frac{N_x - N_{min}}{N_{max} - N_{min}} \qquad (3)$$

Table 2. The relative value calculation formula interpretation.

Type	Interpretation
A(r)	The relative interval value of the absolute value
N	The number
Nx	The absolute sequence number of the corresponding
Nx	Nmin ≤ Nx ≤ Nmax
Min	The minimum value of the sequence number
Max	The maximum value of the sequence number

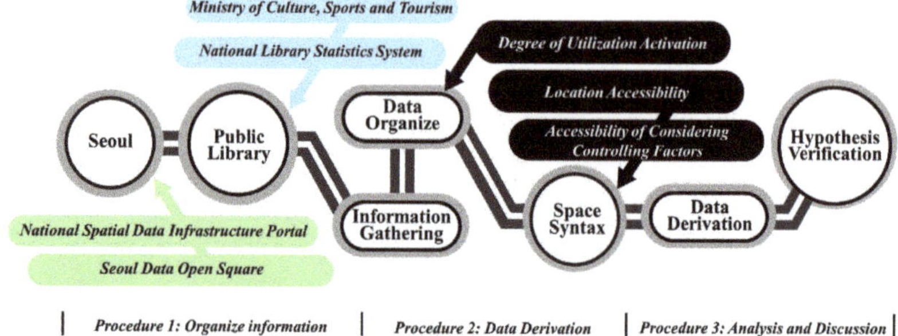

Figure 5. Research process.

3. Analysis and Discussion

3.1. Distribution and Topological Status of Public Libraries in Seoul

Pertaining to the distribution of 783 public libraries by administrative district, Figure 6 shows that Eunpyeong-gu had the greatest number of public libraries, followed by Songpa-gu, Guro-gu, and Gangseo-gu. In contrast, Jung-gu had the lowest number of public libraries. Although a growth in the number of public libraries was far from visible until the 1990s, this number has since rapidly increased due to the greater demand for cultural activities, as previously mentioned. Furthermore, the increasing number of public libraries was a phenomenon that practically defined the framework to meet this cultural demand. In addition, general public libraries accounted for the majority of public libraries that opened before the 1990s, while the number of small libraries was very low. However, when examining all public libraries, 59.6% were public, 40.4% were private, depending on the founding body, 15.4% were general public libraries, and 84.6% were small libraries. This trend highlights that government agencies were the main entities operating public libraries until the 1990s. However, since the 1990s, an increasing number of libraries have been jointly operated by the government and the private sector. In addition, this development trend also demonstrated a clear shift in the administrative pattern of libraries from general public libraries to small libraries. Considering these aspects, including the rapid increase in the supply of public libraries since the 1990s and their functional changes, the dispersed distribution of public libraries in Seoul is notable, with a concurrent large-scale spread of small libraries, as opposed to an intentional absolute concentration.

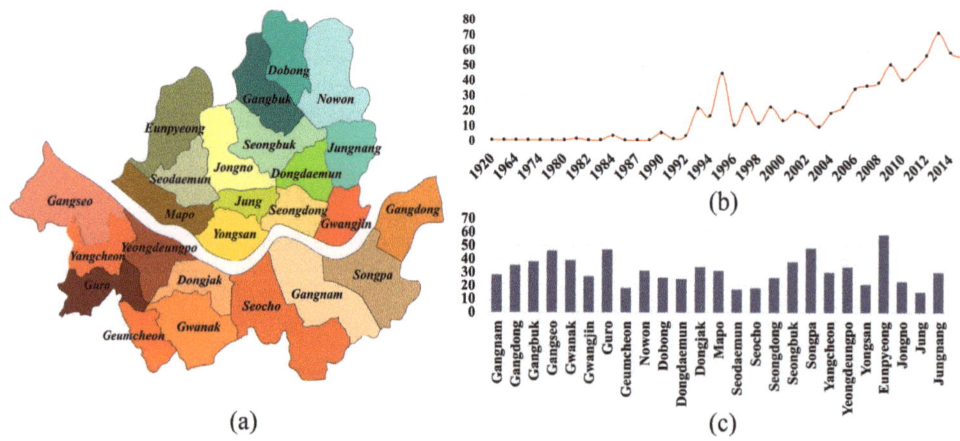

(a) (b) (c)

Figure 6. The current status of public libraries in Seoul: (**a**) administrative districts in Seoul; (**b**) trends in public library growth by year; (**c**) distribution of public libraries by administrative district.

According to Table 3, the maximum, minimum, and average values of the location integration of public libraries in Seoul were 0.569, 0.057, and 0.382, respectively. In terms of the integration of public libraries considering control factors, the maximum, minimum, and average values were 1.494, 0.661, and 1.030, respectively, and the average value of the index expressing DUA was 1.279. Since DUA was close to 1, the total degree of using public libraries in Seoul corresponded to a level of maintaining the utilization rate. However, the significant gap between the maximum (16.124) and minimum (0.290) values indirectly showed the potential of large disparities in DUA among public libraries.

Table 3. Urban hierarchy actuality in Seoul.

Type	Location Accessibility	Integration Accessibility Considering Control Factors	Degree of Utilization Activation
Axial line/Convex space/Total library	167,292	1068	783
Max	0.569	1.494	16.124
Min	0.057	0.661	0.290
Avg	0.382	1.030	1.279

Figure 7 exhibits the integration of public libraries in Seoul by describing the location accessibility of public libraries and accessibility considering the control factors. In terms of the location, the administrative districts of Gangnam-gu, Seocho-gu, Gwangjin-gu, Seongdong-gu, and Dongdaemun-gu revealed a relatively high integration. In contrast, Jung-gu and Jongno-gu districts showed the highest accessibility of public libraries according to the control factors. These results may help explain how control factors affect location accessibility. Figure 8 shows a 3D scatter plot created by converting the location integration of public libraries, integration according to the control factors, and DUA into relative values. In that graph, most public libraries with a relative DUA above 0.6 corresponded to small libraries, and small private libraries were concentrated between 0.8 and 1.0.

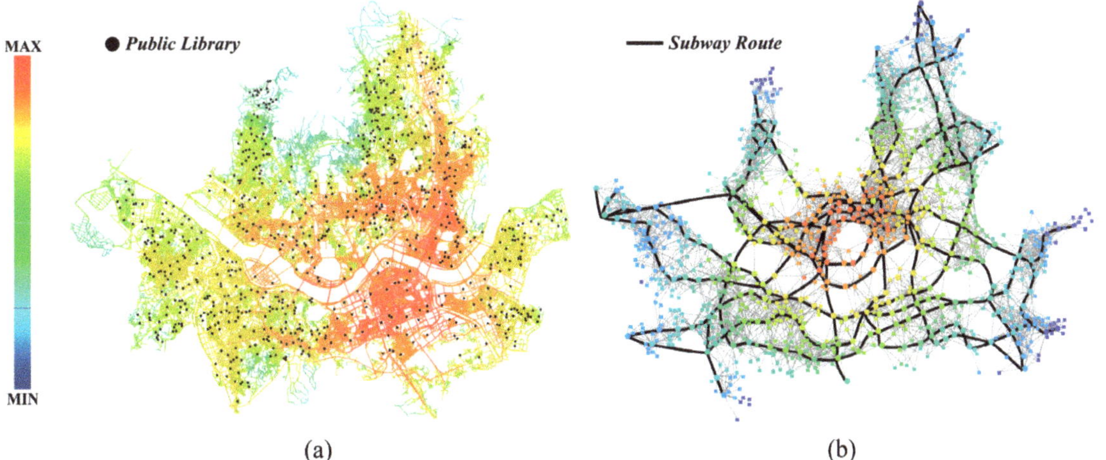

Figure 7. The integration of public libraries in Seoul: (**a**) axial map integration graph; (**b**) convex space integration graph.

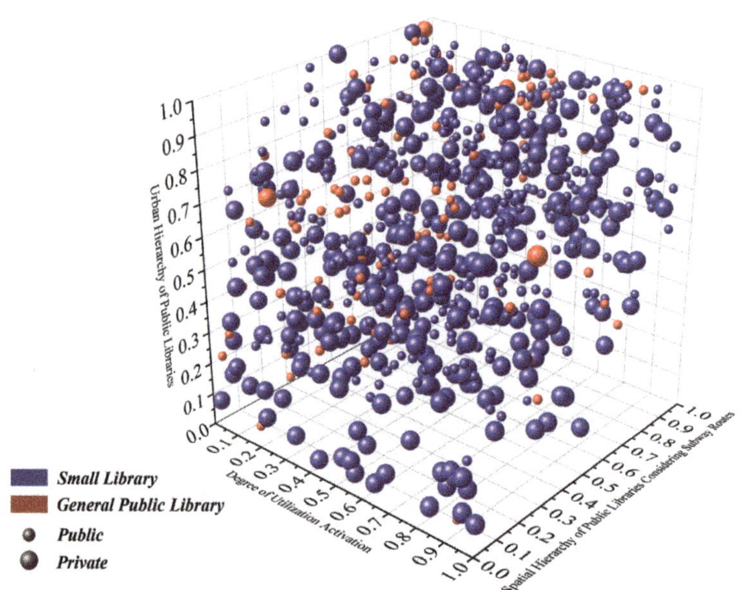

Figure 8. A 3D scatter plot created by the location integration of public libraries, integration according to control factors, and DUA.

3.2. Correlation Analysis of Factors

Prior to validating the correlation of our data, it is necessary to verify the normality and independence of these relative values. Therefore, the normal data distribution was firstly examined using a P-P plot.

According to Figure 9, the location accessibility of public libraries, accessibility considering the control factors, and DUA-converted relative values all fell within a range close to the diagonal of the normal P-P plot. The residuals were distributed above and below the Y = 0 baseline in a ± 0.06 range. Therefore, our data followed a normal distribution. Based on this, the Friedman test was conducted on the derived relative values to determine the independence of the overall distribution of the three indicator variables. As previously mentioned, attributes of sequential data were included when converting absolute values into relative values so as to organize these data in descending order. Table 4 shows the results of the analysis. The chi-square value of the location accessibility of public libraries, accessibility considering the control factors, and DUA was 136.350, corresponding to a significance level of $p < 0.001$. Therefore, there is a clear difference in the overall distribution of the three variables.

Table 4. Descriptive statistics.

Kinds	N	Mean	Std. Deviation	Min	Max	25th	Percentiles 50th (Median)	75th	Mean Rank	Chi-Square	Sig.
Location Accessibility ***		0.620	0.247	0.036	0.999	0.431	0.652	0.819	2.33		
Accessibility Considering Control Factors ***	783	0.470	0.286	0.002	0.999	0.219	0.469	0.706	1.78	136.350	0.000
DUA ***		0.500	0.289	0.001	0.998	0.249	0.500	0.750	1.89		

*** $p < 0.001$.

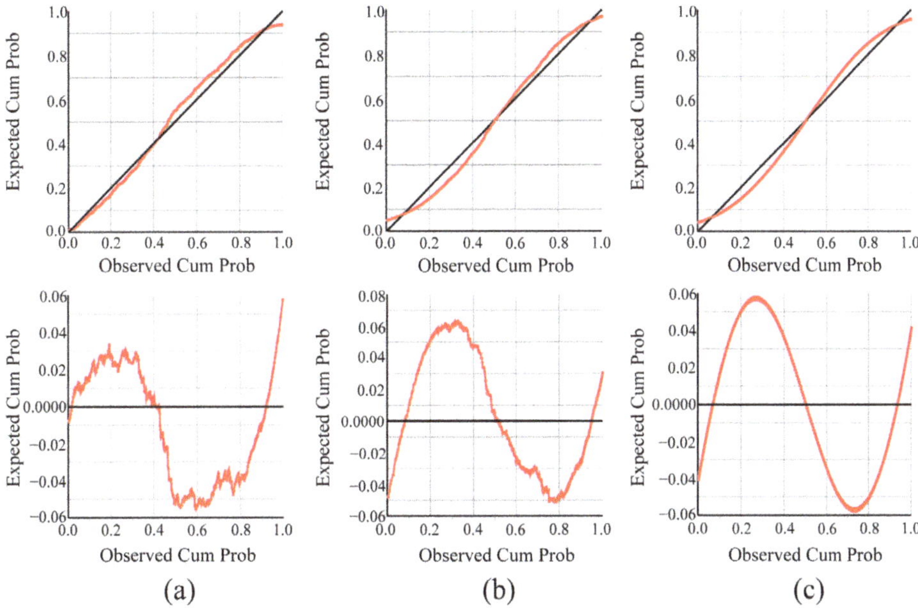

Figure 9. Normal and detrended normal P−P plot: (**a**) urban hierarchy of public libraries; (**b**) spatial hierarchy of public libraries considering the subway routes; and (**c**) degree of utilization activation.

Consequently, a correlation analysis was conducted between the location accessibility, accessibility considering the controlling factors, and the DUA of public libraries in Seoul. However, the correlation between the variables may appear in either linear or nonlinear distributions. Therefore, the correlation analysis was conducted separately using the Pearson correlation coefficient and Spearman's rho correlation coefficient, respectively.

As shown in Table 5, the Pearson's r correlation coefficient was 0.469 between the location accessibility and accessibility considering the control factors and 0.028 between the location accessibility and DUA, and the significance levels were $p < 0.01$ and $p > 0.05$, respectively. The Pearson's r correlation coefficient between the accessibility considering the controlling factors and DUA was −0.002, and the significance level was $p > 0.05$. Considering these results, the location accessibility and accessibility considering the controlling factors (independent variables) may be connected through a linear relationship. However, there was no linear relationship found between the independent and dependent variables.

Table 5. The linear correlation.

	Correlations	Location Accessibility	Accessibility Considering Control Factors	DUA
Location Accessibility	Pearson Correlation	1		
	Sig. (2-tailed)			
	N	783		
Accessibility Considering Control Factors	Pearson Correlation	0.469 **	1	
	Sig. (2-tailed)	0.001		
	N	783	783	
DUA	Pearson Correlation	0.028	−0.002	1
	Sig. (2-tailed)	0.433	0.945	
	N	0783	783	783

** Correlation is significant at the 0.01 level (2-tailed).

Table 6 shows the results of reviewing the nonlinear correlation using Spearman's rho correlation coefficient. In general, the Spearman's rho correlation coefficients between the independent variables (location accessibility and accessibility considering the control factors) and the dependent variable (DUA) were 0.025 and -0.002, respectively, and the significance level was $p > 0.05$. The Spearman's rho correlation coefficient between the location accessibility and accessibility considering the controlling factors was 0.468 and the significance level was $p < 0.01$.

Table 6. Nonlinear correlation.

	Correlations		Location Accessibility	Accessibility Considering Control Factors	DUA
Spearman's rho	Location Accessibility	Correlation Coefficient	1		
		Sig. (2-tailed)			
		N	783		
	Accessibility Considering Control Factors	Correlation Coefficient	0.468 **	1	
		Sig. (2-tailed)	0.001		
		N	783	783	
	DUA	Correlation Coefficient	0.025	-0.003	1
		Sig. (2-tailed)	0.490	0.930	
		N	783	783	783

** Correlation was significant at the 0.01 level (two-tailed).

By performing the correlation analysis using Pearson's r and Spearman's rho correlation coefficients, the location accessibility and the accessibility considering controlling factors (the independent variables) were likely to demonstrate linear or nonlinear distributions and appeared significant in terms of correlation. However, according to the research design, the accessibility considering controlling factors is a result calculated based on location accessibility. In other words, although the independence and normality of the data could be confirmed by a normal P-P plot, the research design corresponded to the two cases of expressing accessibility using the same space syntax theory. Therefore, the independent variable was considered to be the cause of internal correlation in the correlation analysis. In addition, the accessibility of public libraries (the independent variable) in both cases did not correlate with the DUA (the dependent variable), in either linear or nonlinear distributions. Therefore, the null hypothesis proposed in this study was rejected, and the alternative hypothesis was adopted instead.

3.3. Discussion

Our findings revealed that the DUA of public libraries did not correspond to the results obtained from interpreting the location accessibility of public libraries and accessibility considering the control factors. These results proved that there is inadequate evidence for estimating DUA based on the accessibility of public libraries in the present day in Seoul and indirectly explained why the utilization of public libraries in the outskirts of Seoul was higher than the average utilization rate published in Zhao's study. Furthermore, the present study demonstrated that the accessibility of public libraries in Seoul was not a major factor for affecting DUA. Although these findings do not exclude a causal relationship in terms of estimating DUA based on accessibility, as indicated previously, they do confirm that the demand for public libraries is determined by changes in customs and habits, in accordance with David Hume's theory of causation. In addition, it becomes evident that it is extremely difficult to apply causal relationships established in the past directly to the present time. Accordingly, it is possible that the main reason people use public libraries is their personal will to obtain knowledge, regardless of whether physical access is easy or not. In other words, the determinants of public library DUA today are highly likely to be influenced by internal factors, such as user motivation, public library service, and space.

4. Conclusions

This study investigated the development, distribution, and utilization rate of public libraries in Seoul to review whether we could still apply the previously established causal relationship between the accessibility of public libraries and DUA to the present time. The location accessibility of public libraries, accessibility considering subway control factors, and DUA were calculated, and research hypotheses were raised. The results of the study can be summarized as follows.

First, the number of public libraries in Seoul has significantly increased since the 1990s. Until the 1990s, relatively large general public libraries were predominant. Since then, the government and the private sector have collaborated to initiate the rapid growth of small libraries in Seoul. In addition, small private libraries accounted for most of the public libraries with a high DUA. Public libraries in Seoul corresponded to a level of maintaining the utilization rate on average; however, there were large discrepancies in DUA among public libraries.

Second, consideration of the subway control factors revealed a distinct difference in physical accessibility between the location accessibility of public libraries in Seoul and the hierarchical center of accessibility, confirming that the addition of additional controlling factors, such as subway routes, can significantly affect the hierarchical accessibility of public libraries. In addition, the location accessibility of public libraries and accessibility considering subway control factors appeared to have independent topological structures, although correlating with each other.

Third, the recent accessibility of public libraries in Seoul did not have a causal relationship with the DUA. The results of the correlation analysis based on linear and nonlinear distributions showed that the independent variable corresponding to the accessibility of public libraries set in this study did not correlate with the DUA. It should be mentioned that these results do not contradict the causal relationship claimed by previous researchers. However, it can be suggested that the reason for interpreting the usage rate according to the development of public libraries may have shifted from the accessibility of public libraries to the users' motivation or the public libraries themselves.

The results of this study are significant in that they present directions for improving the use of public libraries in densely populated cities with growing cultural demand and provide baseline data for establishing and opening new public libraries. However, the present study has certain limitations. For instance, we did not consider the combined use of private transportation, buses, highways, and public transportation when investigating accessibility considering the control factors; therefore, further research needs to be conducted to supplement the results of the present study by including other transportation means. In addition, it is necessary to shift the focus of this research to the users of public libraries or the space itself to understand the factors that directly affect DUA.

Author Contributions: Conceptualization, K.H. and X.Z.; methodology, X.Z.; writing—original draft preparation, X.Z.; writing—review and editing, K.H.; supervision, K.H. All authors have read and agreed to the published version of the manuscript.

Funding: This research was supported by the BK21 FOUR Service-Design-driven Social Innovation Educational Research Team at Dongseo University.

Institutional Review Board Statement: Not applicable.

Informed Consent Statement: Not applicable.

Data Availability Statement: No additional data are available.

Conflicts of Interest: The authors declare no conflict of interest.

References

1. Korea Promoted to Developed Nation by Unctad. Available online: https://koreajoongangdaily.joins.com/2021/07/04/business/economy/Unctad-developed-country-developing-country/20210704185600398.html (accessed on 1 September 2022).
2. UNCTAD Classifies, S. Korea as Developed Economy. Available online: https://english.hani.co.kr/arti/english_edition/e_international/1002230.html (accessed on 1 September 2022).
3. Wang, C.; Ghadimi, P.; Lim, M.K.; Tseng, M.L. A literature review of sustainable consumption and production: A comparative analysis in developed and developing economies. *J. Clean. Prod.* **2019**, *26*, 741–754. [CrossRef]
4. Guerrero, M.; Liñán, F.; Cáceres-Carrasco, F.R. The influence of ecosystems on the entrepreneurship process: A comparison across developed and developing economies. *Small Bus. Econ.* **2021**, *57*, 1733–1759. [CrossRef]
5. Zhang, C.Q.; Chen, P.Y. Applying the three-stage SBM-DEA model to evaluate energy efficiency and impact factors in RCEP countries. *Energy* **2022**, *241*, 122917. [CrossRef]
6. Dova, E.; Sivitanidou, A.; Anastasi, N.R.; Tzortzi, J.G.N. A mega-event in a small city: Community participation, heritage and scale in the case of Pafos 2017 European Capital of Culture. *Eur. Plan. Stud.* **2022**, *30*, 457–477. [CrossRef]
7. Sanetra-Szeliga, J. Culture and heritage as a means to foster quality of life? The case of Wrocław European Capital of Culture 2016. *Eur. Plan. Stud.* **2022**, *30*, 514–533. [CrossRef]
8. Plaza, B.; Aranburu, I.; Esteban, M. Superstar museums and global media exposure: Mapping the positioning of the Guggenheim Museum Bilbao through networks. *Eur. Plan. Stud.* **2022**, *30*, 50–65. [CrossRef]
9. Sand, J. *Tokyo Vernacular: Common Spaces, Local Histories, Found Objects*; University of California Press: Berkeley, CA, USA, 2013.
10. Bertacchini, E.; Nuccio, M.; Durio, A. Proximity tourism and cultural amenities: Evidence from a regional museum card. *Tour. Econ.* **2021**, *27*, 187–204. [CrossRef]
11. Prentice, R. Experiential cultural tourism: Museums & the marketing of the new romanticism of evoked authenticity. *Mus. Manag. Curatorship* **2001**, *19*, 5–26. [CrossRef]
12. Brida, J.G.; Disegna, M.; Vachkova, T. Visitor satisfaction at the museum: Italian versus foreign visitor. *Tourism. Int. Interdiscip. J.* **2013**, *61*, 167–186.
13. Ahmad, S.; Abbas, M.Y.; Taib, M.Z.M.; Masri, M. Museum Exhibition Design: Communication of meaning and the shaping of knowledge. *Procedia Soc. Behav. Sci.* **2014**, *153*, 254–265. [CrossRef]
14. Blunden, J. The Language with Displayed Art(efacts): Linguistic and Sociological Perspectives on Meaning, Accessibility and Knowledge-Building in Museum Exhibitions. Ph.D. Thesis, University of Technology Sydney, Ultimo, NSW, Australia, 2017.
15. Shen, L. Out of information poverty: Library services for urban marginalized immigrants. *Urban Libr. J.* **2013**, *19*, 4.
16. Begum, D.; Roknuzzaman, M.; Shobhanee, M.E. Public libraries' responses to a global pandemic: Bangladesh perspectives. *IFLA J.* **2022**, *48*, 174–188. [CrossRef]
17. Mushtaq, A.; Arshad, A. Public library use, demographic differences in library use and users' perceptions of library resources, services and place. *Libr. Manag.* **2022**, *43*, 563–576. [CrossRef]
18. Martin, J.; Jennings, M. Tomorrow's museum: Multilingual audiences and the learning Institution. *Mus. Soc. Issues* **2015**, *10*, 83–94. [CrossRef]
19. Cha, J. Multilingual Museums: A Proposal to Increase Linguistic Diversity in Contemporary Art Museums. Master's Thesis, University of San Francisco, San Francisco, CA, USA, 2018.
20. Kwon, N.H.; Song, K.J. A national study explaining the public library use among korean adults: Examining the influence of individual characteristics. local library inputs, and local government investments. *J. Korean Biblia Soc. Libr. Inf. Sci.* **2014**, *25*, 291–312. [CrossRef]
21. Han, Y.O.; Cho, M.A.; Kim, S.K. A study on the current states and problems for multi-cultural families in libraries. *J. Korean Soc. Libr. Inf. Sci.* **2009**, *43*, 135–160. [CrossRef]
22. Shim, J.Y. Identifying information needs of public library users based on circulation data: Focusing on public libraries in Seoul. *J. KOSIM* **2021**, *38*, 173–199. [CrossRef]
23. Lee, K.J. The influence factors on the numbers of visitors and reference room users of public libraries: Based on the national libraries statistical data 2018. *J. Korean Soc. Libr. Inf. Sci.* **2020**, *54*, 105–125. [CrossRef]
24. Kim, M.H.; Moon, J.M. A Study on the correlation between viewing behavior and exhibiting methods in museums-focusing on viewing behavior on weekdays and weekends in medium sized history museums in Korea. *J. Asian Archit. Build. Eng.* **2013**, *12*, 173–180. [CrossRef]
25. Zhao, X.L. A Study on Urban Hierarchical Characteristics of Public Libraries: Focused on the Libraries in the Boundary District(gu) of Seoul, Korea. Ph.D. Thesis, Sangmyung University, Seoul, Republic of Korea, 2022.
26. Ko, H.K.; Lim, C.Z.; Lim, H.K. An analysis of the planning guideline of space size for public library. *J. Archit. Inst. Korea Plan. Des.* **2012**, *28*, 139–146.
27. Yoon, H.Y. Current status and policy tasks of public library in Seoul. *J. Korean Libr. Inf. Sci. Soc.* **2018**, *49*, 1–23. [CrossRef]
28. Seoul Infographics. Available online: https://www.si.re.kr/node/54729 (accessed on 10 September 2022).
29. Song, S.E.; Kim, S.T. Research on changes of spatial configuration due to complexation of public library. *Korean Inst. Inter. Des. J* **2011**, *20*, 311–320.
30. Choi, Y.J.; Yoon, D.S. A study on the mixed-use planning of public libraries for community revitalization. *Korean Inst. Inter. Des. J.* **2020**, *29*, 116–125. [CrossRef]

31. Chang, Y.K. A study on the concepts and programs of 'Makerspaces' at public libraries. *J. Korean Soc. Libr. Inf. Sci.* **2017**, *51*, 289–306. [CrossRef]
32. Park, S.J.; Lee, J.Y. A study on the site selection of public libraries using analytic hierarchy process technique and geographic information system. *J. Korean Soc. Inf. Manag.* **2005**, *22*, 65–85. [CrossRef]
33. Yim, S.J.; Kim, C.H. Analysis of selecting an appropriate location for infrastructure considering the range of service area and accessibility: Focusing on public libraries. *J. Korean Reg. Dev. Assoc.* **2020**, *32*, 45–66. [CrossRef]
34. Kim, H.J.; Lee, J.G.; Yeo, K.H. Regional disparities and determinants of spatial accessibility of public libraries in Seoul. *Seoul Stud.* **2015**, *16*, 109–127. [CrossRef]
35. Won, J.J.; Ahn, K.H. The effect of locational and facility characteristics on public library use: Focused on public libraries in Seoul. *J. Archit. Inst. Korea Plan. Des.* **2010**, *26*, 79–86.
36. Park, J.S.; Lee, K.S. Development of integrated accessibility measurement algorithm for the seoul metropolitan public transportation system. *J. Korean Reg. Sci. Assoc.* **2017**, *33*, 29–41, UCI: G704-000510.2017.33.1.001.
37. Lee, K.T.; Kim, I.K.; Shim, J.Y.; Kim, J.H. Improvement of Zone-based regional transportation mode choice model reflecting behavioral difference by travel distance. *J. Korean Soc. Transp.* **2020**, *38*, 346–360. [CrossRef]
38. Kim, Y.; Park, J.H. Factors influencing the intention of knowledge sharing in public Libraries: Based on self-determination theory. *J. Korean Biblia Soc. Libr. Inf. Sci.* **2021**, *32*, 247–265. [CrossRef]
39. Han, S.W.; Chang, W.K. A study on user's perception survey for establishment of regional base library: Centered on gwangju metropolitan library. *J. Korean Libr. Inf. Sci. Soc.* **2022**, *53*, 267–292. [CrossRef]
40. Kang, Y.J.; Lee, J.Y. Study on the place image factors for place branding of public libraries. *J. Korean Soc. Libr. Inf. Sci.* **2022**, *56*, 129–159. [CrossRef]
41. Jeong, T.K.; Seo, H. A study on the openness of reading space in the public libraries: Focused on the public libraries in Seoul. *J. Archit. Inst. Korea Plan. Des.* **2022**, *38*, 55–63. [CrossRef]
42. Song, M.S.; Chang, I.H.; Hoang, G.S. A study on perceptions of users and specialists for establishing mid-to long term development plan for libraries in anyang city. *J. Korean Soc. Libr. Inf. Sci.* **2022**, *56*, 67–93. [CrossRef]
43. Lee, S.S.; Lee, T.S.; Joo, S.H. Content analysis of online book curation services in korean public libraries. *J. Korean Libr. Inf. Sci. Soc.* **2022**, *53*, 189–209. [CrossRef]
44. Yoon, S.Y. Big data platform for public library users: Focusing on the cultural programs and community service. *J. Korean Biblia Soc. Libr. Inf. Sci.* **2022**, *33*, 347–370. [CrossRef]
45. Jones, S. Optimizing public library resources in a post COVID-19 world. *J. Libr. Adm.* **2020**, *60*, 951–957. [CrossRef]
46. Alajmi, B.M.; Albudaiwi, D. Response to COVID-19 pandemic: Where do public libraries stand? *Public Libr. Q.* **2021**, *40*, 540–556. [CrossRef]
47. Smith, J. Information in crisis: Analysing the future roles of public libraries during and post-COVID-19. *J. Aust. Libr. Inf. Assoc.* **2020**, *69*, 422–429. [CrossRef]
48. Binks, L.; Braithwaite, E.; Hogarth, L.; Logan, A.; Wilson, S. Tomorrow's green public library. *Aust. Libr. J.* **2014**, *63*, 301–312. [CrossRef]
49. Hvenegaard Rasmussen, C. The participatory public library: The Nordic experience. *New. Lib. World.* **2016**, *117*, 546–556. [CrossRef]
50. Michele Moorefield-Lang, H. Makers in the library: Case studies of 3D printers and maker spaces in library settings. *Libr. Hi Tech.* **2014**, *32*, 583–593. [CrossRef]
51. Kang, M.H. Analysis on the trends in researches of public service of the library. *J. Korean Libr. Inf. Sci. Soc.* **2014**, *45*, 361–394. [CrossRef]
52. Lee, S.S. Public library service positioning strategy. *J. Korean Libr. Inf. Sci. Soc.* **2013**, *44*, 279–303. [CrossRef]
53. Jeong, D.K.; Noh, Y.H. A study on users' perception of specialized services through service quality evaluation of public libraries. *J. Korean Soc. Inf. Manag.* **2018**, *35*, 51–75. [CrossRef]
54. Chun, B.A. Estimating spatial accessibility to public libraries on the regional scale: A case study on gangwon province. *J. Korean Cartogr. Assoc.* **2014**, *14*, 93–105, UCI: G704-SER000008969.2014.14.1.008.
55. Jung, D.E. Analyzing Supply Characteristics of Local Living Service Facilities in Seoul: Focused on the Quantitative Supply and Walking Access. Master's Thesis, University of Seoul, Seoul, Republic of Korea, 2020.
56. Yim, S.J. Analysis of Selecting an Appropriate Location for Infrastructure Considering the Range of Service Area and Accessibility: Focusing on Public Libraries. Master's Thesis, Chung-Ang University, Seoul, Republic of Korea, 2020.
57. Kim, Y.W.; Piao, G.S.; Kim, M.S. A study on the change of urban spatial networks in Gwangju and Naju due to Gwangju-Jeonnam innovation city. *J. Archit. Inst. Korea Plan. Des.* **2020**, *36*, 99–108. [CrossRef]
58. Lee, W.H. Accessibility to the central city and the development of backward regions: Towards a new spatial strategy. *J. Korean Assoc. Reg. Geogr.* **2013**, *19*, 436–445.
59. Bang, Y.C.; Ahn, Y.J. Does the development of Daegu innovation city affect nearby housing prices in respect of urban spatial structure?: The evidence from Hedonic price models by the comparison of distance-measurements. *J. Korean Reg. Dve. Assoc.* **2016**, *28*, 131–146.
60. Jeon, B.Y.; Liu, S.P.; Hong, Y.K.; Lee, M.H. Analysis of urban infrastructure service areas in the mid-living areas of Cheongju city considering the characteristics of local traffic: Focusing on the network analysis method. *J. Korea Acad. Ind. Coop. Soc.* **2022**, *23*, 704–716. [CrossRef]

61. Jung, S.E. A study on the urban spatial structure of Ulsan based on the Space-Syntax. *Des. Converg. Stud.* **2014**, *13*, 195–208, UCI: G704-SER000008947.2014.13.1.006.
62. Kang, C.D. A study on impacts of street network on land-use density in Seoul, Korea. *Korea Spat. Plan. Rev.* **2013**, *76*, 129–148. [CrossRef]
63. Joo, Y.J. A study on the movement of street-based urban morphology using analysis of integrated land use-transportation. *Spat. Inf. Res.* **2011**, *19*, 63–72, UCI: G704-000574.2011.19.3.003.
64. Kim, J.H. A Study on Changing Urban Spatial Structure of Seoul by GIS and Space Syntax: 1970's~2000's. Ph.D. Thesis, Korea University, Seoul, Republic of Korea, 2009.
65. Republic of Korea Library Act. Available online: https://www.law.go.kr/LSW/eng/engMain.do?eventGubun=060124 (accessed on 1 September 2022).
66. Republic of Korea Ministry of Health and Welfare. COVID-19 Quarantine Policy 2020. Available online: http://www.mohw.go.kr/eng/index.jsp (accessed on 10 September 2022).
67. Peng, Y.S. Causal analysis in the social sciences. *Sociol. Stud.* **2011**, *3*, 1–32. [CrossRef]
68. Moon, C.O. Causal relation and experience in Hume and Whitehead. *Stud. Philos. East West.* **2012**, *65*, 77–94. [CrossRef]
69. Bae, J.H. Kant's proof of the causal principle. *J. Korean Philos. Soc.* **2018**, *147*, 215–237. [CrossRef]
70. Bak, J.C. An interpretative model for the Hume's causality claims. *Korean J. Philos. Sci.* **2014**, *17*, 45–72.
71. Kim, B.J. A new interpretation of the problem of induction in Hume's philosophy. *Mod. Philos.* **2021**, *17*, 37–62. [CrossRef]
72. Jakeway, I. Mental Causation: The Happy Autonomy of Psychology and Physics. Ph.D. Thesis, King's College London, London, UK, 2022.
73. Sánchez-Cañizares, J. Integrated information theory as testing ground for causation: Why nested hylomorphism overcomes physicalism and panpsychism. *J. Conscious. Stud.* **2022**, *29*, 56–78. [CrossRef]
74. Hillier, B.; Hanson, J. *The Social Logic of Space*; Cambridge University Press: Cambridge, UK, 1984.
75. Zhao, X.L.; Moon, J.M. Analysis of urban spatial accessibility of museums within the scope of Seoul. *Buildings* **2022**, *12*, 1749. [CrossRef]
76. Pan, M.Y.; Shen, Y.F.; Jiang, Q.C.; Zhou, Q.; Li, Y.H. Reshaping publicness: Research on correlation between public participation and spatial form in urban space based on Space Syntax: A case study on Nanjing Xinjiekou. *Buildings* **2022**, *12*, 1492. [CrossRef]
77. Lee, J.H.; Kim, Y.S. Rethinking art museum spaces and investigating how auxiliary paths work differently. *Buildings* **2022**, *12*, 248. [CrossRef]
78. Kim, Y.O.; Kim, J.Y.; Yum, H.Y.; Lee, J.K. A study on Mega-Shelter layout planning based on user behavior. *Buildings* **2022**, *12*, 1630. [CrossRef]
79. Rashid, M. On spatial mechanisms of social equity: Exploring the associations between street networks, urban compactness, and social equity. *Urban Sci.* **2022**, *6*, 52. [CrossRef]
80. Kim, J.Y.; Kim, Y.O. Residents' Spatial-Usage behavior and interaction according to the spatial configuration of a social housing complex: A comparison between High-Rise apartments and perimeter block housing. *Sustainability* **2022**, *14*, 1138. [CrossRef]
81. Wang, M.; Yang, J.; Hsu, W.L.; Zhang, C.; Liu, H.L. Service facilities in heritage tourism: Identification and planning based on space syntax. *Information* **2021**, *12*, 504. [CrossRef]
82. Suchoń, F.; Olesiak, J. Historical analysis of the example of nowy sącz in space syntax perspective. guidelines for future development of urban matrix in medium-sized cities. *Sustainability* **2021**, *13*, 11071. [CrossRef]
83. Huang, B.X.; Chiou, S.C.; Li, W.Y. Landscape pattern and ecological network structure in urban green space planning: A case study of Fuzhou city. *Land* **2021**, *10*, 769. [CrossRef]
84. Altafini, D.; Cutini, V. Towards a spatial approach to territorialize economic data in urban areas' Industrial agglomerations. In Proceedings of the ICCSA 2022: Computational Science and Its Applications—ICCSA 2022 Workshops, Malaga, Spain, 4–7 July 2022; pp. 370–386. [CrossRef]
85. Kim, S.T. A Study on the Multi-Dimensional Space Analysis Model Using 3D Graphic Engine. Ph.D. Thesis, Hanyang University, Seoul, Republic of Korea, 2008.
86. Ministry of Land, Transport and Maritime Affairs. Available online: http://www.ltm.or.kr/2009/board.php?board=kkknewsmain&page=58&category=3&command=body&no=1210&PHPSESSID=277b59ab67120a43fa1ee757b443b0ce (accessed on 10 September 2022).
87. 2021 Overview of National Cultural Infrastructure, Ministry of Culture, Sports and Tourism. Available online: https://www.mcst.go.kr/kor/s_policy/dept/deptView.jsp?pSeq=1563&pDataCD=0417000000&pType=02 (accessed on 12 September 2022).
88. National Library Statistics System. Available online: https://www.libsta.go.kr/ (accessed on 15 September 2022).
89. Seoul Subway Operation Status Statistics. Available online: https://data.seoul.go.kr/dataList/247/S/2/datasetView.do (accessed on 15 September 2022).
90. National Spatial Information Portal. Available online: http://www.nsdi.go.kr/lxportal/?menuno=3085 (accessed on 15 September 2022).

Article

Analysis of Pedestrian Behaviors in Subway Station Using Agent-Based Model: Case of Gangnam Station, Seoul, Korea

Joo Young Kim and Young Ook Kim *

Department of Architecture, Sejong University, Seoul 05006, Republic of Korea
* Correspondence: yokim@sejong.ac.kr; Tel.: +82-2-3408-3762; Fax: +82-2-3408-4331

Abstract: Numerous pedestrians interact with the subway station space by finding entrances into this closed area to use the subway system; further, they may use transfer transportation facilities or the complex functions nearby, such as commercial. Many studies examine pedestrian behaviors in subway stations, but most focus on special situations such as disasters and evacuation. Because it is important to analyze gait patterns in everyday situations, this study aims to verify the explanatory power of actual gait behavior by using space syntax theory in constructing an optimal agent-based model. To this end, first, pedestrian characteristics and space types are classified using pedestrian data from Gangnam Station. Second, the depthmapX program is used to develop an appropriate agent-based model for stations. Third, a simulation is run to calculate the frequency of the agent movement at each gate, which is matched with the observed pedestrian volume. Fourth, the relationship between the frequency of the agent movement and pedestrian volume is analyzed using Statistical Package for the Social Sciences. The results show that although agent-based models have limitations in explaining pedestrian patterns in the entire subway station, they are capable of explaining these patterns along the shortest paths between ticket gates and station entrances.

Keywords: agent-based model; space syntax; subway station; pedestrians' behaviors

Citation: Kim, J.Y.; Kim, Y.O. Analysis of Pedestrian Behaviors in Subway Station Using Agent-Based Model: Case of Gangnam Station, Seoul, Korea. *Buildings* **2023**, *13*, 537. https://doi.org/10.3390/buildings13020537

Academic Editors: Michael J. Ostwald and Ju Hyun Lee

Received: 24 January 2023
Revised: 11 February 2023
Accepted: 12 February 2023
Published: 15 February 2023

1. Introduction

1.1. Background

As urban areas expand, subways are constructed to alleviate traffic congestion. However, many subway stations in the city center are overcrowded during peak hours [1]. Therefore, predicting people's space use patterns in stations is important to both their convenience of movement and safety. In addition, many people find it challenging to navigate the station space, given that it is a closed underground space that has many multifunction facilities such as transfer transportation and commercial facilities [2]. Because the behaviors of pedestrians in station space contribute to increased congestion and unused space, it is important to understand how they move from the platforms to the exits after passing through the ticket gates.

In this regard, many commercial software programs support agent-based models to analyze and predict pedestrian behaviors—for example, Building EXODUS, Simulex, Pathfinder, and Unity 3D [3]. However, these programs are typically used to analyze crowd diversion mechanisms for use in emergency evacuation rather than in the daily environment. Further, they set complex conditions for individual agents and are hence suitable for applications in a limited closed space and under special circumstances [4]. In addition, these simulation programs are not used for large and the diverse elements of the surrounding environment cannot be incorporated into them [5]. Therefore, researchers have been seeking ways to overcome the limitations of such simulations and have called for new approaches [6].

In this context, pedestrian behaviors in subway stations under normal circumstances differ from those during an emergency evacuation, when only the shortest distance paths are used. In reality, pedestrians do not calculate the most optimal path to the destination with every step. In other words, they choose paths continuously by using available visual information [7]. Hence, a visual perception-based simulation of space is required to predict their behaviors under normal circumstances. To this end, this study examines subway station space using the Visual Graph Analysis (from now "VGA") method of space syntax theory [8–11]. Nevertheless, given that in an analysis of space, VGA cannot be used to examine micro-space use behaviors, an agent-based model was developed by employing space syntax theory. This model can be used to analyze pedestrian behaviors under normal circumstances by incorporating the field of view for agents and the visibility characteristics of the station [12]. However, it needs to be verified in various spatial environments to confirm that it is completely reliable [13].

In sum, this study seeks to apply and verify an agent-based model by examining its correlation with the observed pedestrian volume in a major subway station in which the pedestrian environment is important. The study provides insights into pedestrian behavior patterns according to the characteristics of the subway station space and discusses potential applications for the presented agent-based model.

1.2. Literature Review

1.2.1. Studies on the Effect Factors on Pedestrian Behavior in Subway Stations

Modern subway stations not only transport passengers but also serve as transport hubs and multiuse facilities. Consequently, these are perceived as key urban facilities that significantly influence the development of surrounding areas [14]. Proposals to develop subway stations as multiuse facilities that reflect local identity are often presented [15,16]. Many studies have focused on user satisfaction with the station's physical environment or the appropriateness of station size [17–19]. The key focus area of studies that examine stations as multiuse facilities is often the design of an efficient transit center or the development of an integrated information system [17,20,21]. Thus, subway stations are studied as important spaces that attract people and generate complex behaviors in the urban environment.

Meanwhile, various studies have also considered the characteristics of the spatial configuration of subway stations since it affects human behavior [22]. For instance, Durmisevic and Sariyildiz [23] derived space syntax indicators to assess the quality of underground station space, and Yoon and Kim [2] introduced such indicators as elements of wayfinding in public space within. In addition, Van der Hoeven and Van Nes analyzed design of underground space in two subway stations using space syntax method [24]. These studies have also discussed ways to improve the environment by verifying that the subway station environment is related to spatial configuration indicators.

Other studies have examined the relationship between the spatial configuration of subway stations and pedestrian behaviors in more detail. For example, Kim and Kim [7] showed that pedestrian behaviors in the space between the platform and the ticket gates are related to the field of view, which is dependent on a spatial configuration, but that there was little correlation in other areas. Moreover, Ueno [10] analyzed the spatial characteristics of multiuse commercial facilities connected to Shibuya Station and their relationship to pedestrian volume. This study showed that although the correlation was low overall, there was a positive correlation at the ticket gates and station entrances. Thus, these studies have commonly identified a specific subway station space where pedestrian behavior is determined by spatial configuration. Notably, these studies suffer from a limitation in that they have not analyzed how various pedestrian behaviors are influenced by the spatial configuration of subway stations.

Another stream of studies has investigated the effects on pedestrian behavior of environmental factors other than the spatial configuration in order to compensate for this limitation. In this regard, Okamoto et al. [9] studied an underground market in Nagoya, Japan, and argued that elements other than spatial configuration—such as the visual depth from the corridor and the number of store tenants—significantly affect pedestrian behavior.

Further, Xu and Chen [11] measured urban spatial vitality in underground spaces connected to subways by analyzing the relationship between hourly pedestrian volumes and spatial characteristics. They found that vitality is related not only to the accessibility to space syntax elements and the field of view but also to the size of the space, the range of functions, and transportation elements—and that the effects differ according to the time of the day. In other words, since pedestrian behaviors at a station are influenced by factors such as the surrounding function, scale, and environment, it is difficult to explain their behavior by considering existing spatial configuration indicators alone.

In sum, pedestrian behaviors inside subway stations differ from general pedestrian behaviors that depend on spatial configuration.

1.2.2. Studies of Pedestrian Behavior Using Agent-Based Models

Many studies on pedestrian behaviors have used the space syntax approach [25], including that of Turner and Penn [13], who proposed the agent-based model and simulated pedestrian movements using agents with a defined field of view as they moved in space. This approach is based on Gibson's ecological theory of perception [26]. This model does not depend on learned paths or destinations, but instead follows movement rules based on the spatial configuration of buildings [13,27]. The model can be used in both a closed internal space [28] and a large open space [29,30]. Some key studies that have used this model are as follows.

Turner and Penn [13], who studied the Tate Britain gallery, are the first to have verified the correlation between the agent-based model and the characteristics of pedestrian behaviors. They showed that when agents were programmed to change direction after 3 steps, the model showed the highest correlation with observed pedestrian patterns ($R^2 = 0.76$). Above all, it is meaningful that they confirmed the validity of the agent-based model. Hu et al. [31], who analyzed pedestrian patterns in a plaza, based their simulations on scenarios of various conditions experienced by pedestrians. They tested for the scenario most similar to the current gait pattern, thus demonstrating the possibility of combining spatial organization and multiple agents. Omer and Kaplan [32] examined pedestrian behavior in the urban environment using an agent-based model, that incorporated a street network and land use. They showed that the agent-based model is more effective than a multiple regression analysis (from now "MRA") model in explaining pedestrian behaviors. Thus, these studies have confirmed that the agent-based model can be used to explain urban pedestrian behaviors. Jiang and Jia [33], who analyzed a road network through agent-based simulations, found no difference in behavior between pedestrians with and without a final destination. This result is significant for controlling crowd flow. Cheliotis [34] tested agent-based models of space design and human behaviors and asserted that these models explain the complexities of human and crowd behavior in space. Thus, this study confirmed the validity of the agent-based model for analyzing pedestrian behavior in spaces ranging from the interior space of a building to open spaces in a city.

The following studies have analyzed rail stations using agent-based models. Castle et al. [35] argued for applying these models in rail stations—pointing out the need for crowd evacuation and exit simulations—but did not conduct empirical research. Tang and Hu [36] used an agent-based model to study Sihui Station in Beijing, China, focusing on examining the spatial design of the station's public space and parking space and of the retail space outside the station. They proposed a solution to address congestion at the station entrance and strategies to improve pedestrian circulation patterns. Although agent-based model studies for subway stations are insufficient, the importance of the movement path connected to the entrance has been confirmed. Meanwhile, the analyses in these studies are limited to the station's surrounding areas and overlook its internal space. Therefore, the present study is meaningful in that it analyzes pedestrian behaviors in the indoor space connected to the subway entrance by using the agent-based model validated in previous studies.

2. Methods and Data

2.1. Methods

This study explores how well an agent-based model can explain observed pedestrian behavior in the underground space connected to subway stations. The hypothesis in this study is as follows: "An agent-based model is capable of explaining pedestrian behavior in the underground space connected to subway stations". To verify this statement, the study proceeds in five steps.

First, pedestrian behavior and subway station space types were analyzed, given that the underground space connected to station is influenced not only by ticket gates and the external environment of the station but also by visual accessibility. Therefore, pedestrian data were collected to analyze actual pedestrian behaviors, and space types were defined according to pedestrian behaviors. Second, an appropriate agent-based model that reflects the characteristics of the underground space connected to subway stations was created. Using depthmapX program—was used to set values for the agent analysis tool—including the analysis length, the number of agents and locations, and the agent set parameters. Spatial configuration characteristics and Turner and Penn's study [13] were considered because these authors applied an agent-based model in real space to study visitor behaviors in a gallery and verified their results. Although present study Turner and Penn's results [13] were initially considered, some parameters have changed the model to fit observed pedestrian behaviors. Third, an agent-based simulation was run for comparison with the observed pedestrian volume at each observation location. The data included the number of agents who moved through the cells, which can be compared with pedestrian behavior. Fourth, the study's hypothesis was supported by explaining the observed pedestrian volume using an agent-based simulation. To this end, correlation and regression analyses were conducted using Statistical Package for the Social Sciences (from now "SPSS"). The correlation analysis correlation coefficients and significance for each category. In the simple linear regression analysis, the adjusted R^2, Durbin–Watson values, and significance was calculated to ascertain the extent to which the agent-based model can explain the observed pedestrian volume. A section with conclusions and a discussion follows the results section. Figure 1 summarizes the research method and process.

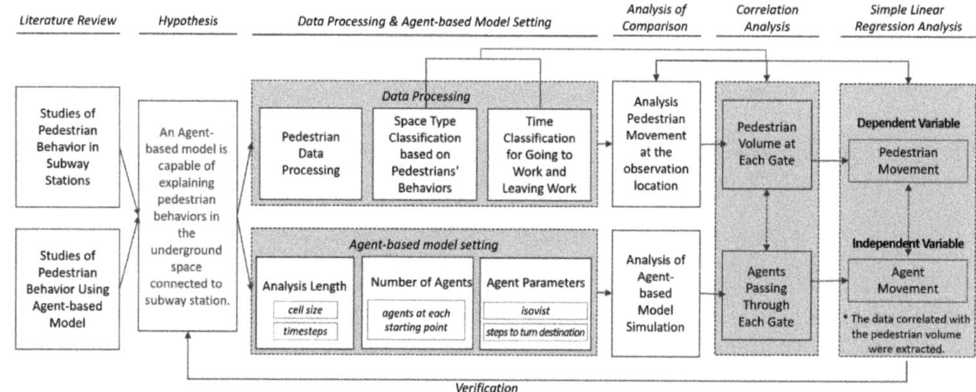

Figure 1. Research method and process.

2.2. Study Area Characteristics

The study area selected was Gangnam Station, Seoul, Korea. Specifically, the study area was limited to the station's first basement (B1) level, which has ticket gates and entrances. Gangnam Station is among the busiest subway stations in Korea and is connected to many multiuse facilities. Table 1 shows the five subway stations with the highest annual ridership in 2021. The first is Sinnonhyeon, followed by Gangnam. Gangnam and Jamsil Stations are both transportation hubs connected to the subway and other modes of transportation. They are also connected to underground commercial areas. Gangnam Station was selected for this study since it has both a high level of ridership and a wide range of space use behaviors.

Table 1. The five subway stations with the highest annual ridership in 2021.

Rank	Station Name	Line	Annual Ridership
1	Sinnonhyeon	Line 9	18,762,507
2	Gangnam	Line 2	17,563,866
3	Noryangjin	Line 9	15,582,731
4	Jamsil	Line 2	14,544,907
5	Yeouido	Line 9	14,312,939

Note: Open Government Data (https://www.data.go.kr/data/15099330/fileData.do (accessed on 5 January 2023).

In Gangnam Station (Line 2), the platform is located in the second basement (B2) level, and the ticket gates and the station entrances are located in the level (Figure 2). Pedestrian behaviors on the platform depend on the transfer locations in the previous stations rather than the spatial configuration of this station itself [8]. Therefore, this study excluded the B2 level and focused on the B1 level, where pedestrian behaviors are related to the spatial configuration of the station space. In the B1 level, a corridor links the underground commercial area to the Sinbundang Line for people who wish to transfer. Gangnam Station (Line 2) has 8 entrances. Since the road intersection above this station does not have crosswalks, people sometimes use the B1 level to cross the intersection. More than 89 city and inter-city bus routes connect Gangnam Station to Seoul's other centers and to satellite cities around Seoul [37].

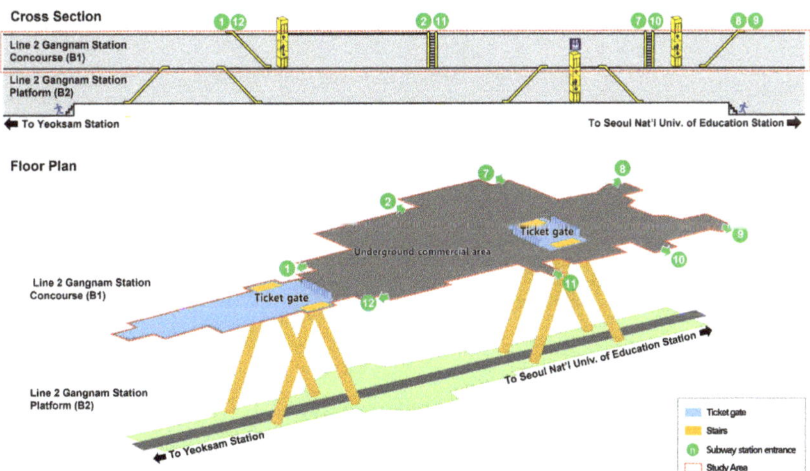

Figure 2. Study area (first basement level (B1), Line 2 Gangnam Station).

2.3. Components and Path Selection of Agent-Based Model

The agent-based model based on space syntax was simulated in the depthmapX program. The components of the environment were the analysis length, the number of agents, and the starting points.

The analysis length was the time the agent enters and travels. The number of agents was set for each starting point. The agent parameters included the agent stride length, walking speed, isovist, and steps before changing the direction to the destination. The agent stride length and walking speed are related and set the cell size of 1 step. The isovist is the angular extent the agent is heading toward the destination. Therefore, when the agent changes direction, the agent chooses a new random destination within the isovist [12]. This model is not for setting a final destination and move, but for continuously changing the destination according to visibility.

Figure 3 shows the agent's path selection process in the model [12,38]. First, the scope of analysis space, environment, and agent parameters is set. Then, the number of agents at the starting point is entered. Each agent establishes a destination at the starting point and takes steps before changing direction. Upon arrival, the agent determines a temporary new destination using the isovist. It is based on Poisson distribution. If the agent's field of view is unclear, they take a step to the left or right to set a new destination. When an agent arrives at a new destination, the agent moves iteratively according to the same rules.

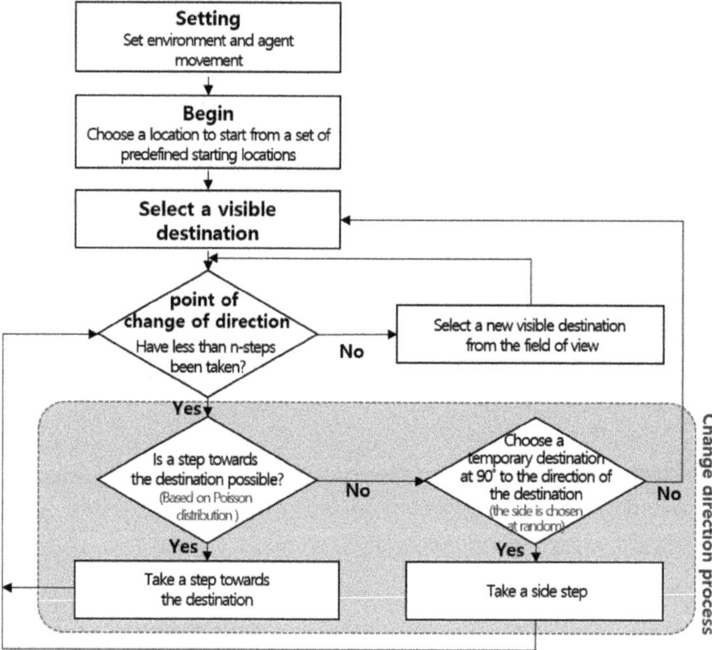

Figure 3. Flow chart of the agent decision process.

2.4. Pedestrian Data and Behaviors

2.4.1. Pedestrian Data

In this study, pedestrian data from Kim et al. [8] was used for analysis. Specifically, they divided the study into convex spaces and counted the number of pedestrians at the "gates" defined at the boundaries. They collected data for 5 min between 8–9 a.m. and 6–7 p.m. on a Friday. Although Kim et al. [8] included data on 48 gates, this study used data collected from 28 gates only since the study area was restricted to the B1 level and excluded the B2 level (Figure 4). Gangnam Station area is a major office district where the pedestrian volume is highest during commute hours and reaches the highest level during Friday evening commute hours. Therefore, it was considered appropriate to use these data as representing the busiest times.

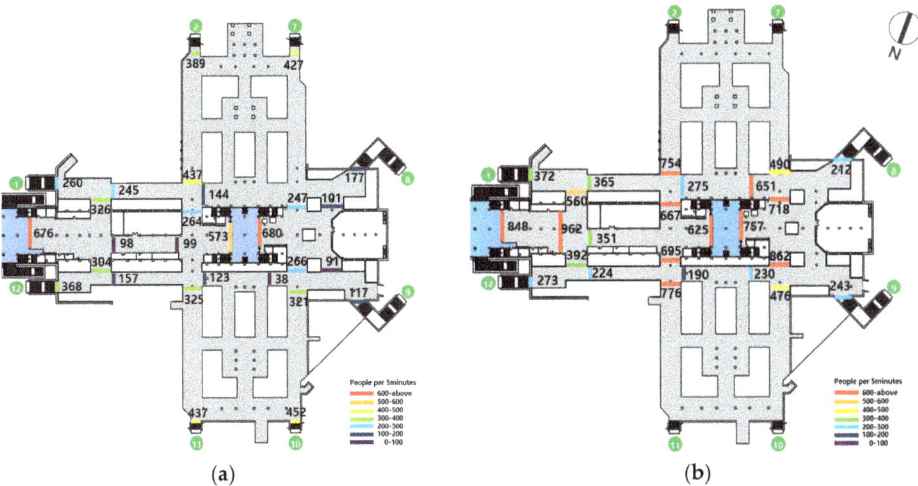

Figure 4. Pedestrian volume count in first basement level (B1), Gangnam Station: (**a**) morning commute hours (8–9 a.m.); (**b**) evening commute hours (6–7 p.m.). (Maps created by the author from Kim et al. [8]).

The data collected from the 28 gates during morning and evening commute hours are shown in Figure 4a,b, respectively. The areas filled in blue are separated by ticket gates, and the numbers in green circles refer to station exit numbers. Pedestrian volume is denoted with colored lines—the red lines denote higher volumes, and the purple lines denote lower volumes. The numbers next to the colored lines are the number of pedestrians who passed the gates during the 5 min observations.

The highest pedestrian volume during the morning commute was observed at the ticket gates, followed by station exits 10(452), 11(437), and 7(427), and the southbound corridor (437). Meanwhile, station exits 8 and 9 had relatively low pedestrian volumes, as did the western section of the station. Similarly, the highest pedestrian volume during the evening commute was observed at the ticket gates and relatively low pedestrian volumes were observed at exits 8 and 9. However, the corridors connecting the ticket gates in the center to exits 2, 7, 10, and 11 had very high pedestrian volume levels. It was observed that there were more people in the corridors, given that the pedestrian count at the ticket gates was approximately 700.

A comparison of the pedestrian volumes during the morning and evening commute hours showed that the pedestrian volume was higher in the evening. In the morning, pedestrians passed the ticket gates, and then immediately dispersed via corridors or exits. In the evening, in addition to subway riders, there were pedestrians who remained in the underground space or passed through the station and left via other exits. This can be observed in the pedestrian volume difference between the ticket gates and nearby corridors. For example, for the northeast ticket gate, the number of pedestrians in the morning was 676. The number of pedestrians in the three connected corridors was 326, 304, and 98, respectively. The difference between ticket gate count and corridor count is 52 and shows the ratio of pedestrians who travel through the connected corridors. However, in the evening, the number of pedestrians at the ticket gate and at a nearby point were 848 and 962, respectively. The pedestrian counts in the three connected corridors was 560, 351, and 392. The difference exceeds 100 even close to the ticket gate and exceeds 340 in connected corridors. This result indicates that pedestrians not only use the ticket gates but also either stay in the station space or pass through it to exit the station.

2.4.2. Space Type Classification Based on Pedestrian Behaviors

Among the pedestrians inside Gangnam Station, workers comprise the largest group [37]. This indicates that during commute hours, most pedestrians have a definite destination and are well aware of its location and the shortest paths to reach it. They tend to travel along the shortest path to their destination, from the subway platform to the ticket gates to the station exits [8,39]. Since pedestrian behaviors differ depending on whether they take the shortest path or not, this study examined pedestrian behaviors separately for different sections of travel. To this end, "shortest path space" was defined as in Figure 5, by directly connecting station exits to the nearest ticket gates (Figure 5a). Spaces outside the "shortest path space" were categorized simply as the corridors of the underground commercial areas.

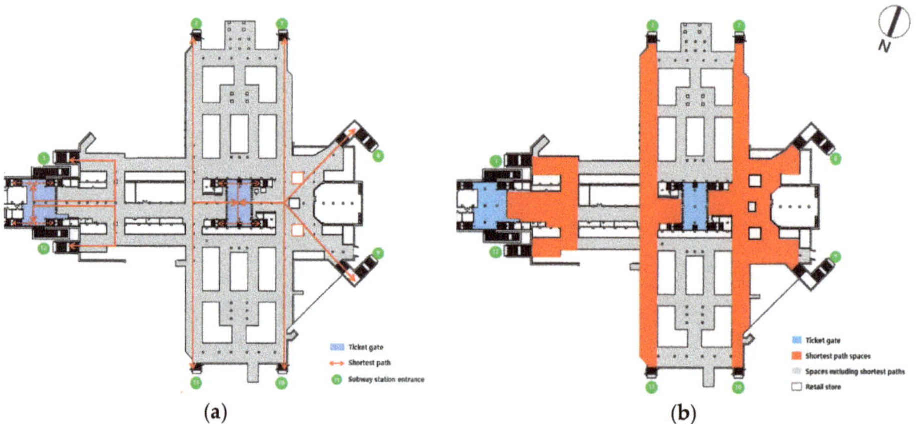

Figure 5. Definition of shortest path space between ticket gates and subway station entrances: (**a**) shortest paths; (**b**) shortest path space.

3. Verification of Pedestrian Volume Using Agent-Based Model

3.1. Definition of Agent-Based Model and Calibration

3.1.1. Analysis Length

The analysis length was set to 5 min after an agent is released. This corresponds to the actual observation period of 5 min and enables direct comparison of agent simulation results with observed pedestrian volume. In addition, it takes approximately 5 min for people to travel from the subway platform to a station exit through a ticket gate.

In the depthmapX program, "timestep" refers to the analysis period. The study area is divided into cells (unit space), and the cells represent the travel time. Sutherland et al. [40] defined the cell size at 0.75 m × 0.75 m and assumed a walking speed of 1.5 m/s. Therefore, 1 s is equivalent to 2 timesteps.

Given that the width of ticket gates in the study area is 0.5 m at most, the cell size cannot be set as 0.75 m × 0.75 m because the space around the ticket gates would be "closed." VGA, which provides the basis for conducting agent-based modeling, requires a grid setting so that all spaces are connected and accessible [38]. Hence, the cell size was set at 0.375 m × 0.375 m, which is equivalent to a half step and also enables agents to pass through the ticket gates. In addition, the timesteps were set to 4 per second, or 1200 since the analysis length was 5 min (300 s).

3.1.2. Number of Agents

To determine the number of agents, public ridership data for Gangnam Station in October 2007 was applied to enable a direct comparison with the observed pedestrian volume. The average ridership between 8–9 a.m. was 3627 entries and 19,408 exits, corresponding to 1919 in 5 min. The average ridership between 6–7 p.m. was 14,765 entries and 11,726 exits, corresponding to 2207 in 5 min.

The number of agents to be introduced depends on the cell size, and the number of agents to be released per 1/4 s was entered for starting points. For the starting points, the stairs to the subway platform (eight sections) reflect the number of exits, and the station's eight exits/entrances reflect the number of entries. The number of agents introduced for the analysis length (1200 timesteps) is 1.58 for the morning commute and 1.85 for the evening commute. The ratio between entries/exits is approximately 1:53 during the morning commute, and 1:0.79 during the evening commute, which is similar to the actual pedestrian volume in the station (1 in the equation).

$$\begin{aligned}
&\text{Morning Commute (8-9 a.m.): } 1200 \text{ [timesteps]} \times 1.59 \text{ [agents]} = 1908 \text{ agents} \\
&\text{Evening Commute (6-7 p.m.): } 1200 \text{ [timesteps]} \times 1.84 \text{ [agents]} = 2208 \text{ agents}
\end{aligned} \tag{1}$$

3.1.3. Agent Parameters

The agent parameters were adopted from Turner and Penn's study [13] on Tate Britain Gallery in London. In this study, agents were programmed such that their field of view was 170° and they changed direction after three steps. The highest correlation between the predicted and observed pedestrian volume ($R^2 = 0.76$) was observed in this setting, which confirmed the model's validity.

The field of view angle in the present study was also set to 170° (15 bins), and "steps before turn decision" was initially set to 6 steps or 1.5 s in line with Turner and Penn's setting of 1.5 s or 3 steps [13]. However, these parameters did not match the actual pedestrian behaviors, and after several trials, it was discovered the setting of 12 steps best fit the actual pedestrian behavior. It can be surmised that in Turner and Penn's study [13], people changed direction frequently to view the artworks in the gallery, whereas in the subway space, during commute hours people tend to move in a relatively straight way to their destinations.

3.2. Comparison of Agent-Based Model Simulation and Observed Pedestrian Volume

3.2.1. Analysis of Agent-Based Model Simulation

The gate count results from the movement of agents through the cells can be compared with the observed pedestrian volume. As agents move during the 5 min period, cells record the frequency of their movement. In Figure 6, colors closer to red (blue) indicate cells with higher (lower) frequency values.

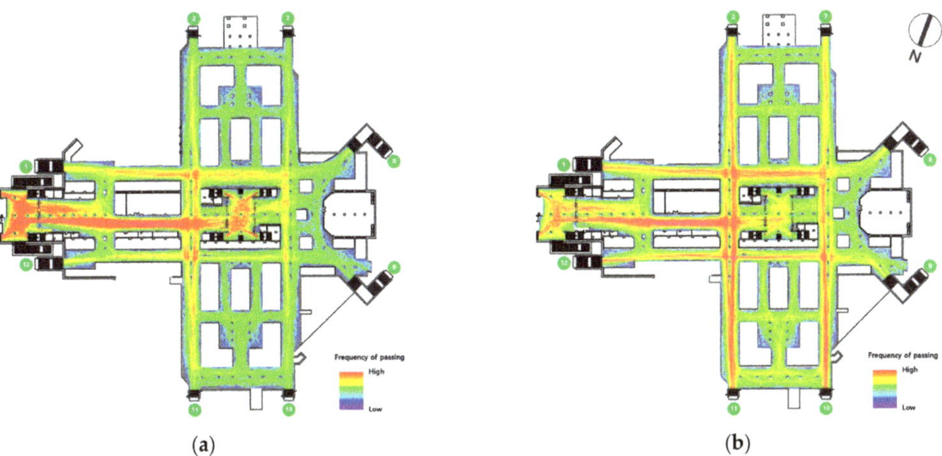

Figure 6. Agent-based simulation results: (**a**) morning commute hours (**b**) evening commute hours.

The morning commute results showed that the section with the highest frequency is the ticket gates in the station's east section, followed by the corridor linking those ticket gates in the east to the ones in the center. The corridor linking station exits 2 and 11 had the next highest frequency, and the corridors around the underground commercial areas showed relatively low frequency. The evening commute results differ—the corridors near the station exits showed higher frequency values than the ticket gate sections. This finding reflects the fact that the proportion of 'exits (entries)' is higher during the morning (evening) commute.

3.2.2. Comparison of Simulated and Observed Pedestrian Volume

The simulation result includes frequency values, which indicate how many times agents passed through each cell. Table 2 presents the frequency values from the cells corresponding to the locations in the station where the actual number of pedestrians was counted. For observation locations without data, the frequency value of the cells was excluded. In sum, a database containing data from 28 locations during the morning commute and 25 locations during the evening commute was prepared. All 53 locations were included in the statistical analysis. In addition, given that pedestrians tend to travel along the shortest paths [7], gates were classified into those along the shortest paths and those that are not, in order to examine their spatial characteristics. In all, 35 gates were along the shortest paths (17 for the morning commute, and 18 for the evening commute), and 18 gates were not (11 for the morning commute, and 7 for the evening commute).

Table 2. Data on pedestrian movement and agent movement at the observation location.

Location	Morning Commute Hours		Evening Commute Hours		Gate Type
	Pedestrian	Agent	Pedestrian	Agent	
1	676	1128	848	771	
2	-	-	962	1355	
3	260	287	372	369	
4	368	216	273	325	
5	326	368	560	304	
6	304	380	392	312	
7	437	363	754	600	
8	264	630	667	794	
9	-	-	695	931	Gates located on the shortest paths
10	325	508	776	752	
11	573	765	625	609	
12	680	558	757	421	
13	-	-	490	432	
14	247	530	718	631	
15	266	403	862	648	
16	321	327	476	638	
17	101	158	-	-	
18	91	145	-	-	
19	177	216	212	329	
20	117	132	243	274	
21	389	125	-	-	
22	427	102	-	-	
23	437	147	-	-	
24	452	110	-	-	
25	245	463	365	442	
26	98	999	351	767	Gates not located on the shortest paths
27	157	318	224	357	
28	99	984	-	-	
29	123	332	-	-	
30	144	705	275	812	
31	-	-	190	405	
32	-	-	651	812	
33	38	298	230	420	

3.3. Investigation of Observed Pedestrian Volume Using Agent-Based Model

3.3.1. Correlation Analysis

Table 3 shows the correlations between the observed pedestrian volume and the total number of agents by observation period and gate type. Following Turner and Penn's approach to verification of the simulation results [13], the pedestrian volume and the frequency value of agents were both log-transformed so that the distribution is close to normal distribution.

Table 3. Correlation between observed pedestrian volume and agent number.

		Total	Morning Commute Hours	Evening Commute Hours
All gates	Pearson correlation coefficient	0.346	0.035	0.636
	p-value	0.006 *	0.429	0.000 *
	N	53	28	25
Gates located on the shortest paths	Pearson correlation coefficient	0.833	0.827	0.789
	p-value	0.000 *	0.000 *	0.000 *
	N	35	17	18
Gates not located on the shortest paths	Pearson correlation coefficient	−0.321	−0.690	0.671
	p-value	0.097	0.009 *	0.050 *
	N	18	11	7

* Correlation coefficient is significant at the 0.01 (99%) level (both sides).

The results in Table 3 show that the observed pedestrian volume and simulation frequency values are positively correlated at 0.35 for all gates. For gates located on the shortest paths, the correlation is very high at 0.83. For gates not located on the shortest paths, the correlation is negative but not statistically significant. Regarding the morning commute hours, there is no correlation for all gates—but for those along the shortest paths, it is 0.83. During evening commute hours, there is a positive correlation of 0.64, and for the shortest paths, it is 0.79. In other words, the observed pedestrian volume and the total number of agents show high correlation in the shortest path spaces. In contrast, the observed pedestrian volume during the morning commute hours differs from that revealed by the agent simulation. For gates not located on the shortest paths, there is only a weak correlation, both during morning and evening commute hours, at the 0.05 confidence level, but the correlation is negative for the morning commute and positive for the evening commute. This difference likely stems from the limitation of a small sample size.

3.3.2. Simple Linear Regression Analysis

After extracting simulation data that had a positive correlation with the observed pedestrian volume, a simple linear regression analysis was conducted to examine agent attributes that explain the observed pedestrian volume clearly. The observed pedestrian volume was the dependent variable, and the frequency value of agents is the independent variable. Both R^2 and Durbin–Watson values were checked to determine how well the model fits the data. For the Durbin–Watson test, a value close to 2 indicates that the residuals from the regression analysis are not autocorrelated. An analysis of variance (ANOVA) was used to evaluate whether the results are significant at the $p < 0.01$ level (99% confidence interval).

Table 4 shows that for the entire study area including all gates, the adjusted R^2 is very low at 0.103 and the regression is not statistically significant at the 99% confidence level. However, for gates on the shortest paths, the regression is statistically significant and R^2 is 0.693. Table 5, which presents results from morning commute hours, shows that the regression is statistically significant only for gates on the shortest paths, and the adjusted R^2 is 0.663.

Table 4. Simple linear regression analysis for the entire study area.

Model Summary					
Model	**R**	**R^2**	**Adj. R^2**	**Std. Error of Estimate**	**Durbin–Watson**
All gates	0.346 [a]	0.120	0.103	0.663	1.009
Gates on the shortest paths	0.833 [a]	0.693	0.684	0.350	1.570

ANOVA [b]						
Model		**Sum of Squares**	**df**	**Mean Square**	**F**	**Sig.**
All gates	Regression	3.062	1	3.062	6.955	0.011 [c]
	Residual	22.450	51	0.440	-	-
	Total	25.511	52	-	-	-
Gates on the shortest paths	Regression	9.140	1	9.140	74.530	0.000 [c]
	Residual	4.047	33	0.123	-	-
	Total	13.187	34	-	-	-

Coefficients [d]						
Model		**Unstandardized**		**Standardized**	**t**	**Sig.**
		B	**Std. Error**	**Beta**		
All gates	Constant	3.440	0.893	-	3.853	0.000
	Log agent	0.388	0.147	0.346	2.637	0.011
Gates on the shortest paths	Constant	0.339	0.655	-	0.518	0.608
	Log agent	0.927	0.107	0.833	8.633	0.000

[a] Predicted value: (constant), log agent; [b] dependent variable: log pedestrian; [c] predicted value: (constant), log agent; [d] dependent variable: log pedestrian.

Table 5. Simple linear regression analysis for morning commute hours.

Model Summary					
Model	**R**	**R^2**	**Adj. R^2**	**Std. Error of Estimate**	**Durbin–Watson**
Gates on the shortest paths	0.827 [a]	0.684	0.663	0.348	1.903

ANOVA [b]						
Model		**Sum of Squares**	**df**	**Mean Square**	**F**	**Sig.**
Gates on the shortest paths	Regression	3.935	1	3.935	32.488	0.000 [c]
	Residual	1.817	15	0.121	-	-
	Total	5.752	16	-	-	-

Coefficients [d]						
Model		**Unstandardized**		**Standardized**	**t**	**Sig.**
		B	**Std. Error**	**Beta**		
Gates on the shortest paths	Constant	1.201	0.824	-	1.458	0.165
	Log agent	0.829	0.145	0.827	5.700	0.000

Table 6 presents the results for the evening commute hours, and although the regression is statistically significant for all gates (adjusted R^2 = 0.378), the Durbin–Watson value is less than 1 and the regression does not have adequate explanatory power. For gates on the shortest paths, the regression is statistically significant and the adjusted R^2 is 0.599.

Table 6. Simple linear regression analysis for evening commute hours.

Model Summary					
Model	R	R^2	**Adj. R^2**	**Std. Error of Estimate**	**Durbin–Watson**
All gates	0.636 [a]	0.404	0.378	0.406	0.867
Gates on the shortest paths	0.789 [a]	0.622	0.599	0.290	1.554

ANOVA [b]						
Model		**Sum of Squares**	**df**	**Mean Square**	**F**	**Sig.**
All gates	Regression	2.573	1	2.573	15.610	0.001 [c]
	Residual	3.792	23	0.165	-	-
	Total	6.365	24	-	-	-
Gates on the shortest paths	Regression	2.221	1	2.221	26.374	0.000 [c]
	Residual	1.347	16	0.084	-	-
	Total	3.568	17	-	-	-

Coefficients [d]						
Model		**Unstandardized**		**Standardized**	**t**	**Sig.**
		B	**Std. Error**	**Beta**		
All gates	Constant	1.276	1.232	-	1.036	0.311
	Log agent	0.774	0.196	0.636	3.951	0.001
Gates on the shortest paths	Constant	1.292	0.977	-	1.321	0.205
	Log agent	0.799	0.156	0.789	5.136	0.000

[a] Predicted value: (constant), log agent; [b] dependent variable: log pedestrian; [c] predicted value: (constant), log agent; [d] dependent variable: log pedestrian.

In sum, the regression is statistically significant and can explain the observed pedestrian volume for only the shortest path spaces. The two regression equations are presented below (two of an equation), as shown in Equation (2) and Figure 7. The agent simulation has the best explanatory power when the morning and evening commute data are both included (68.4%). Specifically, the pedestrian volume for the morning commute is slightly better explained. This study's hypothesis that "An agent-based model is capable of explaining pedestrian behavior in the underground space connected to subway stations" is therefore rejected for the entire study area. However, it is not rejected when applied only to the shortest path spaces.

$$\text{Shortest path gate during morning and evening commute hours: } y = 0.684x + 0.339$$
$$\text{Shortest path gate during morning commute hours: } y = 0.663x + 1.201 \tag{2}$$
$$\text{Shortest path gate during evening commute hours: } y = 0.599x + 1.292$$

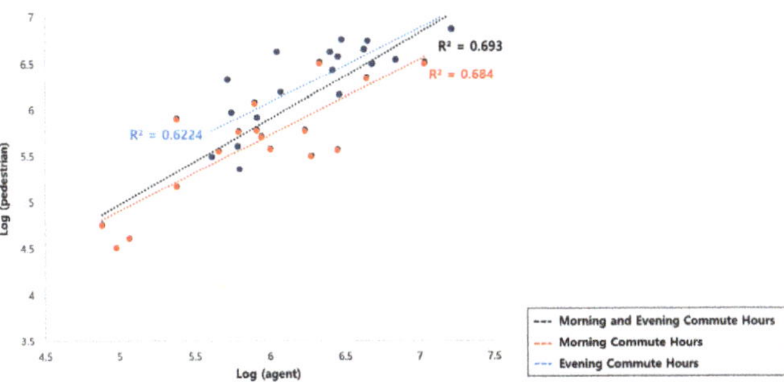

Figure 7. Simple linear regression analysis between the observed pedestrian volume and the number of agents for gates on the shortest path.

4. Discussion

This study analyzed how an agent-based model can explain real pedestrian behaviors in a subway station. The correlation between the frequency of agent movement from the agent-based model and the observed pedestrian volume was weak and did not have explanatory power ($R = 0.346$). However, when only the shortest path spaces were analyzed separately, the correlation was high and had strong explanatory power ($R = 0.833$). In other words, the main result of this study is that it verified that the agent-based model can explain the behavior of pedestrians only for the shortest path within the subway station. Based on previous studies and the results from this study, the following four points need to be discussed.

First, it should be ensured that pedestrians have the appropriate viewing angle in the space corresponding to the shortest path connecting the ticket gate and the station exit to ensure smooth pedestrian flows. In this study, the pedestrian behavior patterns show that the people who use subway stations during commute hours depend on their field of view and follow the shortest paths between the ticket gates and the station exits. In particular, the explanatory power for the morning commute hours was 66% higher than for the evening commute hours (60%). Notably, prior studies have reported this pattern and have also revealed that subway users are well aware of the shortest path to their destinations and choose to travel along these paths [8,39]. In contrast, the correlation was low for spaces outside the shortest path spaces. While this result is not statistically significant ($R = -0.321$. p-value $= 0.097$), it indicates that the spaces outside the shortest path spaces are used as travel space to destinations—indicating that these serve as shortcuts based on the experiences of pedestrians rather than on their field of view. This result is related to Penn's argument that the pedestrian behavior patterns for the spaces along the destination path differ from those determined by spatial configuration.

Second, the agent-based model used to analyze the subway station space should reflect the factors that affect movement by considering the various walking purposes of pedestrians, given that the results of the model used in this study varied depending on the characteristics and space use behavior of the agent. The correlation between the results from the agent-based model and the observed pedestrian volume was low for the subway station space because pedestrians engage in many types of behaviors simultaneously. The observation of pedestrians revealed that some subway riders are aware of the shortest paths from the subway platform to station exits and travel with a clear purpose, but others either look around the underground commercial area or remain in one location. In Turner and Penn's analysis of the Tate Britain Gallery [13], the explanatory power of agents was high at 76%, because most pedestrians had a common goal of artwork appreciation. Therefore, it is necessary to experiment with different model parameters, for example, by modifying agent conditions or introducing a space-type variable, depending on pedestrian behavior.

Third, subway stations need a physical environment plan to respond to the changing pedestrian behaviors by the time of day. This study analyzed data from a station's pedestrian peak hours, that is, the morning and evening commute hours on a Friday. The correlation between the frequency of agent movement and the observed pedestrian volume was very high ($R = 0.827$) along the shortest paths during the morning commute hours, and slightly lower ($R = 0.789$) during the evening commute hours. During the evening commute hours, the correlation was low ($R = 0.636$) for the entire station as well. These results are in accordance with Xu and Chen's argument [11] that spatial vitality varies in different periods of the day depending on environmental variables. In addition to subway riders, there were pedestrians who did not use the subway but stayed in one space during the evening commute hours. This finding indicates that although pedestrian behavior during morning and evening commute hours differs, a multiple regression analysis that includes factors that likely influence such behaviors at other times of the day must be conducted.

Fourth, future studies should adjust the environment and agent settings of the agent-based model according to the characteristics of the actual physical environment and pedestrian behaviors. For the agent program parameter in the present study, "steps before turn decision" was initially set to the default value of three steps [13], but this was not appropriate as a predictive model of pedestrian volume in a subway station. Other values were tested, and it was determined that 12 steps were the most optimal for this study. Therefore, when applying an agent-based model for a structure, it is necessary to apply appropriate parameters for each space type. If researchers modified it to better match actual pedestrian behaviors, the agent-based model would become an effective tool capable of predicting pedestrian behaviors accurately.

5. Conclusions

This study contributes to the literature in two ways. From the academic perspective, it developed an agent-based model—capable of predicting general pedestrian behaviors—and applied it not to a general building structure but, rather, to a subway station, which is a unique space. From the methodological perspective, the study showed that although this model has shortcomings, it has the potential to predict natural pedestrian behaviors and can be applied to diverse types of architectural space.

Regardless, this study has some limitations. First, the pedestrian data size was not large enough, since data from observation locations in only one station were used. Therefore, this study attempted to reveal the model's validity by testing the residuals and the significance of the regression equation. Second, this study used only an agent-based model to explain actual pedestrian patterns using simple linear regression. Since the goal of the study was to verify the explanatory power of the agent-based model other factors were not considered. Therefore, future studies need to build a more meaningful predictive model and test it using data collected from several subway stations. In addition, multiple regression analysis models should be verified by examining various factors influencing gait behavior.

Therefore, in the future, it is necessary to build a more meaningful predictive model with comprehensively approached data and data from several subway stations. In addition, the multiple regression analysis models should be verified by examining various gait behavior influencing factors.

Despite these limitations, this study is meaningful because it presents an agent-based model that reflects the reality of the subway station space where pedestrians engage in various pedestrian behaviors. The analysis results can provide primary data for developing agent-based models for built and urban environments. In addition, this study can be used as a guide when evaluating and planning a pedestrian-centered subway station space.

Author Contributions: Conceptualization, J.Y.K. and Y.O.K.; methodology, Y.O.K.; software, J.Y.K.; validation, J.Y.K.; formal analysis, J.Y.K. and Y.O.K.; resources, Y.O.K.; data curation, J.Y.K. and Y.O.K.; writing—original draft preparation, J.Y.K.; writing—review and editing, J.Y.K. and Y.O.K.; visualization, J.Y.K.; supervision, Y.O.K. All authors have read and agreed to the published version of the manuscript.

Funding: This research received no external funding.

Data Availability Statement: Data derived from the current study can be provided to readers upon request.

Conflicts of Interest: The authors declare no conflict of interest.

References

1. Gupta, S. Facilities for Passenger Movement to Decongest Underground Stations. *J. Inst. Eng. Ser. A* **2014**, *95*, 269–276. [CrossRef]
2. Yoon, H.J.; Kim, Y.O. Analytic Study on the Spatial Configuration of Public Space in Subway: With Focus to the Wayfinding Elements. *J. Archit. Inst. Korea Plan Des.* **2002**, *18*, 187–194.
3. Liu, J.; Zhang, R.; Sun, L.; Yan, W. Fire Evacuation in Complex Underground Space of Personnel. Proceedings of IOP Conference Series: Earth and Environmental Science, Chengdu, China, 23–25 April 2021; IOP Publishing: Bristol, UK, 2021.
4. Moussaïd, M.; Helbing, D.; Theraulaz, G. How Simple Rules Determine Pedestrian Behavior and Crowd Disasters. *Proc. Natl. Acad. Sci. USA* **2011**, *108*, 6884–6888. [CrossRef] [PubMed]
5. Bellomo, N.; Dogbe, C. On the Modeling of Traffic and Crowds: A Survey of Models, Speculations, and Perspectives. *SIAM Rev.* **2011**, *53*, 409–463. [CrossRef]

6. Thalmann, D.; Musse, S.R. *Crowd Simulation*, 2nd ed.; Springer Science & Business Media: London, UK, 2012.
7. Kim, C.J.; Kim, Y.O. A Study on the Movement Pattern of Passengers in Subway Station. *J. Archit. Inst. Korea Plan Des.* **2007**, *23*, 71–78.
8. Kim, Y.O.; Kim, C.J.; Cho, I.O.; Kim, A.H.; Nam, S.W. The Effect of Spatial Layout on Space Use Pattern in Subway Station. Proceedings of KSR Conference, Jeju, Korea, 17–18 May 2007; 2007.
9. Okamoto, K.; Kaneda, T.; Ota, A.; Meziani, R. Correlation Analyses between Underground Spatial Configuration and Pedestrian Flows by Space Syntax Measures: A Case Study of Underground Mall Complex in Nagoya Station. *Natl. Libr. Aust. Cat. Publ. Entry* **2014**, *1*, 116–128.
10. Ueno, J.; Nakazawa, A.; Kishimoto, T. An Analysis of Pedestrian Movement in Multilevel Complex by Space Syntax Theory-in the Case of Shibuya Station. *J. Civ. Eng. Archit.* **2009**, *118*, 1–12.
11. Xu, Y.; Chen, X. The Spatial Vitality and Spatial Environments of Urban Underground Space (Uus) in Metro Area Based on the Spatiotemporal Analysis. *Tunn. Undergr. Space Technol.* **2022**, *123*, 104401. [CrossRef]
12. Penn, A.; Turner, A. Space Syntax Based Agent Simulation. In *Pedestrian and Evacuation Dynamics*; Schreckenberg, M., Sharma, S.D., Eds.; Springer-Verlag: Berlin, Germany, 2002; pp. 99–114.
13. Turner, A.; Penn, A. Encoding Natural Movement as an Agent-Based System: An Investigation into Human Pedestrian Behaviour in the Built Environment. *Environ. Plan. B Plan. Des.* **2002**, *29*, 473–490. [CrossRef]
14. Bae, W.K.; Ahn, H.J.; Jeong, M.K. An Implication Study for Making the Station-Oriented Neighborhood Resulting from the Analysis of Land Use and Pedestrian Environment on Newly Designated Station Area—Focused on the No.9 Subway Line, Second Section (Samjung Station-Coex Station). *J. Urban Des. Inst. Korea Urban Des.* **2010**, *11*, 113–128.
15. Won, J.Y. "Compound" Program Proposal for Railway Station Reflecting Regional Context: Focused on the Case of Shinchon Station. Master's Thesis, Ewha University, Seoul, Korea, 2005.
16. Lee, S.K. Bujeon Complex Transfer Station, Developing a Creative Transfer Center That Combines Various Functions Is Emerging as a Task. *Busan Dev. Forum* **2017**, *165*, 20–25.
17. Kim, J.H.; Kim, S.G.; Lee, K.N. Assessment of the New Capacity and Los of Transfer Facilities in the High-Speed Railway Stations. *J. Korean Soc. Civ. Eng. D* **2008**, *28*, 735–740.
18. The Korea Transport Institute. Evaluation of Service Quality of Rail Station on Demand-Side Perspective and Application. 2014. Available online: http://www.dbpia.co.kr/journal/articleDetail?nodeId=NODE06335902 (accessed on 27 October 2022).
19. Jang, S.Y.; Han, S.Y.; Kim, S.G. A Study on Level of Service of Pedestrian Facility in Transfer Stations at Urban Railroad. *J. Korean Soc. Railw.* **2010**, *13*, 339–348.
20. Choi, Y.K.; Lee, J.Y.; Cho, Y.S. A Study on the Planning of the Mixed-Use Complex for Urban Revitalization. *J. Archit. Inst. Korea Plan. Des.* **2009**, *25*, 183–190.
21. The Korea Transport Institute. A Study on Planning Guidelines for Transit Center Complex. 2010. Available online: http://www.dbpia.co.kr/journal/articleDetail?nodeId=NODE02416776 (accessed on 29 October 2022).
22. Hillier, B.; Hanson, J. *The Social Logic of Space*; Cambridge University Press: Cambridge/London, UK; New York/New Rochelle, NY, USA; Melbourne/Sydney, Australia, 1984.
23. Durmisevic, S.; Sariyildiz, S. A Systematic Quality Assessment of Underground Spaces–Public Transport Stations. *Cities* **2001**, *18*, 13–23. [CrossRef]
24. van der Hoeven, F.; van Nes, A. Improving the Design of Urban Underground Space in Metro Stations Using the Space Syntax Methodology. *Tunn. Undergr. Space Technol.* **2014**, *40*, 64–74. [CrossRef]
25. Hillier, B.; Penn, A.; Hanson, J.; Grajewski, T.; Xu, J. Natural Movement: Or, Configuration and Attraction in Urban Pedestrian Movement. *Environ. Plan. B Plan. Des.* **1993**, *20*, 29–66. [CrossRef]
26. Gibson, J.J. *The Ecological Approach to Visual Perception: Classic Edition*; Psychology press: New York, NY, USA; London, UK, 2014.
27. Turner, A. From Axial to Road-Centre Lines: A New Representation for Space Syntax and a New Model of Route Choice for Transport Network Analysis. *Environ. Plan. B Plan. Des.* **2007**, *34*, 539–555. [CrossRef]
28. Pelechano, N.; Allbeck, J.M.; Badler, N.I. Controlling Individual Agents in High-Density Crowd Simulation. In Proceedings of the 2007 ACM SIGGRAPH/Eurographics Symposium on Computer Animation, San Diego, CA, USA, 2–4 August 2007.
29. Batty, M. Agent-Based Pedestrian Modelling. *Adv. Spat. Anal. CASA Book GIS* **2003**, *81*, 81–106.
30. Torrens, P.M. Intertwining Agents and Environments. *Environ. Earth Sci.* **2015**, *74*, 7117–7131. [CrossRef]
31. Hu, H.; Luo, Z.; Chen, Y.; Bian, Q.; Tong, Z. Integration of Space Syntax into Agent-Based Pedestrian Simulation in Urban Open Space. In *Proceedings of 22nd International Conference on Computer-Aided Architectural Design Research in Asia(CAADRIA:) Protocols Flows and Glitches, Suzhou, China, 5-8 April 2017*; The Association for Computer-Aided Architectural Design Research in Asia: Hong Kong.
32. Omer, I.; Kaplan, N. Using Space Syntax and Agent-Based Approaches for Modeling Pedestrian Volume at the Urban Scale. *Comput. Environ. Urban Syst.* **2017**, *64*, 57–67. [CrossRef]
33. Jiang, B.; Jia, T. Agent-Based Simulation of Human Movement Shaped by the Underlying Street Structure. *Int. J. Geogr. Inf. Sci.* **2011**, *25*, 51–64. [CrossRef]
34. Cheliotis, K. An Agent-Based Model of Public Space Use. *Comput. Environ. Urban Syst.* **2020**, *81*, 101476. [CrossRef]
35. Castle, C.; Waterson, N.; Pellissier, E.; Le Bail, S. A Comparison of Grid-Based and Continuous Space Pedestrian Modelling Software: Analysis of Two Uk Train Stations. In *Pedestrian and Evacuation Dynamics*; Springer: Berlin/Heidelberg, Germany, 2011.

36. Tang, M.; Hu, Y. Pedestrian Simulation in Transit Stations Using Agent-Based Analysis. *Urban Rail Transit* **2017**, *3*, 54–60. [CrossRef]
37. Chae, S.K. Remodeling Design Directions on the Gangnam Station Underground Public Pedestrian Facility. *J. Archit. Inst. Korea Plan. Des.* **2012**, *56*, 48–52.
38. Turner, A.; Doxa, M.; O'sullivan, D.; Penn, A. From Isovists to Visibility Graphs: A Methodology for the Analysis of Architectural Space. *Environ. Plan. B Plan. Des.* **2001**, *28*, 103–121. [CrossRef]
39. Jung, R.H.; Chung, J.H.; You, S.Y. Analysis of Route Choice Behavior in Subway Station: Focusing on Walking Distance. *Korea Spat. Plan. Rev.* **2016**, *88*, 81–100.
40. Sutherland, D.; Kaufman, K.; Moitoza, J. Kinematics of Normal Human Walking. In *Human Walking*; Rose, J., Gamble, J.G., Eds.; Williams and Wilkins: Baltimore, MD, USA, 1994.

 *buildings*

Article

Evaluating Experiential Qualities of Historical Streets in Nanxun Canal Town through a Space Syntax Approach

Yabing Xu [1], John Rollo [2,*] and Yolanda Esteban [2]

[1] School of Architecture and Urban Planning, Shandong Jianzhu University (Shandong Architecture and Engineering University), Jinan 250101, China; xuyabing20@sdjzu.edu.cn
[2] School of Architecture and Built Environment, Deakin University, Melbourne, VIC 3220, Australia; yolanda.esteban@deakin.edu.au
* Correspondence: john.rollo@deakin.edu.au; Tel.: +61-3-522-78329

Abstract: Many studies have been conducted to measure the experiential qualities of historical streets using the standards and principles released by many global organizations. However, little attention has been paid to the effect of spatial characteristics of historical heritage. This study proposes a space syntax-based methodology, first developed by Bill Hillier and Julienne Hanson with colleagues from the Bartlett School of Architecture, while introducing factors such as complexity, coherence, 'mystery', and legibility from the work of environmental psychologist Stephen Kaplan and the urban designer Gordon Cullen. Our intention is to help inform urban designers in understanding people's spatial cognition of historical streets, and thereby assist designers and managers in identifying where cognitive experiences can be improved. The proposed method is applied to Nanxun, which is a developed canal town currently in decline in Zhejiang Province, China. This will be treated as the case study in order to explore the implication of the space syntax analysis. The impact from spatial characteristics on the evaluation is indirect and largely determined by the road-network of the canal town. As for Nanxun, the findings of this research suggest that the government's priority is to solve current negative tourist perception based on a conservation restoration plan. The findings of this research provide a reference for policymakers to better understand the experiential qualities of historical streets in townscapes.

Keywords: space syntax; experiential qualities; spatial analytics; design cognition; computational design

Citation: Xu, Y.; Rollo, J.; Esteban, Y. Evaluating Experiential Qualities of Historical Streets in Nanxun Canal Town through a Space Syntax Approach. *Buildings* **2021**, *11*, 544. https://doi.org/10.3390/buildings11110544

Academic Editors: Pierfrancesco De Paola, Michael J. Ostwald and Ju Hyun Lee

Received: 8 September 2021
Accepted: 8 November 2021
Published: 15 November 2021

Publisher's Note: MDPI stays neutral with regard to jurisdictional claims in published maps and institutional affiliations.

1. Introduction

1.1. Research Aim

The Beijing–Hangzhou Grand Canal is recognized by the United Nations Educational, Scientific and Cultural Organization (UNESCO) as a World Heritage Site (WHS). It is the longest artificial river and one of the oldest canals in the world [1]. Many of the cities along the Grand Canal are undergoing rapid economic growth, and there are increasing numbers of both large and small developments impacting the heritage significance of these areas.

In order to help mitigate the loss of the spatial experience of the historic character of the Grand Canal towns, this research adopts a space syntax-based analysis method to assist in evaluating the experiential qualities of their historic streets. The aim of the research is to help public authorities, developers, urban designers, planners, architects, and landscape architects in possibly extending these experiences in new developed areas through sensitively designed interfaces and extensions to the heritage protected Grand Canal townscapes.

1.2. Research Background and Context

Buildings, open spaces, or other features, which contribute positively to the character of a conservation area in a historic townscape [2] (defined as the visual appearance of a town or urban area), require a systematic analysis of their spatial characteristics in order to help evaluate the influence of these characteristics on user/visitor experience [3]. Experiential qualities of historical streets have been recognized as a major indicator to evaluate whether or not various sites need to be conserved and ways in which this can be realized. Townscapes have been a major concern for many researchers who study architecture, landscape architecture and urban design. The protection of historical and cultural heritage is an important cause to promote national unity, strengthen the national image and improve the quality of life of a nation's residents. However, economic globalization has brought an effect of 'compression' to the cultures of ethnic groups and regions around the world, as described by Yang and Yu [4], and this effect has quietly infiltrated into all aspects of society, especially for the preservation of historical heritage. Current trends in protection of historical heritage testify to increasing attention to the research on evaluating experiential qualities of historical streets which is widely recognized as a significant indicator when exploring experiential qualities of historical streets [2].

Space syntax aims to represent and analyze a diverse range of spatial layouts and to assist in explaining possible relationships between human behaviors and space characteristics [3,5]. Griffiths and Vaughan [6] in their paper, 'Mapping spatial cultures: contributions of space syntax to research in the urban history of the nineteenth-century city', applied historical data from maps to establish a mapping model applying space syntax theory on historical areas of 19th Century industrial cities with the use of HGIS (Historical Geographical Information Systems). Griffiths and Vaughan's [6] paper serves as an example of how the application of Space Syntax may yield informed insights as to why certain sociopolitical patterns of behavior may have been able to occur in different historic urban settings. Hence this paper begins to explore how the structure and morphology of many of China's historic Canal Town's provide a unique and intriguing visitor experience which is gradually being eroded with increasing development pressures.

The following paper is divided into five distinct sections. Following the introduction, the second part, Literature Review, appraises the previous research on space syntax and spatial cognition. The Third section, Methodology, discusses the process of the research which focuses on two subsections: Stage 1-Data Acquisition, the application of six Space Syntax indicators to measure the detailed information of the case study canal town of Nanxun 'Stage 2: Data Inference' reinterprets the data from stage 1, that is, the high and low value results of the six indicators, with low and high cognitive inference, and applies both serial vision theory and visual graph analysis (VGA) to assess the perceptions of Nanxun. The fourth Section Case Study: Nanxun, presents the application of Space Syntax and Visibility Graph Analysis (VGA) to provide an informed understanding of the spatial cognition of the town of Nanxun canal town. It elaborates on the limitations and future research, and presents a range of conserved and rebuilt suggestions to help improve people's spatial cognition in the historic districts of Nanxun. The final section, Conclusion, lists a number of important considerations is drawn in the fifth section.

2. Literature Review

2.1. Overview of Space Syntax

Space syntax was formulated by Bill Hillier and Julienne Hanson with colleagues from the Bartlett School of Architecture, University College London during the late 1970s and early 1980s. It involves a set of concepts and procedures for the study of spatial formations in order to advance understanding regarding the relationship between human spatial behavior and the form of the built environment, be it at an architectural building scale or the morphological structure of a city [7]. Two capstone moments in the development of space syntax occurred with the publication of Hillier's and Hanson's book 'The Social Logic of Space' in 1984 [8], and 'Space is the Machine [9]: A configurational Theory of Architecture', authored by Bill Hiller in 1999 (both were published by Cambridge University Press) [10]. The development of space syntax since the 1980s has undergone numerous advancements and applications, not only within architectural and planning practice submissions, but also in other research fields such as archaeology and human geography [11].

Scholars and researchers in the field of urban design have argued that architectural and urban morphology could be applied to spatial cognition. Zhao and Zou studied spatial characteristics in townscape and applied design cognition [12]. Penn [13] reviewed previous research work on syntax space and environmental cognition from his group, especially multiple doctoral degree dissertations. Penn [13] and his students utilized computer computation modelling to simulate the individual decision-making processes at the level of the individual. In Penn's opinion, cognitive space should be a topological terminology. Montello [14] studied on "thinking about thinking". He explored how significance is the cultural differences in spatial cognition. He reviewed the "Standard Social Science Model" which is often mentioned and implemented in studies related to the differences at that time and held a similar opinion with other scholars that a set of assumptions prevents observing universal aspects of human cognition. Ishikawa and Montello [15] reviewed two theories that hold the opinion that the cultural differences are significant: the carpentered-world hypothesis and the ecological hypothesis. Yu and Liu [16] combined an axis and image map to explore the special forms of Gulangyu of Xiamen in China. Based on their analysis, the authors put forward that ignoring the social and cultural environment cannot be ignored for the spatial pattern of the urban morphology. Wang and Bramwell [17] took the historical city block of China as an example and conducted a comparison using space syntax analysis. It was found that the space syntax analysis performed outstanding in research on the cognitive map due to its high degree of correlation. Additionally, the results of comparison of space syntax analysis and image map analysis indicates that the coherent structure and organization and the continuity are major elements to understanding a city [18].

Previous research results mainly focus on the comparison and analysis of relationships among the space syntax analysis, cognitive maps and space pictures [19,20]. In light of the large number of studies, Boeing [19] proved the relationship between isovist and visibility graphs. Gerald and Jan [20] study was based on established methods and published experimental data. The study investigated space syntax in different scales based on isovist and place graph. Since the architectural theory includes many complex factors which are difficult to numerically represent, the prediction as well as the accuracies of the spatial qualities would be doubtful. Initially, Gerald and Jan [20] discussed theories and methods in translating spatial qualities into isovist measurements. However, some navigation experiments where space properties are well represented by isovist and visibility graph measurements could benefit the responses and navigation behavior prediction. Meanwhile, according to the statistical results in the study, the measurements from a selected single position could be used for prediction purposes. Previous research has been conducted to modify the space syntax model by analyzing tourists' behaviors and big data analysis [20]. For example, Lebendiger and Lerman [21] aimed to find the correlations between urban spatial configurations and human emotions in an urban area to offer initial evidence that certain space sequences do cause positive and negative emotional arousal. However, there are few studies in which the questionnaire survey method, a classical method of acquiring people's space cognition information, corresponded to the spatial characteristics represented by the analysis variables in the space syntax model, or systematically discussed the spatial characteristics of historical city blocks and the interpretation level of the space syntax model.

Space syntax could be applied on historical districts. It also aims to represent and analyze a large range of spatial layouts and to assist in explaining the relation between human behaviors and space characteristics on historical streets [22]. The major contribution of the combination of space syntax analysis and geographic information system lies in improving the efficiency of expressing the competence of space [23–25]. In addition, space syntax mainly focuses on the open space which can express the representation patterns of a certain spaces [26]. Kubat et al. [27] investigated pedestrian and vehicular activity in Sharjah's historical center in order to understand current movement patterns in an to attempt to provide suggestions on the conservation of historical areas based on the understanding of spatial configuration. Space syntax analysis is favored in previous research on historical districts, since this research method provides a visualized and quantitative approach to exploring and researching historical districts in the fields of management, conservation and planning. For example, Li et al. [28] applied space syntax analysis to a historical district in China in order to explore the relationship between the street network and tourism preferences, while providing suggestions on tourism management. Additionally. Sheng et al. [29] considered Chaoyang in China as a case study and analyzed the traffic situation and visual integration of the central block using space syntax in order to provide a basis for the planning and re-design of historical districts. The study presented in this paper will provide a reference for policymakers to better understand the experiential qualities of historical streets in townscapes.

2.2. Main Analysis Factors of Spatial Morphology Information

Space syntax is a calculative method to understand the spatial patterns and configuration in order to assist in capturing user experience and reflecting visual and compositional attributes within a given space [30]. Figure 1 shows the Topological analysis methods of space syntax based on axial lines with additional consideration of shortest physical paths.

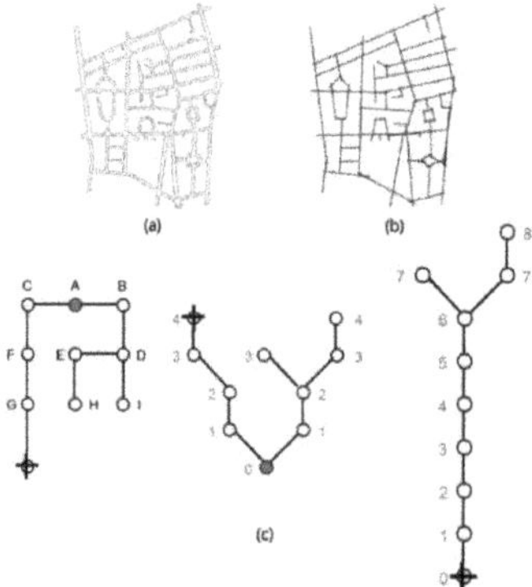

Figure 1. Topological analysis methods of space syntax based on axial lines (**a**), (**b**) with additional consideration of shortest physical paths (**c**).

Topological and geometrical analysis of space applies VGA. In addition, metric analysis which uses metric distance, the distance in metres from one space to another. Different spatial patterns are generated by assessing the three types of distance. In Figure 1, the major definitions used in Space Syntax can be concluded as follows:

(1) Topological (paths with fewest turns)
(2) Geometrical (paths with least angle change)
(3) Metric (shortest physical paths)

Space syntax is one of the computational analytical methods. Metric distance belongs to the wider theoretical and methodological field of space syntax. Figure 1 illustrates three types of distance metrics including topological, geometrical and metric. Topological, geometrical, and metric measures in space syntax draw a theoretical distinction between three types of spatial analysis.

2.3. Main Theories of Experiential Qualities of Historical Streets

Stephen Kaplan [31], writing in the 1970s within the emerging field of environmental psychology, is still considered relevant today, and while his research discussed factors that influence peoples' preferences of landscapes, the authors believe that many of his ideas could offer valuable insight regarding tourist's preferences in the experience of townscapes and historical streets. Kaplan [31] listed certain misunderstandings and misconceptions about preferences and perception and helped researchers, especially young scholars in architecture and urban planning, in thinking about four factors (as shown in Table 1) from two persisting purposes: 'making sense', which denotes the knowing or understanding of the things that happened in the space within a large time and space scale; and 'involvement', which denotes a learning activity. These two purposes are not contradictory since 'making sense' refers to the perceived structure of the environment and peoples' involvement with it, and references a supportive environment as being a place rich in possibility. The four factors are separately presented as blow:

- Complexity belongs to the involvement element at the initial analysis stage. This concept is similar to 'diversity' or 'richness' which relates to how much is 'going on' or 'there is to see' in a specific scene.
- Coherence is the initial analysis stage concept for making sense. This concept would require that all the objects in people's views should keep some kind of harmonic or internal consistency to deliver the message or story of a scene to people.

(Kapan notes that many people often only hold a certain amount of information in their 'working memory' at any one time, and that this appears to between five and seven major units or 'chunks' of information. Working memory is important for reasoning and the guidance of decision-making and behavior as distinct from short term memory [32]. Thus, if a scene possesses a high degree of complexity, it may require more structure to organize visual information, in a way that helps to realize coherence.)

- Mystery is different from surprise or novelty, rather, according to Kaplan [31], it is a spatial factor that represents the promise of new information offered by the scene, as one moves through a place. This would seem to align with the 'journey' oriented experience along a streetscape. The close connection and continuity between the observation and what is expected should bring in a high mystery feeling. Mystery raises curiosity rather than ambiguity or uncertainty [31].
- Legibility, this concept is similar to coherence and deals with the space structure with its variation and readability. The high legibility space will easily minimize the difficulty in finding one's way (Wayfinding). A highly legible scene is simple enough for people to oversee and generate a cognitive map. This factor focuses on the influences on the moving within the space. Kaplan writes "legibility entails a promise, a prediction, but in this case not of the opportunity to learn but to function ... it deals with the structuring of space, with its differentiation, with its readability ... Coherence concerns the conditions for perceiving while legibility concerns the conditions for moving within the space" [31].

Table 1. Four factors of preference matrix (Source: Kaplan, 1979).

Level of Interpretation	Making Sense	Involvement
The Visual Array	Coherence	Complexity
Three Dimensional Space	Legibility	Mystery

Based on the four concepts, Kaplan's research generated a 2 by 2 matrix in terms of the visual array as shown in Table 1, with respect to form, line, color, and texture, and of three-dimensional space with respect to spatial experience. He also discusses confusion in the idiosyncrasy of perception. Since perception and interpretation are combined together, the difficulty in separating these two indicators in the research is anticipated. Hence it is probably more appropriate for the researcher to use 'preference' rather than 'judgment' in their research and when conducting participatory interviews. Kaplan also suggests that the four factors: form, line, color, and texture, which are often used in the landscape analyses may conflict with the realities of human perception since they only work for a two-dimensional picture plane and provide a limitation on the sampling of the properties of the spatial experience. Kaplan's matrix may provide a way to evaluate the townscape experience. The challenge in adopting aspects of Kaplan's work is how to numerically illustrate the issues in the 2 by 2 matrix so as to create a symmetrical rating methodology which can be applied in conjunction with spatial syntax.

3. Methodology

The methodology is divided into two stages which are applied to the case study analysis of the Canal town Nanxun (see Section 4): 'Stage 1—Data Acquisition' and 'Stage 2—Data Inference'.

Stage 1: Data Acquisition, introduces the main analysis factors of spatial morphology information on the topological, geometrical, and metrics of the Space Syntax model [32]. Working with the software Depthmap, which is a multi-platform software designed to perform a set of spatial network analyses, data from six Space Syntax indicators were applied to measure detailed spatial information of Nanxun canal town's street network, and include: 1. connectivity value, 2. control value, 3. mean depth value, 4. local integration value, 5. global integration value, and 6. unintelligibility value.

1. Connectivity is treated as an indicator to explore the degree of the connection of a given space with other spaces.
2. Control value is wildly used to measure the control effect of a given space on other spaces.
3. Mean depth is an important feature to draw the degree of convenience from a given space to other space.
4. Local Integration value is favored in literature in illustrating the centrality of a given space in remaining spaces of a certain region and
5. Global integration relates to the integration of a line towards all the other lines of the axial map
6. Intelligibility value relates to "the degree to which what can be seen and experienced locally in the system allows the large–scale system to be learnt without conscious efforts" [5].

With respect to the relationship between the space syntax model and subjective analysis methods, the research will measure Nanxun's spatial topological relationship under a walking scale (k = 3, as mentioned in Table 2) by employing the connectivity value, control value, mean depth value, integration value and intelligibility value in the space syntax model. If a space has a high connectivity value, this means many other spaces connect with this space [33]. The control value can reflect the spatial connectivity [24]. Furthermore, the mean depth value refers to the shortest topological distance of the space from all other spaces. The higher the integration value, the higher the accessibility and commonality of the space [28]. The intelligibility value is used to measure whether the local spatial structure helps to establish understanding of the entire inter-spatial 46 system, indicating spatial identifiability [34].

'Stage 2: Data Inference', reinterprets the data from stage 1, that is, the high and low value results of the six indicators, with low and high cognitive inference based on the visual experiential qualities of Nanxun's historic street network and morphology derived from the writings of Gordon Cullen [2] and Stephen Kaplan [31]. This is conveyed in Table 2 Analysis variables in space syntax model and indicators of spatial cognition.

While previous studies have been conducted to attempt to evaluate the experiential qualities of spaces via questionnaires, the second stage of the methodology uses both Serial Vision theory and Visual Graph Analysis (VGA) to assess the perceptions of Nanxun canal town. The first stage of the visibility analysis addresses a serial vision approach which was first developed by the urban designer Gorden Cullen in his influential work *Townscape* first published in 1961 [2]. The concept of serial vision is significant to this study as it appears to reflect the rates of perception being experienced by an observer, which can be explained as a sequence of visual images captured by an observer when walking through a town at a uniform speed [2]. This may have a significant correlation to the analysis of space syntax, especially regarding connectivity value, control value and integration value and also appears to share interesting parallels with the work of Kaplan [31].

Table 2. Formula and description of analysis variables in space syntax model (Duan, J. & Hillier, B. 2007. Urban Space 3: Space Syntax and Urban Planning, Nanjing, Southeast University Press).

Variable	Formula	Scale	Description of Variables
Connectivity value	$C_i = k$	The 'i' space	Refers to the number k of spaces connected to the 'i' space; the higher the value, the closer the space is to the surrounding space, the stronger the influence on the surrounding space and the better the space permeability.
Control value	$C_{1i} = \sum_{j=1}^{k} \frac{1}{C_i}$	The 'i' space	Refers to the reciprocal sum of other spatial connection values connected to the 'i' spatial node; the larger the value, the greater the degree of control the 'i' space has over the space it intersects.
Mean depth value	$D_i = \frac{\sum_{j=1}^{n} d_{ij}}{n-1}$	The 'i' space	Refers to the shortest topological distance of the 'i' space from all other spaces; the lower the value, the more convenient the space is.
Integration value	$I_i = \frac{n(log_2((n+2)/3)-1)+1}{(n-1)(D_i-1)}$	The 'i' space	The most commonly used and important analysis variable, refers to the degree of spatial aggregation or dispersion between the 'i' space and other spaces. The higher the value, the higher the accessibility and commonality of the space. It is divided into the global integration value I_i and the local integration value $I_i(k)$. The global integration value refers to the relationship between the 'i' space and all other spaces and the local integration value refers to the relationship between the 'i' space and other spaces within the K steps (usually K = 3 at the walking scale).
Intelligibility value	$R^2 = \frac{\Sigma(C_i-\overline{C})(I_i-\overline{I})^2}{\Sigma(C_i-\overline{C})^2 \Sigma(I_i-\overline{I})^2}$	The whole space	Indicates the degree of correlation between local integration and global integration, used to measure whether the local spatial structure helps to establish understanding of the entire inter-spatial system ($\overline{C}$ is the average of the Connectivity value of all units; $\overline{I}$ is the average value of the global integration of all units in space); the larger the value, the stronger the spatial comprehensibility.

4. Case Study: Nanxun

4.1. Overview of Nanxun

Nanxun is located in Nanxun District, Huzhou, intersecting with Jiangsu, Zhejiang and Shanghai. In the Ming and Qing Dynasty, Nanxun was a famous town producing silk, and it is also an ancient town enriched with diverse people and combining both Eastern and Western architecture. Within the ancient towns in Jiangnan, the reputation of Nanxun is not accounted for the greatest, but the richness of Nanxun residents is remarkable. The rivers and streets in Nanxun are quite similar to other Jiangnan canal towns, as dwellings are crossed with each other, teahouses and restaurants are particularly prosperous. All dwellings are sitting next to rivers by connection of masonry steps, hence people normally conduct daily activities on the bank shore, such as washing clothes and vegetables. If there were boats parking in there, people could also get into the town through the dock. From aesthetic perspective, the white wall and grey tiles are reflected on the surface of river, and the flags hung over teahouses andrestaurantscouldbringpeopletheuniqueexperience-ofJiangnanrivertown. Within the Nanxun planned protection area of the ancient town, at two sides off general traditional building area and important river streets, it is allowed to upgrade, rebuild, demolish and build various architectures and rectify the environment. However, all construction activities shall be continuously under the principle of continuity and logic of urban texture. It is not allowed to implement demolishing and construction engineering in large scale.

4.2. Data Acquisition—The Spatial Syntax Analysis of Nanxun

To measure the spatial characteristics objectively, this research employs space syntax to systematically reflect the spatial characteristics of Nanxun canal town, as outlined in Section 3—Methodology.

Figure 2 displays the road-network and waterway system plan in Nanxun.

Figure 2. Road-network and waterway system plan in Nanxun.

Waking road, restrictive road, open space, main entrance and secondary entrance are shown in the Figure 2. On the whole, the axis structure of the ancient town area in Nanxun is not relatively simple. However, the spaces which have weak connectivity may increase mystery, mystery raises curiosity rather than ambiguity or uncertainty. In addition, historical and cultural blocks are social spaces in which a connection can be established between the urban context and social activity, containing unique emotional experiences and place attachments. Therefore, when studying the spatial characteristics of historical districts in order to optimize their development, we should not only take into account the spatial organization, but also consider the experiences of people's spatial cognition. Table 3 displays the spatial characteristics applied main analysis factors of space syntax of twenty-two major tourist streets in the Nanxun ancient town area.

Table 3. Space syntax analysis variables of ancient town area in Nanxun.

Road Name	Connectivity Value	Control Value	Mean Depth Value	Local Integration Value	Global Integration Value
Nanshan Street	4	1.250	503.820	61.482	144.248
Xiangshan Street	8	2.430	447.064	78.161	177.863
Renrui Road	5	1.726	341.369	75.074	178.014
Naxi Street	8	3.000	529.911	76.449	179.394
Wangshan Road	4	1.250	623.082	51.999	143.830
Wangjin Road	5	1.726	593.174	60.266	144.499
Hongshuo Road	6	2.000	495.372	70.366	157.337
Hongfeng Road	8	2.500	500.074	73.483	161.042
Baoshan Street	5	1.726	475.293	74.050	181.519
Bianmin Road	5	1.726	381.320	74.584	183.228
Hongyang Road	6	2.000	376.238	60.753	163.024
Nanfeng Street	5	1.726	373.797	65.942	183.368
Jiahe Road	5	1.726	355.918	43.901	146.679
Jiaye Road	5	1.726	317.762	51.432	166.219
Xida Street	5	1.726	427.487	69.349	183.058
Fengzhou Road	6	2.000	562.447	68.111	167.891
Dongda Street	5	1.726	610.750	69.860	143.203
Renshou Street	8	2.500	683.029	57.210	119.598
Fengtiao Street	5	1.726	656.531	71.724	141.614
Kaolaowan Street	5	1.726	657.381	73.291	143.335
Average value	5.65	1.900	495.591	73.247	160.448

As shown in Table 3, the integration value in the central historic block is higher than that in other regions of Nanxun town. The local integration value of these streets in this area clearly indicate that Xiangshan Street has the highest value, which is 78.161. It is also worth noting that Xiangshan Street is the main tourist street intersecting in the middle of the block, and is also the street with an elevated global integration value. With significant public exposure Xiangshan Street is the main passageway of the ancient block space organization [35]. Xiangshang street also presents the greatest connectivity value of 8 which is significantly higher than all of the 10 streets in ancient Nanxun, its control value of 2.430 is also much higher than that of other streets. The average local integration value of the main tourist streets of Nanxun is 61.482.

4.3. Data Inference—What May the Data Imply Regarding Perceived Experience?

To better understand the possible relationship between spatial cognition and spatial characteristics, the five indicators employed in this research in Table 2 are correlated with nine spatial cognition indictors in Table 4.

Figure 3a illustrates that the functional streets have significant differences in the connectivity value of the axis, with the average control value being 5.65 in the ancient town area. Xiangshan street, Naxi street, Hongfeng Road and Renshou Street's connectivity value is as high as 8, Nanshan Street and Wangshan Road's connectivity value is as low as 4. Figure 3b shows local integration value, Xiangshan Street is the highest as 78.161, higher than the lowest named Jiahe Road (43.901) ~78%. As shown in Figure 3c, Jiaye Road has the lowest value, 317.762, which is ~58.8% compared to the average value. The global integration value, see Figure 3d, Shenxing street has the lowest global integration value, it means the street has lower coherence and legibility. The street should increase publicness by means of rebuilt public space and tourist center. Renshou street is the main tourist street in Nanxun ancient town area, as we can see that the five values of this street are close to average values.

Table 4. Analysis variables in space syntax model and indicators of spatial cognition.

Analysis Variable in Space Syntax Model	Spatial Characteristic	Indicators of Spatial Cognition
Connectivity value	Spatial permeability	Easy understanding for road-network structure
		Intriguing journey through the town
Control value	Spatial connectivity	Few dead-end roads
		Well interconnected between lane ways and streets
		Few repeated-visit
Mean depth value	Spatial compactness	High tour efficiency
Integration value	Spatial accessibility	A clear main tour road
	Space publicity	A significant gathering centre
Intelligibility value	Spatial identifiability	Accurate determined the current location
		Efficient way finding to the destinations
		Clear and easy identified spatial functional zoning of the block

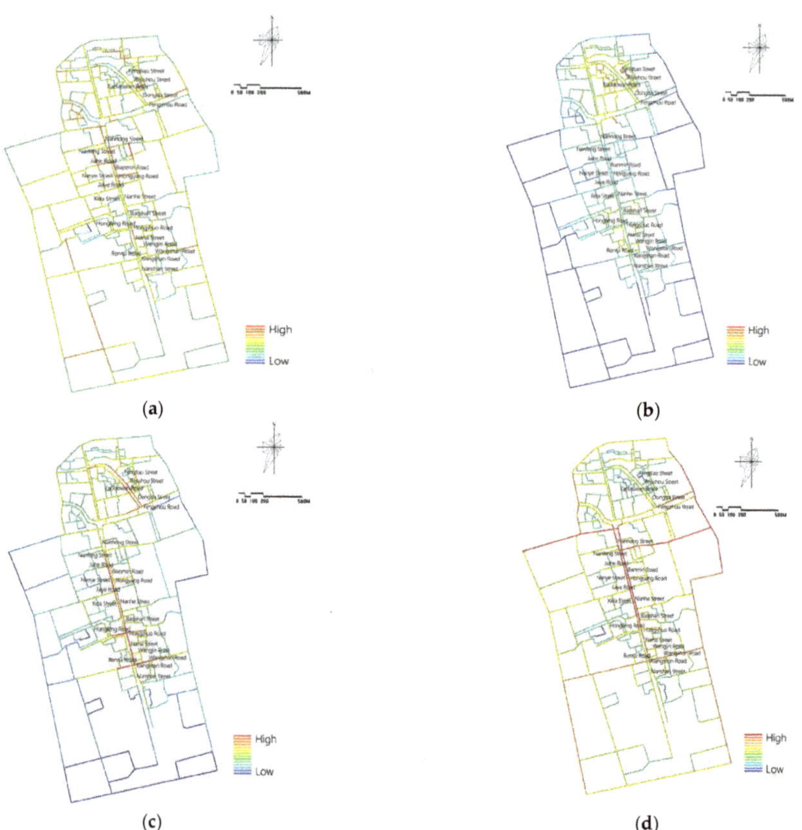

Figure 3. The space syntax model analysis result on Nanxun town: (**a**) connectivity value, (**b**) local integration value, (**c**) mean depth value, and (**d**) global integration value.

In Figure 3a, there are three streets and one road named Xiangshan Street, Naxi Street, Renshou Street, and Hongfeng Road with the highest connectivity value. The level of accessibility of the streets near the Grand Canal is relatively high, such as Bianmin Road and Xida Street on the west of the Canal, and Hongyang Street on the east of the Canal. Figure 3b shows that Renshou Street has the highest mean depth value in the whole town. It is obvious in Figure 3c that Nanfeng Street has the highest value in local integration, which is 183.368. The higher global integration value is concentrated in the middle of the ancient town area as presented in Figure 3d. Table 3 shows the spatial characteristics of major roads in Nanxun town (Xu et al. 2020). As the color contours in Figure 3d and values in 1, there are four streets with the highest global integration values, namely, Xida Street (183.058), Nanfeng Street (183.368) and Bianmin Road (183.228). Xu et al. [35] suggested that a stronger global integration value makes it easier to accumulate pedestrian flow. Therefore, Xida Street, Nanfeng Street and Bianmin Road are the most public, and convenient spaces for tourists to gather. Local integration value of streets in the ancient town area indicate that Xiangshan Road has the highest value, which is 78.161. It is also the street with the higher global integration value and hence receives the highest public exposure being the main passageway of the block space organization. Xiangshan street is the greatest in connectivity and its control value is much higher than that of other streets.

Figure 4 presents the intelligibility value of Nanxun town. If a spatial group has a high degree of intelligibility, it means that its overall spatial layout is more easily recognized and understood by people [35]. Generally, the range of the intelligence value R^2 can be evaluated according to being either 'weak', 'good' or 'strong' [35]. The intelligibility value of Nanxun ancient town area is calculated as 0.489, the intelligence value R^2 can be 0–0.5, which indicates that the spatial identifiability is weak. The high legibility space will easily minimize the difficulty in finding one's way. A highly legible scene is simple enough for people to over-see and generate a cognitive map. This term focused on the influences on moving within the space. Therefore, the cognitive degree of the Nanxun's spatial group could be improved, however by doing so might interfere with the perceived Mystery of the townscape when considering the four factors of Kaplan's [31]. Preference Matrix. In other words, rather than having an urban morphology where one is able to readily identify their location from any position within the whole town structure, having a more complex street pattern may slow their cognition and thereby enhance their experince of place [10].

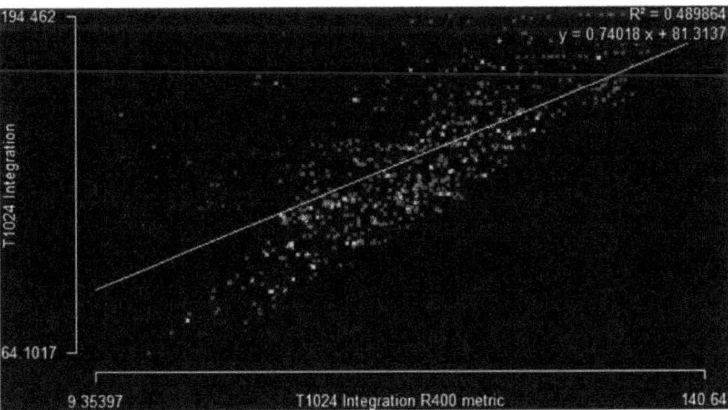

Figure 4. The space syntax model analysis result on Nanxun: Intelligibility value.

4.4. Focused Visibility Analysis of Baoshan Historical Block in Nanxun Ancient Town Area

Nanxun has large volume of people moving through the ancient town area than other historic canal towns, the efficient management pattern of Nanxun is probably a significant reason for more positive rather than negative visitor feedback. Therefore, an analysis of the benefits obtained from the promotion of management efficiency from Nanxun could assist towns such as Nanyang. Figure 5 displays the serial vision of Baoshan historical block in Nanxun ancient town area.

Figure 5. Serial vision of Baoshan historical block in Nanxun ancient town area.

Compared to other canal towns, Nanxun has more notable landmarks as shown in Figure 5, the stone arch bridge in photo 1, green trees in photo 7, and eaves along with white walls and black tiles in the other photos. The white walls are not well-maintained and part of them are peeling off. Therefore, the serial vision in Nanxun is still lacking certain characteristics. In contrast to the often-dramatic serial version examples of the historic European towns evaluated by Gordon Cullen [2], the successful Chinese ancient town, such as Wuzhen, provides the tourist a quiet and peaceful feeling when moving within the environment, whilst, none of the images from ancient towns includes cultural heritage monuments or historic buildings. They reflect the normal life of ordinary Chinese people in the past hundreds of years. The "standard" serial images include certain landmarks, such as a monument in photo 2, a sculpture in photo 3 and a great temple in photos 4 to 5. The images are significantly different from one to another in that the route the captured slowly unfolds, not all being revealed, but they are continuous showing spatial prominence through the intrigue that is revealed with the slight shift in geometry.

Topological and geometrical analysis of VGA or access graph analysis, metric analysis which uses metric distance, the distance in meters from one space to another. According to the main tourist routes include Baoshan street, Xida street and representative space of the town, the pedestrian flow simulation on VGA model analysis scope including part of the tour path and the space of the square and open space is selected. However, compared with the VGA, the analysis scope of the model is reduced, and the water area on the visible layer is not apparent, so the analysis scope of the model is different.

When viewed from the water space on the south side of the Baoshan historical block to the north side as shown in Figure 6, the street space and bridgehead open space are captured in the first step of sight depth, and the space is transparent and unobstructed. At the second step of sight depth, the main street space and east-west street space connected with the bridgehead open space can be seen, and the northeast space has the largest sight depth. When the line of sight is viewed from the courtyard space on the east side of the study area to the western side, due to the block of the building, the area of the first step of sight depth is the smallest, mainly inside the courtyard space. However, in the second step of sight depth, the main street space connecting the north and the south is completely visible, but the line of sight is difficult to penetrate further to the west side. When the sight is viewed from the main street space on the northern side of the study area to the southern side, the main street space is in the first sight depth range. However, due to the block of buildings, only part of the open space at the bridgehead can be seen. In the second step of the line-of-sight depth, the global view of the other spaces can essentially be seen, which is the point of view that needs the least line depth to comprehend the global view. When the line of sight is viewed from the main street space in the west of the study area to the east, most of the east-west main street spaces and the west open space can be seen in the first step of line-of-sight depth. In the second step, the north-south main street space is also fully visible, but the east side space visibility is weak.

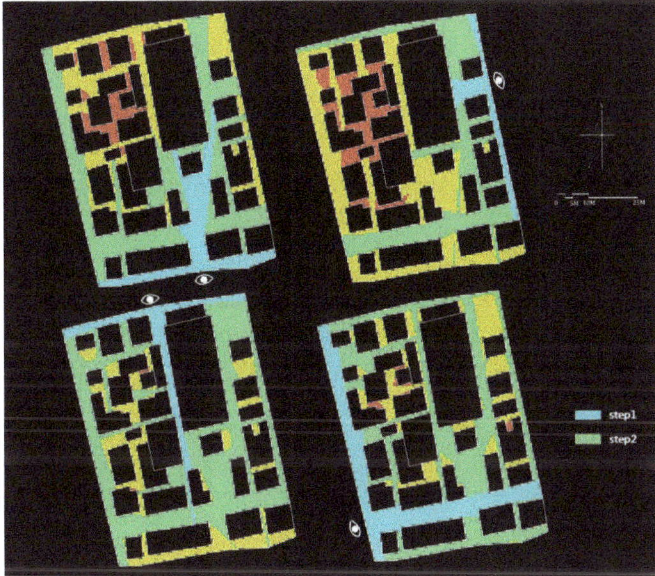

Figure 6. Visibility Graph Analysis of the highlighted block in Figure 5 (The step1 denotes the most likely selected space for a tourist at the location of the eye symbol and step2 denotes the most likely selected space for tourist after step1).

The analysis is from the four directional viewpoint of the Baoshan historical block as shown in Figure 6. When viewed from the water space on the southern side of the Baoshan historical block to the northern side, the street and public space in front of the ancient stage is completely in the first sight depth, and the space is transparent and unobstructed. Through the second sight depth, the space in the east and west connected with the bridgehead open space can be seen and further extended to the interior. On the whole of the Baoshan historical block, the line-of-sight depth of the northwest space is the largest. When the line of sight is viewed from the east space of the Baoshan historical block to the west side, the three sides are blocked by buildings, resulting in the minimum depth of sight range in the first step. However, the main street space connecting north and south in the second line of sight depth is completely visible, but the line of sight is difficult to penetrate further to the eastern side of the block. When the line of sight is viewed from the main street space on the northern side of the Baoshan historical block to the southern side, although the range of the first line of sight depth is not large, but the involved space is wider. In the second step of line-of-sight depth, the people can basically see other spaces of the whole world, and the point of view of the whole space can be seen with the least depth of sight among the four viewpoints. When the line of sight is viewed from the main street space on the western side of the Baoshan historical block to the east side, the main street space and the street space on the south side can be seen in the first step of sight depth, and penetrate to the western side. In the second step of line-of-sight depth, the bridgehead space and the public space in front of the stage can be seen, but the space visibility in the northeast part of the Baoshan historical block can also be seen. Figure 7 presents the pedestrian flow simulation on the VGA analysis of Baoshan historical block.

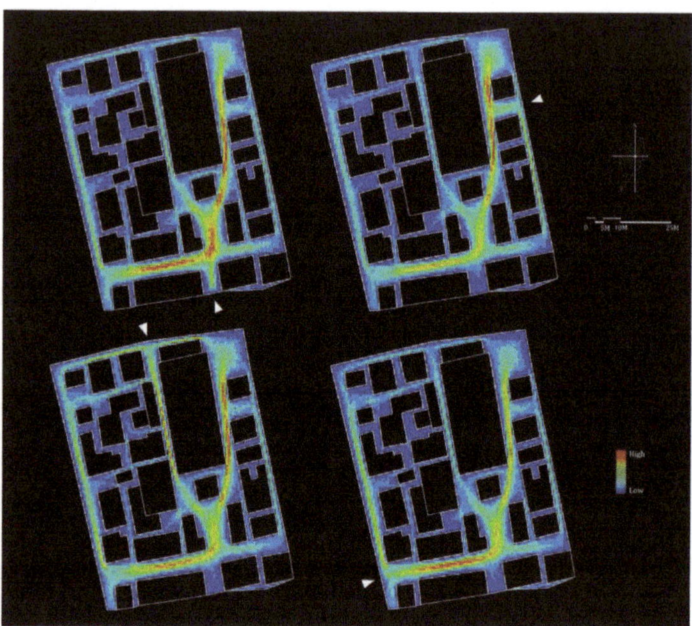

Figure 7. Pedestrian flow simulation on VGA of Baoshan historical block in Figure 5 (The red and the blue denotes the high and low tourist intensity, respectively).

In the simulation of pedestrian flow analysis as shown in Figure 7, when the flow of people is released on the northern side of the historical block, the main space forms an obvious centre in the public space in front of the stage, and it extends to the eastern side and the northern side and gradually decreases. When the flow of people is released in the eastern side of the space, the main space is distributed on the adjacent north-south street, and there is also high flow of people in the open space on the north side. When the flow of people is released in the space on the north side of the space, the main spatial distribution is in the north-south street of the east side. Although the passenger flow of the adjacent north-south street is increased, it is still lower than that of the adjacent street. At the same time, the east-west street space on the south side still attracts higher pedestrian flow. Therefore, the main stream of people is distributed in a mirror-image of 'L-shaped' street space. Through comparative analysis, the public space in front of the ancient stage is in the core get-together space centre, and the space reached by high pedestrian flow is always connected with the open large space regularly. Figure 8 presents the integration of the pedestrian flow simulation on VGA of the Baoshan historical block.

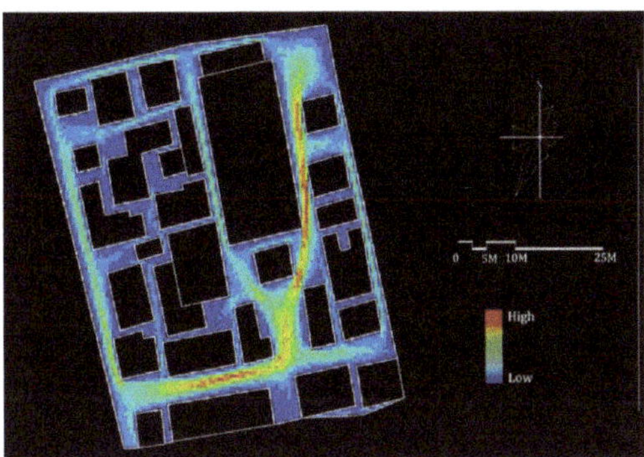

Figure 8. Integration of pedestrian flow simulation on VGA of Baoshan historical block.

Through the global pedestrian flow simulation as shown in Figure 8, it can be seen that the human activity track forms a spatial distribution structure of a mirror-image 'L'. At the same time, it intersects with another street space with weak pedestrian flow, which also forms a more obvious secondary public center. Due to the bottom-up organization deduction of eastern building texture in disorder, the arrival of pedestrian flow is weakened. Therefore, the core position of the public space in front of the ancient stage and beside Xida street could be further extended, and the organizational relationship with the bridgehead space could be integrated to further enhance its spatial function and value.

4.5. Suggestions for Nanxun and Future Research

Exploring the relevance experiential qualities of historical streets in townscape aims to provide a reference for policymakers to better understand the experiential qualities of historical streets in townscapes [36,37].

Based on the space syntax analysis results of the streets of ancient Nanxun, combined with the focused visibility analysis on a historic block near the Baoshan Bridge, several suggestions for improving peoples' experiential qualities in historic districts of Nanxun are proposed. In this research, Nanshan Street and Wangshan Road in Nanxun have the lowest connectivity value and control value in the ancient town area, and Jiaye Road has the lowest mean depth value. Considering the road network and town planning, the findings suggest a conservative restoration to solve current negative experiential qualities of Nanxun. For example, if the main tourist street (Baoshan Street) is rebuilt to raise its connectivity, the visiting time will be extended and the complexity and coherence of the space will be increased; two potential gathering centers may also improve integration values of the town so as to prevent periodic overcrowded and way-losing situations.

While Nanxun's overall block structure is relatively clear and regular compared with other historic canal towns such as Wuzhen, where the sense of perceived mystery is evoked on the first impression appears to be reinforced by a low intelligibility value. The opportunities of applying serial vision studies along with VGA at a micro-scale or sub-block level, revealed slight changes in the geometry of streets and lanes which presents tourists with an intriguing experience. Working with these slight shifts of design intervention by further scaffolding opportunities of landscape treatments to improve experiential qualities of historical streets.

5. Conclusions

This paper presents on-going research that attempts to explore the potential of working with space syntax analysis while introducing factors such as complexity, coherence, 'mystery', and legibility from the work of environmental psychologist Stephen Kaplan [31] and the urban designer Gordon Cullen [33]. The study investigated the spatial form and distribution of the ancient area of the Nanxun canal town. The purpose of the research is to assist in evaluating the spatial and experiential qualities of its historic streets in order to help mitigate the loss of these qualities with increasing development pressures. Reflecting on the computational design analysis for evaluating the relevance of these experiential qualities, further research is needed to explore the application of various environmental psychology factors. The following lists a number of important considerations:

A more detailed study regarding specific physical, social, and economic impacts on spatial cognition, reflecting people's perception of the relevance of the experiential qualities of historical streets, could provide valuable insight to further manage sensitive design outcomes with respect to future development pressures.

Drawing on the methodology developed by Yu, Behbahani, Ostwald and Gu [38], in their paper '*Wayfinding in Traditional Chinese Private Gardens: a Spatial Analysis of the Yuyuan Garden*' [38], further research could focus on a space syntax analysis of the waterways in the canal towns in order to study how various locations, such bridges and access points to the water and canal boats, may have added to their spatial and experiential qualities;

To conclude, an analysis of experiential qualities of historical streets based on a questionnaire or interview survey will be conducted to see if a correlation exists between local residents and tourist perceptions, and the interpretations of the results drawn from the combined space syntax, serial vision, and visibility analysis presented in the paper.

Author Contributions: Conceptualization, Y.X. and J.R.; methodology, Y.X. and J.R.; software, Y.X.; validation, Y.X. and J.R.; formal analysis, Y.X.; investigation, Y.X.; resources, Y.X.; data curation, Y.X.; writ-ing—original draft preparation, Y.X.; writing—review and editing, Y.X., J.R. and Y.E.; visualiza-tion, Y.X.; supervision, J.R. and Y.E.; project administration, Y.X.; funding acquisition, Y.X. All authors have read and agreed to the published version of the manuscript.

Funding: This research was funded by Yabing Xu Doctoral Scholars Grant Program of Shandong Jianzhu University (X21109Z) and the Natural Science Foundation of Shandong Province (ZR202103020684).

Institutional Review Board Statement: Not applicable.

Informed Consent Statement: Not applicable.

Data Availability Statement: The data presented in this study are available within this article.

Acknowledgments: The authors thank the support from Deakin University and Shandong Jianzhu University to conduct this research.

Conflicts of Interest: The authors declare no conflict of interest.

References

1. ICOMOS China. *Principles for the Conservation of Heritage Sites in China*; ICOMOS China: Beijing, China, 2015.
2. Cullen, G. *The Concise Townscape*; The Architectural Press: Oxford, UK, 1961.
3. Turner, A. From axial to road-centre lines: A new representation for space syntax and a new model of route choice for transport network analysis. *Environ. Plan. B Plan. Des.* **2007**, *34*, 539–555. [CrossRef]
4. Yang, L.; Yu, X. A summary of Chinas researches on the protection and utilization of cultural heritage. *Tour. Trib.* **2004**, *4*, 21–26.
5. Hillier, B. *Space Is the Machine: A Configurational Theory of Architecture*; Space Syntax: London, UK, 2007.
6. Griffiths, S.; Vaughan, L. Mapping spatial cultures: Contributions of space syntax to research in the urban history of the nineteenth-century city. *Urban Hist.* **2020**, *47*, 488–511. [CrossRef]
7. Malhis, S. Narratives in mamluk architecture: Spatial and percetual analyses of the madrassas and their mausoleums. *Front. Archit. Res.* **2016**, *5*, 74–90. [CrossRef]
8. Hillier, B.; Hanson, J. *The Social Logic of Space*; Cambridge University Press: Cambridge, UK, 1984.
9. Rana, S.; Batty, M. Visualizing the structure of architectural open spaces based on shape analysis. *Int. J. Archit. Comput.* **2004**, *2*, 123–132.
10. Mishra, S.A.; Pandit, R.K. Space syntax spproach for analyzing crime preventive urban design: Concept review. *J. Adv. Res. Constr. Urban Archit.* **2016**, *1*, 30–35.
11. Usui, H. Statistical distribution of building lot depth: Theoretical and empirical investigation of downtown districts in Tokyo. *Environ. Plan. B Urban Anal. City Sci.* **2019**, *46*, 1499–1516. [CrossRef]
12. Zhao, W.; Zou, Y. Creating a makerspace in a characteristic town: The case of Dream Town in Hangzhou. *Habitat Int.* **2021**, *114*. [CrossRef]

13. Penn, A. Spaces Syntax and Spatial Cognition: Or Why the Axial Line? *Environ. Behav.* **2003**, *35*, 30–65. [CrossRef]
14. Montello, D.R. *A New Framework for Understanding the Acquisition of Spatial Knowledge in Large-Scale Environments*; Oxford University Press: New York, NY, USA, 1998.
15. Ishikawa, T.; Montello, D.R. Spatial knowledge acquisition from direct experience in the environment: Individual diVerences in the development of metric knowledge and the integration of separately learned places. *Cogn. Psychol.* **2006**, *52*, 93–129. [CrossRef]
16. Yu, L.; Liu, Y. Cultures and community in the process of urban development and rehabilitation: Analysis of Gulangyu model Xiamen. *Urban Plan. Int.* **2011**, *25*, 108–112.
17. Wang, Y.; Bramwell, B. Heritage protection and tourism development priorities in Hangzhou, China: A political economy and governance perspective. *Tour. Manag.* **2012**, *33*, 988–998. [CrossRef]
18. Tao, Y. Digital City And Space Syntax: A Digital Planning Approach. *Planners* **2012**, *28*, 24–29.
19. Boeing, G. A multi-scale analysis of 27,000 urban street networks: Every US city, town, urbanized area, and Zillow neighborhood. *Environ. Plan. B Urban Anal. City Sci.* **2020**, *47*, 590–608. [CrossRef]
20. Gerald, F.; Jan, M.W. From space syntax to space semantics: A behaviorally and perceptually oriented methodology for the efficient description of the geometry and topology of environments. *Environ. Plan. B Plan. Des.* **2008**, *35*, 574–592.
21. Lebendiger, Y.; Lerman, Y. Applying space syntax for surface rapid transit planning. *Transp. Res. Part A Policy Pract.* **2019**, *128*, 59–72. [CrossRef]
22. Suchoń, F.; Olesiak, J. Historical Analysis of the Example of Nowy Sącz in Space Syntax Perspective. Guidelines for Future Development of Urban Matrix in Medium-Sized Cities. *Sustainability* **2021**, *13*, 11071. [CrossRef]
23. Lynch, K. *The Image of the City*; MIT Press: Cambridge, MA, USA, 1960; Volume 11.
24. Heyman, A.; Manum, B. Distance, accessibilities and attractiveness; urban form correlates of willingness to pay for dwellings examined by space syntax based measurements in GIS. *J. Space Syntax* **2016**, *6*, 213–224.
25. Rollo, J.; Barker, S. In Perceptions of Place-evaluating experiential qualities of streetscapes. In Proceedings of the 2013 6th State of Australian Cities Conference, Sydney, NSW, Australia, 26–29 November 2013; pp. 1–11.
26. Hillier, B. Studying cities to learn about minds: Some possible implications of space syntax for spatial cognition. *Environ. Plan. B Plan. Des.* **2012**, *39*, 12–32. [CrossRef]
27. Kubat, A.; Rab, S.; Guney, Y.I.; Ozer, O.; Kaya, S. In Application of space syntax in sevelopings: A regeneration framework for sharjahs heritage area. In Proceedings of the 8th International Space Syntax Symposium, Santiago De Chile, Chile, 3–6 January 2012.
28. Li, X.; Lv, Z.; Zheng, Z.; Zhong, C.; Hijazi, I.H.; Cheng, S. Assessment of lively street network based on geographic information system and space syntax. *Multimed. Tools Appl.* **2017**, *76*, 17801–17819. [CrossRef]
29. Sheng, Q.; Zhou, C.; Karimi, K.; Lu, A.; Shao, M. The application of space syntax modeling in data-based urban desian: An example of Chaoyang square renewal in Jilin city. *Landsc. Archit. Front.* **2018**, *6*, 103–113. [CrossRef]
30. Ellard, C. Neuroscience, wellbeing, and urban design: Our universal attraction to vitality. *Psychol. Res. Urban Soc.* **2020**, *3*, 52–54. [CrossRef]
31. Kaplan, S. In Perception and landscape: Conceptions and misconceptions. In Proceedings of the Our National Landscape: A Conference on Applied Techniques for Analysis and Management of the Visual Resource, Incline Village, NV, USA, 23–25 April 1979; Elsner, G.H., Smardon, R.C., technical coordinators, Eds.; Gen. Tech. Rep. (PSW-GTR-35); Pacific Southwest Forest and Range Exp. Stn. Forest Service, USA Department of Agriculture: Berkeley, CA, USA, 1979; pp. 241–248.
32. Miyake, A.; Shah, P. *Models of Working Memory Mechanisms of Active Maintenance and Executive Control*; Cambridge University Press: Cambridge, UK, 1999.
33. Duan, J.; Hillier, B.; Shao, S.; Dai, X. *Space Syntax and Urban Planning*; Southeast University: Nanjing, China, 2007.
34. Shahbazi, M.; Bemanian, M.R.; Lotfi, A. A comparative analysis of spatial configuration in designing residential houses using space dyntax method (Case Studies: Houses of isfahan and modern architecture styles). *Int. J. Appl. Arts Stud. (IJAPAS)* **2018**, *3*, 81–104.
35. Xu, Y.; Rollo, J.; Jones, D.S.; Esteban, Y.; Tong, H.; Mu, Q. Towards sustainable heritage tourism: A space syntax-based analysis method to improve tourists spatial cognition in Chinese historic districts. *Buildings* **2020**, *10*, 29. [CrossRef]
36. Lin, C.-H.; Morais, D.B.; Kerstetter, D.L.; Hou, J.-S. Examining the role of cognitive and affective image in predicting choice across natural, developed, and theme-park destinations. *J. Travel Res.* **2007**, *46*, 183–194. [CrossRef]
37. Berghauser, P.M.; Stavroulaki, G.; Marcus, L. Development of urban types based on network centrality, built density and their impact on pedestrian movement. *Environ. Plan. B Urban. Anal. City Sci.* **2019**, *46*, 1549–1564. [CrossRef]
38. Rongrong, Y.; Behbahani, P.; Ostwald, M.; Ning, G. Wayfinding in traditional Chinese private gardens:a spatial analysis of the Yuyuan garden. In Proceedings of the 49th International Conference of the Architectural Science Association. Living and Learning: Research for a Better Built Environment, Melbourne, VIC, Australia, 2–4 December 2015; pp. 931–939.

Article

In-Site Phenotype of the Settlement Space along China's Grand Canal Tianjin Section: GIS-sDNA-Based Model Analysis

Yan Zhao [1,2], Jian-Wei Yan [1], Yan Li [1], Guang-Meng Bian [1,*] and Yi-Zhao Du [3]

[1] School of Architecture, Tianjin University, Tianjin 300072, China; zhaoyanhit@sina.com (Y.Z.); yanjw22@126.com (J.-W.Y.); liyan1@yeah.net (Y.L.)
[2] Department of Architecture, Tianjin Ren'ai College, Tianjin 301636, China
[3] Department of Urbanism, Faculty of Architecture, Urbanism, and Building Science, Delft University of Technology, 2628 BL Delft, The Netherlands; y.du-4@tudelft.nl
*   Correspondence: bian_guangmeng81@tju.edu.cn; Tel.: +86-158-2272-9656

Abstract: The settlement space along China's Grand Canal composes an important part of the Canal heritage, has a close bearing on the production and life of the residents there, nourishes rich culture and wisdom and boasts vital value of conservation and inheritance. Due to China's rapid urbanization and industrialization, the settlements along the canal have been destroyed to some extent and their in-site characteristics urgently need excavation and conservation. Through field investigation, space syntax and GIS analysis, this paper performs quantitative analysis of the in-site characteristics of 18 typical rural settlements there. The findings show that: (1) The settlement space of industry dominant type for commerce and trade is comparatively dynamic and the capacity of topology and integration and the attractive force of the settlement center are stronger. (2) The dynamic scope of the citizens' everyday traveling in the settlements has the closest correlation with the data of public-service facilities. (3) The settlements along the canal boast multiple, causal and blended in-site phenotype. The research findings provide new standards to categorize the settlements along China's Grand Canal, paths and methods to explore the characteristics of the settlements and new cognitive perspectives to conserve and renew the settlements along China's Grand Canal Tianjin Section.

Keywords: China's Grand Canal; Tianjin Section; settlement space; in-site phenotype; GIS-sDNA

Citation: Zhao, Y.; Yan, J.-W.; Li, Y.; Bian, G.-M.; Du, Y.-Z. In-Site Phenotype of the Settlement Space along China's Grand Canal Tianjin Section: GIS-sDNA-Based Model Analysis. *Buildings* **2022**, *12*, 394. https://doi.org/10.3390/buildings12040394

Academic Editors: Michael J. Ostwald and Ju Hyun Lee

Received: 18 February 2022
Accepted: 21 March 2022
Published: 23 March 2022

Publisher's Note: MDPI stays neutral with regard to jurisdictional claims in published maps and institutional affiliations.

1. Introduction

China's Grand Canal, created during Ancient China in 486 B.C., is a great water conservancy project and the world-earliest canal residing on China's eastern plain. China's Grand Canal consists of three parts: Jing-Hang Canal, Sui-Tang Canal, and East-Zhejiang Canal, which makes it the longest canal in the world. China's Grand Canal spans over 10 Earth latitudes related to 8 Chinese provincial administrations, and served as the main artery of North–South transportation in Ancient China. In June 2014, China's Grand Canal was approved to be included in the World Heritage List as the 46th World Heritage item of China. Since 2019, the state has successively issued the *Guideline for Conserving and Inheriting the Grand Canal Culture*, determining the planning system to conserve, inherit and utilize the canal culture, and effectively guided the conservation, inheritance and utilization of canal culture. China's Grand Canal nourishes towns and rural settlements and enriches the abundant wealth and brilliant culture there. Since China's reform and opening-up, especially in the last 30 years of rapid urbanization and industrialization, large-scale urbanization has caused damage to settlements along the canal, and the local characteristics have been gradually lost and the in-site characteristics of the settlements badly require excavation, conservation and inheritance.

The research on canal heritage keeps up with the times. With continuous improvement to the content, a research system has been gradually formed which includes water resources management [1], ecological environment management [2,3], canal region [4,5], and canal projects [6]. Recently, some scholars on canal village development have discussed the development potential [7] of canal village tourism by means of rhythm analysis. Most of their research on China's Grand Canal has focused on ecological management [8,9] and canal projects, while research on rural settlements was much more limited. The research on conserving and developing settlements along the canal centered more on settlements in cities and towns, from the perspectives of the history of urban development [10], special investigation [11], spatial evolution [12], spatial distribution [13], value cognition [14], and conservation and utilization [15]. Although some scholars have carried out research on these rural settlements in recent years, and their research has mostly selected famous historical and cultural villages and traditional villages as targets [16], failing to classify the in-site phenotype of different rural settlements from a holistic view.

China's Grand Canal Tianjin Section (hereinafter referred to as "Tianjin Section") plays a pivotal role in the development of commerce and trade and the social economy of North China. "In-site" is a natural and cultural attribute particular to an area, boasting multi-scale, multi-dimensional characteristics according to the natural features, history, culture, and customs native to the area [17]. The present research applied such methods as field investigations, spatial syntax, and GIS to quantitatively analyze the in-site characteristics of typical settlements along the Tianjin Section. It proposed the types of settlements along the canal and the paths and methods for researching their in-site characteristics. It adopted the GIS-sDNA syntax model to perform modeling analyses of different types for rural settlements to compare and analyze their configuration characteristics. Meanwhile, it also compared the relevance between the configuration characteristics of the settlement space and the diachronic and synchronic data, brought forth the in-site characteristics of different types of settlement space, and revealed the inherent social logic of the settlement space along the Tianjin Section.

First, the research primarily selected 18 typical settlements along Tianjin Section to undertake field investigations and interview villagers; we acquired data on cultural heritage, tourist resources, public service facilities, and industrial resources and the resource point data were entered into a GIS database. Second, spatial modeling was applied to Google Maps data and was rearranged into ArcMap to create a spatial configuration model to determine the best possible model, including the canal, surrounding water system, settlement space forms, and land for road transport and production. Third, GIS-sDNA was applied to analyze the NQPDA, TPBt, and nuclear density of the spatial configuration. Finally, correlated analysis of the result of spatial configuration characteristics and the diachronic and synchronic investigation data was conducted and summarized, as was the in-site phenotype and law.

2. Research Area and Type Classification

Tianjin lies on the North China Plain border, borders on the Bohai Sea to the east and the west of Beijing, the capital city of China. According to official records, the government of the Ming Dynasty renamed "Haijin Town", a small town along the canal, into "Tianjin" in 1404, where the defense facilities were constructed (Figure 1). With the prosperity of grain transport and a garrison of troops, and thanks to its advantageous military and geographical position, Tianjin gradually became an important city along the canal and a portal for Beijing. China's Grand Canal drove the exchange of goods and materials between North and South China, encouraged cultural exchange, and stimulated vigorous development of settlements along the canal.

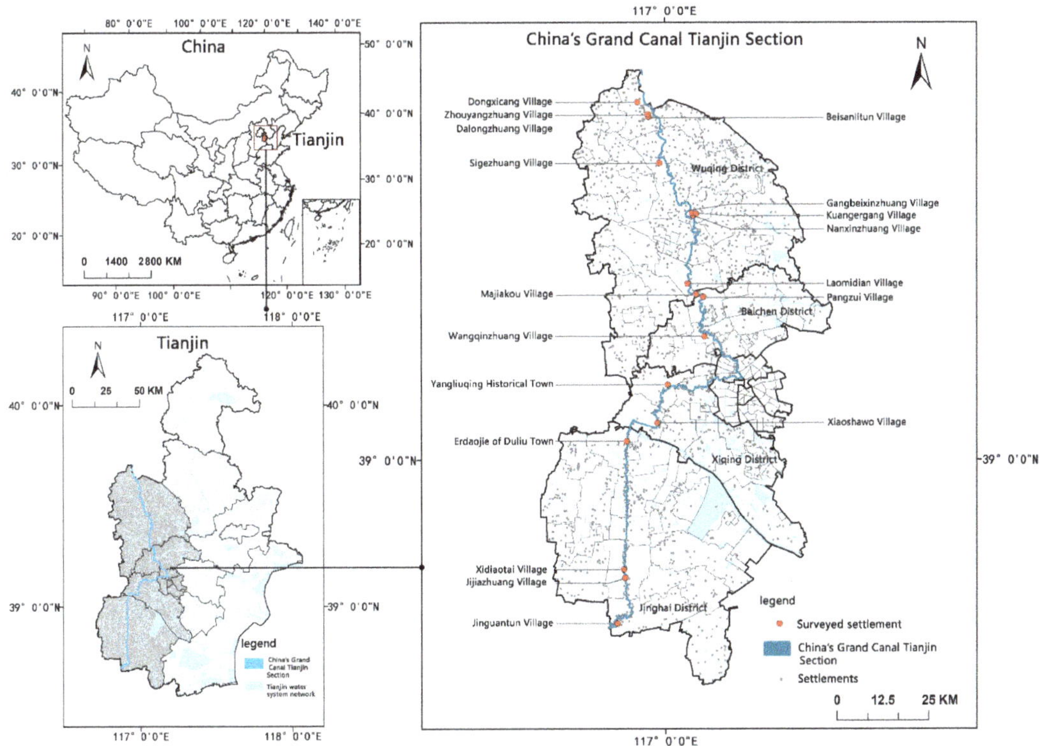

Figure 1. Spatial distribution of the sample settlements.

China's Grand Canal stretches about 182.6 km in Tianjin. The settlements along the canal boast typical features of villages in North China and are typical along the canal. Wuqing District, Beichen District, Qingxi District, and Jinghai District, apart from the city proper, were defined to be in the scope of the settlements along the Tianjin Section to guarantee the diversity and timeliness of the research objects. The selection of typical rural settlements along the Tianjin Section followed the standards of diversity and balance to ensure the typicality and representativeness of the research cases. The settlements along Tianjin Section were divided into four major types, with 16 sub-types for grain-transporting dominant type, industry dominant type, history dominant type, and folk customs dominant type, referring to the historical facts of the canal and the annals of local areas and after a field investigation (Figure 2). The settlements of the grain transporting dominant type refer to the settlements along the canal built for canal digging and dredging, grain transportation and military demands of the country, including the following five sub-types, channel topography, ferries and wharves, dredging intake channels, warehouses, and garrisons. The settlements of the industry dominant type refer to those along the canal for industries such as business, fishery, agriculture, and the handicraft industry, formed depending on grain transportation. The settlements of the history dominant type refer to those along the canal boasting important historical elements, such as typical historical relics, ancient architecture and historical figures. The settlements of folk custom dominant type refer to those built for unique traditional music, traditional dance, traditional handicraft and folklore with the development of grain transporting, and include traditional performance, traditional handicraft and oral traditions. In total, 18 typical sample settlements along the Tianjin Section were selected for in-depth research according to the above categorization (Figures 1 and 2, Table 1).

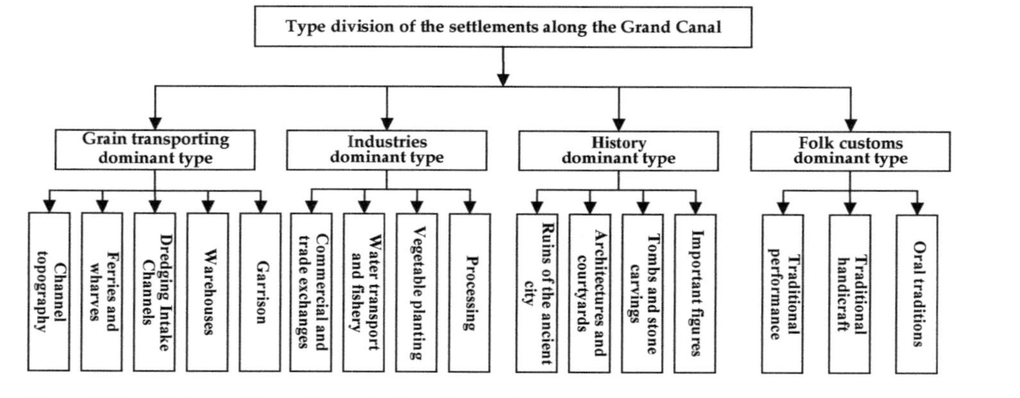

Figure 2. Type division of the settlements along the canal.

Table 1. Basic attributes of the sample settlements.

Type		Name	Area	Location	Population	Year of Village Establishment	Type Traits	Settlement Pattern
Grain transporting dominant type	Channel Topography	Pangzui Village	Shuangjie Town, Beichen District	West bank of the north canal and bordering on the Yongding River in the west	2024	The reign of Emperor Chengzu of Ming Dynasty	Like a mouth according to the natural form of the watercourse of the canal	Pickaxe-shaped
	Ferry and wharf	Laomidian Village	Huangzhuang Street, Wuqing District	West bank of the north canal, bordering on the Longfeng River in the north	3861	Yuan Dynasty	"Shilipo Laomidian Village", grain transporting ferry	Bar-shape with "two rivers separated by a dyke"
	Dredging intake channel	Majiakou Village	Huangzhuang Street, Wuqing District	North of the village, surrounded by the east and north canal and bordering on the Yongding River in the south	2271	Ming Dynasty	Digging for irrigation in Majiakou, discharge of water from the south	Square
	Warehouse	Dongxicang Village	Hexiwu Town, Wuqing District	West bank of the north canal	800	Yuan Dynasty	Fourteen warehouses in Hexiwu set up in the Yuan Dynasty	West–east rectangular
	Garrison	Jinguantun Town	Tangguantun Town, Jinghai District	Intersection of the south canal and the Machangjian River	1118	Ming Dynasty	Jiuxuan Watergate	Trapezoid

Table 1. *Cont.*

Type		Name	Area	Location	Population	Year of Village Establishment	Type Traits	Settlement Pattern
Industries dominant type	Commercial and trade exchanges	Erdaojie of Duliu Town	Duliu Town, Jinghai District	West bank of the south canal	34,413	The reign of Emperor Chengzu of Ming Dynasty	Famous historical and cultural town, a gathering place of merchants and a town of wine vinegar	South-north long, east-west narrow, long ribbon pattern
	Water transport and fishery	Wangqinzhuang Village	Beicang Town, Beichen District	South bank of the north canal	1554	The reign of Emperor Chengzu of Ming Dynasty	Developed water transport and fishery in the Qing Dynasty	West–east rectangular
	Vegetable planting	Xiaoshawo Village	Xinkou Town, Xiqing District	West bank of the south canal and bordering on Lianhuadian in the west	3000	The reign of Emperor Chengzu of Ming Dynasty	The garrison troops of the Yuan Dynasty set up "Rice Affairs Office" in Xiaoshawo Village.	Irregular
	Manufacturing	Jizhuangzi Village	Tangguantun Town, Jinghai District	West bank of the south canal	2421	Qing/Ming Dynasty	Abundant preserved, dried cabbage	Square
History dominant type	Ruins of ancient city	Xidiaotai Village	Chenguantun Town, Jinghai District	West bank of the south canal	1687	Western Han Dynasty	The ancient city of Dongpingshu in Western Han Dynasty and the legend of Jiang Taigong Fishing	Square
	Architectures and courtyards	Zhouyangzhuang Village	Hexiwu Town, Wuqing District	West bank of the north canal	1590	Ming Dynasty	Mosque	Irregular
	Tombs and stone carvings ①	Kuangergang Village	Dajianchang Town, Wuqing District	West bank of the North Canal	418	Ming Dynasty	The imperial monument "Dao Liu Ji Yun" written by Emperor Kangxi	Trapezoid
		Nanxinzhuang Village	Dajianchang Town, Wuqing District	West bank of the north canal	740	Ming Dynasty		Trapezoid
		Gangbeixinzhuang Village	Dajianchang Town, Wuqing District	West bank of the north canal	233	Ming Dynasty		Rectangular
	Important figures	Dalongzhuang Village	Hexiwu Town, Wuqing District	West bank of the north canal	875	Ming Dynasty	Generals of the Yang Family in Yingxi	Trapezoid
Folk customs dominant type	Traditional performance	Sigezhuang Village	Damengzhuang Town, Wuqing District	West bank of the north canal	1892	Ming Dynasty	Bamboo Hobbyhorse Society in Sigezhuang Village, a heritage site of Tianjin	Square
	Traditional handicraft	Yangliuqing Historical Town	Yangliuqing Town, Xiqing District	West bank of the south canal	161,247	Jin Dynasty	New Year picture drawing in Yangliuqing Town	Fan-shaped
	Oral traditions	Beisanlitun Village	Hexiwu Town, Wuqing District	West bank of the north canal	347	Ming Dynasty	Male and Female Brothers of Liu Xiaoguan in the Lasting Words to Awaken the World	Triangular

Data sources: Toponymic network: www.tcmap.com.cn/, accessed on 8 February 2022. Geographical names of Wuqing District. ① Since Kuangergang-Gangbeixinzhuang-Nanxinzhuang cluster on the north bank of the northern canal, the spatial configuration is under strong mutual influence. Therefore, quantitative analysis of the spatial configuration was undertaken as a group.

3. Research Methods

3.1. Theory and Practice of Space Syntax

Space syntax is the paradigm of theoretical research, systematically analyzing architectures and urban space, as proposed by Bill Hillier [18]. The theory of space syntax reveals the essential property of space and society, promotes the importance of "spatial ontology" to new heights, and provides brand new theories and tools to research architecture, urban and rural planning, and human geography. The core connotation of the space syntax theory includes spatial configuration, space division, and natural through-movement [19], specifically shown by the following three aspects: (1) Spatial configuration stresses the holistic relation, refers to a series of correlations of local space (like urban roads), and depends on the entire system [20]. The coinventor of space syntax, Julienne Hanson, proposed decoding the space of a house with the purpose of exploring and explaining abundant social and symbol information [21]. (2) Space division adopts the quantitative method of geometry and topology to better understand people's behaviors and perceptions in a space. For instance, analysis of a street network can help perceive the evolution of land use and further influence the mode of life in a city on the whole [22]. (3) Natural through-movement boasts high value for researching large urban spaces. "The main attribute of the city grid is to make some space superior to other space through movement. The structure of the city grid itself is the main producer of the mode of movement" [23].

The proposal of the space syntax theory provided new methods and techniques to understand and research settlement spaces [24–26], specifically shown from the following perspectives (Figure 3): (1) The evolutionary mechanism of the form of a settlement space. Through the quantitative analysis of the spatial configuration, it put forth a dynamic relation between social functions and space [27] and explored the factors and rules of settlement space that formed their evolution [28]. (2) The features of the phenotypes of settlement spaces. Analysis methods, such as the axis model and segment model were applied to reveal rules and characteristics, as well as the influencing factors, of the phenotype of settlement spaces [29]. (3) The actual testing, investigation, and structural analyses, on the one hand, explained the relationship between spatial form and citizens' perceptions of a space and inspired future planning for developments [30]. On the other hand, it discussed the influence of spatial forms on the travel model, explained the guidance of different travel scales for spatial evolution, and discerned the underlying rule of a settlement space form [31]. (4) The cultural genes of a settlement space; measures, such as J-shaped graph, visual analysis, and testing analysis were applied to perform a quantitative analysis of the settlement architectures, streets, and lanes to explore their social and cultural logic [32]. In brief, space syntax started from spatial ontology, performs quantitative analysis of the features of a space form, the rule of evolution, the social functions, and the cultural logic, and provided a feasible technique to quantitatively research on the "in-site characteristics" of a rural settlement space along China's Grand Canal.

3.2. GIS-sDNA Model Parameters

Spatial design network analysis (sDNA) is a new space syntax modeling tool that can enhance computational stability, draw accurate line models, and undertake various spatial analyses and visualized expressions using GIS geostatistical analysis [33]. This paper performed quantitative analyses of segment models of typical settlements along the Tianjin Section using GIS-sDNA software. Important parameters of GIS-sDNA selected for the present research include:

① Proximity ($NQPDA$): This represents the difficulty degree of other road networks within the search radius of a certain road network. A road network with a higher proximity boasts higher topological integration capacity and centrality and has more attraction for traffic flow traveling through this area.

$$NQPDA_x = \sum_{y \in R_x} \frac{p(y)}{d(x,y)}$$

② Through-movement degree (*TPBt*): Used to measure the probability of traffic flow within the search radius of a road network. The higher it is, the stronger the trafficability characteristic is and correspondingly the more passenger and vehicle flows it bears (In the formula, $OD(y, z, x)$ indicates the total amount of the nodes through node x within search radius R, $Link(y)$ represents the total amount of the nodes of each node y within search radius R and $p(z)$ indicates the weight of node z).

$$OD y, z, x = \begin{cases} 1, x \text{ lies on the shortest path from } y \text{ to } z \\ \frac{1}{2}, x \equiv yz \\ \frac{1}{2}, xy \equiv z \\ \frac{1}{3}, x \equiv y \equiv z \\ 0, \text{ other cases} \end{cases}$$

$$TPBt(x) = \Sigma_{y \in N} \Sigma_{y \in N} OD(y, z, x) \frac{P(z)}{Links(y)}$$

Note: $OD(y, z, x)$ is the shortest topological path between node y passing nodes x and z within radius R; $TPBt(x)$ is the result of the degree of movement through node x.

③ The kernel density analytical approach: The kernel density analytical approach calculates the unit area according to the characteristics of the dots or broken lines, applying a kernel function to match a smooth tapered curve to each dot or broken line. It can improve the visualization quality of the results of the space syntax model and differentiate and analyze the core and evolutionary tendency of a road network.

$$\hat{f}(x) = \frac{1}{mh} \sum_{i=1}^{n} k\left(\frac{x - x_i}{h}\right) \tag{1}$$

m is the number of observations, h is the smoothing parameter or bandwidth ($h > 0$), K is the kernel function, and $(x - x_i)$ is the distance between the estimated point x and the event point x_i.

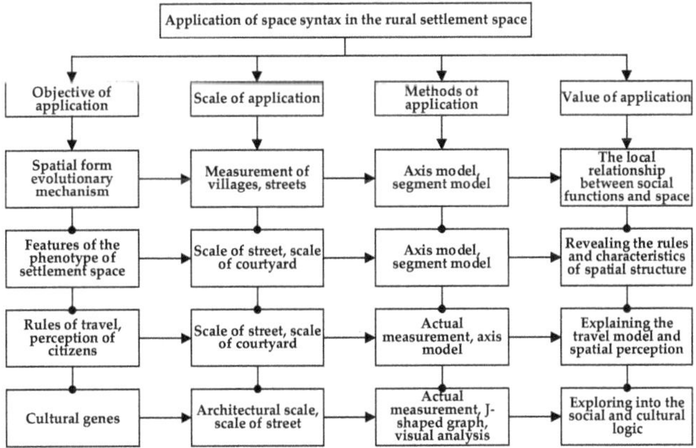

Figure 3. Application of space syntax in the rural settlement space.

At present, sDNA is academically applied to traffic network flow forecasting and model choice in the discipline of transportation [34,35]. Some scholars have applied sDNA and POI data to research the association between city forms and urban vitality and summarized the positive influence of blocks and belt types on urban vitality [36]. Ye and Yeh et al. [37] proposed combining the formal grammar tool "Form Syntax" with space syntax, spatial matrix, and blending to measure street topology, building density, and functional grouping, respectively. The three components can be quantized and combined to represent the phenotype of a city so as to categorize the degree of urbanization.

3.3. Data Sources and Processing

3.3.1. Basic Data Collection

Data collection for the settlements along the Tianjin Section included collection from diachronic and synchronic dimensions. The diachronic dimension is aimed at investigating unique resources formed during the historical evolution of settlements along the canal. Diachronic data chiefly include data of the tangible and intangible cultural heritage of the settlements along the canal. The data on cultural heritage were acquired from the *Detailed Rules and Regulations on Controlling the Space of National Land in the Core Monitoring Area of the Grand Canal Tianjin Section*, the Tianjin Intangible Cultural Heritage Website, and field investigations. The synchronic dimension is aimed at investigating the particular resources in settlements along the canal for the time being. Synchronic data primarily include data regarding tourist resources, public-service facilities, and industrial resources of the settlements there. First-hand data were acquired by the authors through field investigations into rural settlements and through interviews with villagers from February to July, 2021, and the resource point data were entered into a GIS database (Table 2).

Table 2. Data collection for the settlements along Tianjin Section.

Dimension of Division	Main Type	Sub-Type	Basic Type
Diachronic dimension	Cultural heritage	Tangible cultural heritage resources	Water heritage, ancient building, ancient site, ancient stone carving, ancient tomb, traditional residence
		Intangible cultural heritage resources	Oral tradition, traditional performance, folk customs and etiquette, traditional knowledge, traditional practice, traditional handicraft
Synchronic dimension	Tourist resources	Knowledge type	Science and technology, natural wonders, literary art
		Experience type	Picking garden, agronomic experience, experience of festivals
		Recreation type	Farm house, family stay, holiday and recuperation
		Tourism type	Countryside park, canal track, characteristic landscape
	Data of public-service facilities	Administration and service	Village community, party member activity room, rural police office, water company
		Educational institutions	Kindergarten, middle and primary school, training institutions
		Cultural and sports facilities	Cultural square, fitness field, sports ground
		Medical treatment and public health	Village clinic, private clinic, nursing institution, water closet
		Business service	Pedlars' market, convenience store, life service facilities, catering and food
	Industrial resources	The primary industry	Cultivated land, plantation, forestry and husbandry, facility agriculture
		The secondary industry	Production and warehousing land, manufacturing industry, village-run industry, technology culture

3.3.2. Space Modeling Data

Quantitative descriptions were made using the segment model based on GIS-sDNA to analyze the phenotype of spatial configurations of the 18 typical rural settlements along the Tianjin Section. The Google Map data, with a resolution ratio of 0.3 m, were adopted to draw a line model in ArcMap to reduce the model errors, ensure analytical effectiveness, and appropriately expand the model [38]. The modeling data included the canal, the surrounding water system, the form of the settlements, the land for roads and transportation, and the ecological space. The characteristics of the centrality of the spatial configuration of rural settlements of diverse scales were obtained by analyzing the configuration phenotype of settlements using the overall and local scales with GIS-sDNA. The research took the mostly frequented public settlement spaces, namely, a square or village committee, as the radius of daily travel to reveal the core zone, marginal zone, and activity scope of the settlement space. The standard of local travel radius was set to be the mean radius with the public space of the rural settlements as the center. Therefore, the local core scopes of different rural settlements were different according to their sizes. Generally speaking, the local core radius of rural settlements along the Tianjin Section ranged between 200–500 m.

3.4. Framework of Research on the In-Site Phenotype of the Settlements along the Canal

"In-site phenotype" is a natural and cultural attribute of a local area, boasting the characteristics of multiple scales and dimensions thanks to unique natural features, historical concerns, culture, and folk customs [17]. It is a unique in-site feature of an area and the external expression of "in-site genes" [39], and is comparatively stable, measurable, analyzable, and beneficial for clarifying the reasons for, and tendency of, the development of characteristics. "In-site gene" is the formation mechanism of the unique spatial features of an area and the fundamental reason for regional differences and diversity. Figure 4 shows the technical path of the in-site phenotypes of settlements. The in-site phenotype of settlements is composed of a configuration phenotype and in-site characteristics. The former is the dominant characteristic. Modeling of the macroscopic data of the form of the settlements in the spatial dimension was undertaken using the GIS-sDNA software for quantitative analysis to obtain the centrality characteristics of settlements at multiple scales. This is a recessive characteristic and quantitative analysis can be performed in spatial, diachronic, and synchronic dimensions using the following steps: (1) apply the kernel density analytical method to visualize the expression of the overall proximity and the overall TPBt of the configuration phenotype of the settlements; (2) incorporate the data of the cultural heritage in the diachronic dimension, tourist resources in the synchronic dimension, public-service facilities, and industrial resources into the dimension data for quantitative analysis; (3) compare the data dependency of the configuration phenotype of settlements and diachronic and synchronic data, put forward the in-site characteristics of settlements of different types and reveal the inherent social logic of settlements along the Tianjin Section (Figure 4).

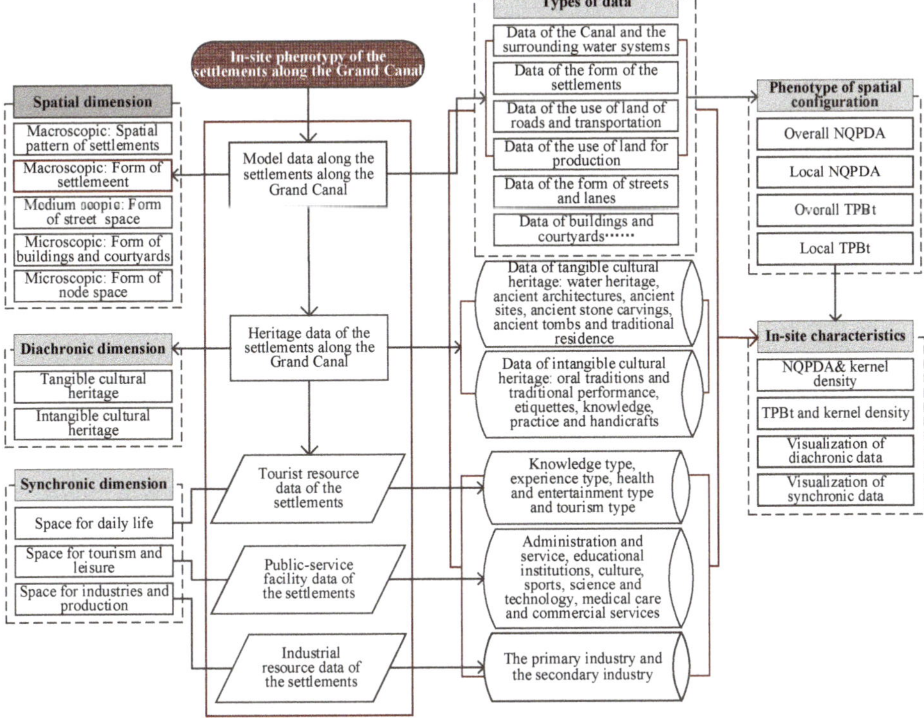

Figure 4. Technical path of in-site phenotype of the settlement space along China's Grand Canal.

4. In-Site Phenotype of the Settlement Space along Tianjin Section

4.1. In-Site Phenotype of the Settlement Space in Grain Transporting Dominant Type

4.1.1. Spatial Configuration Phenotype of the Settlement Space in Grain Transporting Dominant Type

Settlements of the grain transporting dominant type included Pangzui, Laomidian, Majiakou, Jinguantun, and Dongxicang villages (Table 3 and Figure 5). (1) In terms of the overall NQPDA, the mean value of the settlements of Laomidian and Jinguantun was relatively higher, 1.55 and 1.06, respectively, compared to that of Pangzui and Majiakou, which were 0.84 and 0.95, respectively, and that of Dongxicang was the lowest, 0.60. The settlements of Laomidian and Jinguantun boasted strong accessibility, the core space of Pangzui and Majiakou lay inside settlements influenced by river terrain and the spatial structure of Dongxicang was simple, while its core roads were linked with Caizhuang in the north. (2) As to the local NQPDA, the distribution of the five high-value settlements was connected with the spatial form. Pangzui and Laomidian displayed a linear structure parallel with the canal, Maojiakou had a cohesive axis-shaped structure, and Jinguantun and Dongxicang had a dual core cohesive structure vertical to the canal. (3) As to the overall TPBt, the structures of the vital communication lines of trafficability characteristics and the roads of high overall TPBt were consistent in the five settlements, representing the main communication lines to the settlements. (4) As to the local TPBt, high-value streets and lanes were scattered in the core zone of settlements, which were the main functional places of daily activities of villagers.

Table 3. GIS-SDNA syntactic parameters of the settlement space in grain transporting dominant type.

Grain Transporting Dominant Type	Basic Parameters					Parameters of Space Syntax			
						NQPDA		TPBt	
	Core Scope of Settlement Space	Analytical Range of Modeling Space	Local Core Radius ②	Amount of Segments	Mean Length/m	Overall Mean Value	Local Mean Value	Overall Mean Value	Local Mean Value
Pangzui	46.26 ha	70.50 ha	384 m	419	84.40	0.84	0.30	18.07	7.45
Laomidian	33.66 ha	249.49 ha	327 m	713	42.78	1.55	0.53	36.66	11.81
Majiakou	19.18 ha	156.68 ha	247 m	546	56.65	0.95	0.34	16.46	6.84
Jinguantun	53.21 ha	379.06 ha	411 m	575	81.29	1.06	0.26	18.16	6.18
Dongxicang	33.04 ha	304.40 ha	324 m	287	99.23	0.60	0.14	11.09	3.64

② The criterion of the local radius of travel is set to be the mean radius with the public space of the rural settlements as the center. Therefore, the size of the settlements must be different in the local core scope and the local core radius of the rural settlements along the Tianjin Section ranged between 200–500 m.

4.1.2. In-Site Characteristics of the Settlement Space in Grain Transporting Dominant Type

The analysis of the in-site characteristics of the settlement space for the grain transporting type shows that (Figure 6): (1) Pangzui was built according to the river terrain. In the shape of a mouth, it developed a unique settlement form. The overall proximity to Pangzui has a high relevance for data regarding public-service facilities and tourist resources. It was determined through the field investigation that Pangzui invited some artists to the village and built an art gallery and an exhibition center, and family remains to drive the development of the rural settlements. (2) Since ancient times, the pick-shaped Laomidian has been a ferry location for grain transportation. The topological central area and the areas of high-value trafficability characteristics have a high relevancy with public-service facilities and a low relevancy with cultural heritage and industrial resources. (3) As a typical case of a dredging intake channel, the overall proximity of Majiakou shows a spotty distribution and high-value space, and the public-service facility resources have high relevancy. High-value road networks comprise the main roads in the western and southern parts of the settlement. (4) The trapezoid-shaped Jinguantun used to house garrison troops. There are only a few public-service facilities in the high-value zones of the settlement. The proximity to Jiuxuanzha heritage and tourist resources are close. The in-site characteristics of high TPBt show the high trafficability characteristics of road networks in the settlement; the TPBt of the bridge in the west of the village and other high-value road networks have some relevance to tourist resources, public-service facilities, and Jiuxuanzha heritage. (5) In Dongxicang, as one of the fourteen "cang" villages, the high centrality of the area has the highest relevance to public-service facilities, followed by those of cultural heritage and industrial resources.

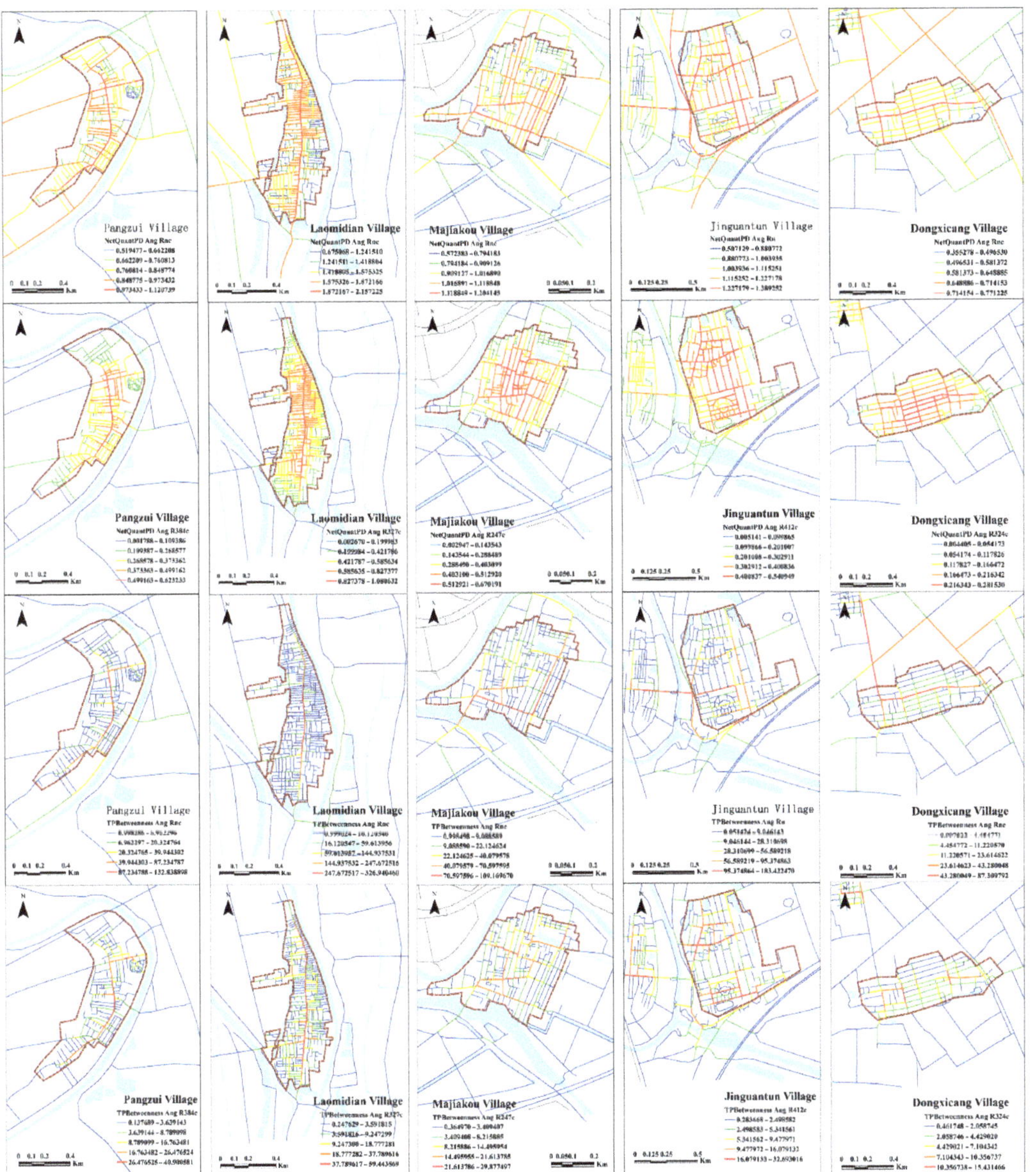

Figure 5. Spatial configuration analytical result of the settlement space in grain transporting dominant type.

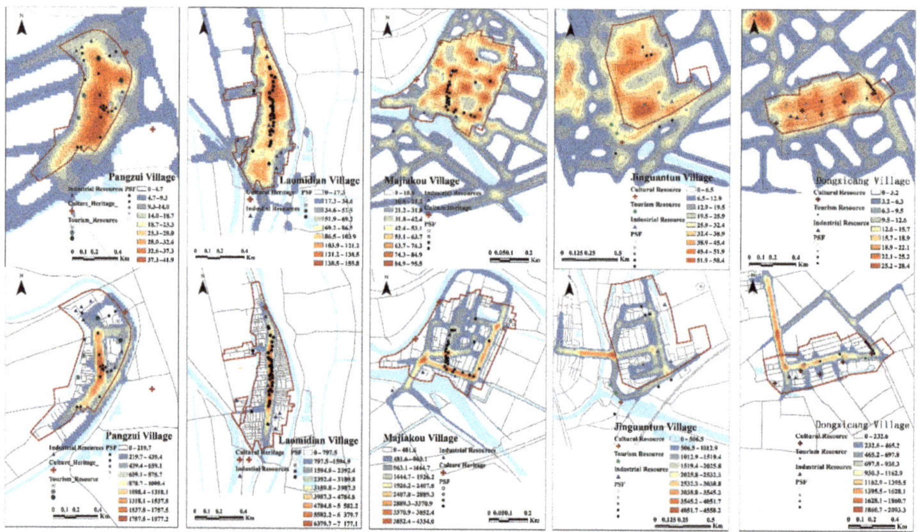

Figure 6. GIS-sDNA analytical result of the in-site characteristics of the settlement space of the grain transporting dominant type.

4.2. In-Site Phenotype of the Settlement Space in Industry Dominant Type

4.2.1. Spatial Configuration Phenotype of the Settlement Space in Industry Dominant Type

The settlements of an industry dominant type included the Erdaojie of Duliu Town, Wangqinzhuang, Xiaoshawo, and Jizhuangzi (Figure 7 and Table 4). (1) The analytical results of the overall proximity showed that the overall mean values for the four rural settlements were Erdaojie of Duliu Town (of 3.80) > Wangqinzhuang (of 1.56) > Jizhuangzi (of 1.27) > Xiaoshawo (of 1.15), indicating that the accessibility of the entire space of typical rural settlements relating to industries was quite strong. Erdaojie of Duliu Town lies in Duliu Town and is densely dotted with road networks. The road networks with a high proximity are in a +-shaped distribution; the core zone of Wangqinzhuang could be divided into southern and northern parts with the core zone in the shape of a fishing net in the north and a grid in the south; Jizhuangzi and Xiaoshawo core zones were in a grid-shaped distribution. (2) The local proximity indicates that the local core zones of the four rural settlements were all in the center of the settlements and featured multi-core distributions. (3) The overall proximity displayed that the structure of the vital communication line with trafficability characteristics and that of some roads with high overall integration degree are basically consistent. The roads with trafficability characteristics in Erdaojie of Duliu Town, Xiaoshawo, and Jizhuangzi span over the canal, representing a high opening degree of space. (4) The structure of four streets with high local TPBt is like that of the local integration degree, displaying the characteristics of a discrete distribution.

Table 4. SDNA syntactic parameters of the settlement space in the industry dominant type.

Industry Dominant Type	Basic Parameters					Parameters of Space Syntax			
						NQPDA		TPBt	
	Core Scope of Settlements	Analytical Range of Modeling Space	Local Core Radius	Amount of Segments	Mean Length/m	Overall Mean Value	Local Mean Value	Overall Mean Value	Local Mean Value
Erdaojie of Duliu Town	50.63 ha	605.97 ha	401 m	2293	48.43	3.80	0.51	39.80	8.71
Wangqinzhuang	40.78 ha	160.28 ha	360 m	798	44.39	1.56	0.58	26.58	10.06
Xiaoshawo	18.42 ha	106.43 ha	242 m	529	44.21	1.15	0.39	22.34	7.70
Jizhuangzi	56.27 ha	30.27 ha	423 m	589	62.47	1.27	0.45	18.03	7.36

Figure 7. Spatial configuration analytical result of the settlement space in the industry dominant type.

4.2.2. In-Site Characteristics of the Settlement Space in Industry Dominant Type

The analysis of the in-site characteristics of the settlement space of the industry dominant type shows that (Figure 8): (1) Duliu Town has been a gathering place of merchants and has had wine and vinegar breweries since ancient times. As to the in-site characteristics of the proximity of the Erdaojie of Duliu Town, high-value zones are highly relevant, with cultural heritage resources such as the Zhang Family Residence, the Liu Family Residence, and the Tankou Historic Site. The phenotype of TPBt indicates that high-value areas have a high relevancy for public-service facilities. (2) Since ancient times, Wangqinzhuang has boasted advanced water transport and fisheries. Now, industries are mainly fruit-planting for orchards, vegetable planting, and processing. The in-site characteristics of proximity exhibit a central area with a high integration degree in the north of the village that has a high relevancy for public-service facilities. Meanwhile, there are also relations with three cultural heritage resources, namely, the Temple of the King of Medicine, the former residence of Huang Jinxiang, and Zhongyingshaolian Martial Art Society, manifesting that cultural and artistic activities are also in-site characteristics of Wangqinzhuang, showing that cultural and art activities are also in-site characteristics of the settlement. The high-value road networks are highly relevant to public-service facilities. (3) Xiaoshawo is rich in turnips and boasts many planting and marketing bases and advanced agriculture. The area of high overall proximity is of high relevance to public-service facility resources, low relevance to tourist resources and has the lowest relevance for industrial resources. The high-value road network is linked to the canal bridge and the rural highway and is the place where most people and vehicles travel. (4) Jizhuangzi is rich in "preserved, dried vegetables". Pickles are the main source of finances of villagers. The in-site characteristics of the proximity shows that the high-value center is evenly distributed among the settlements. Two ancient tombs and former dwellings are located in the center of the area and public-service facility resources were in a linear distribution near the south–north road.

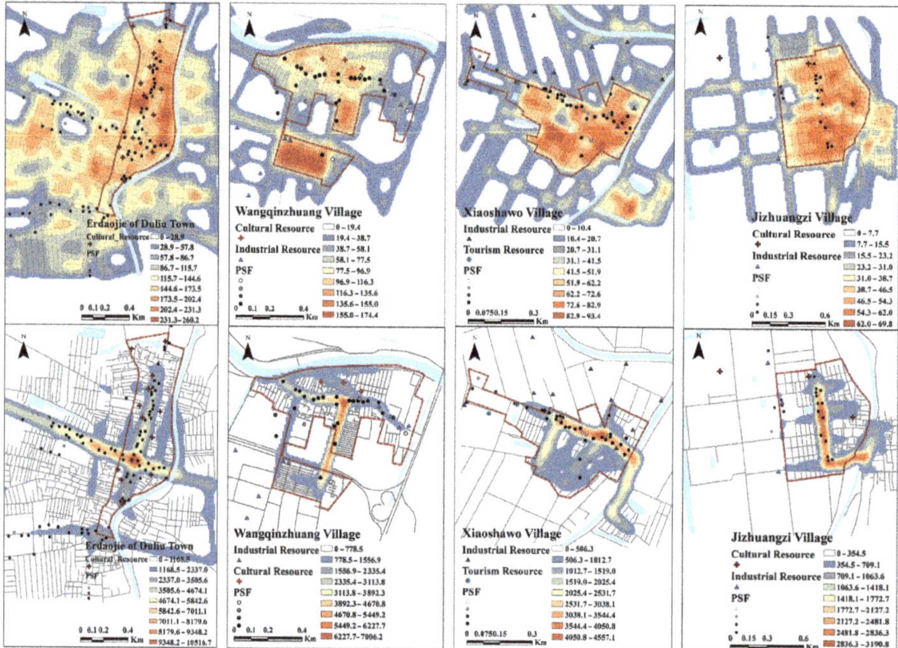

Figure 8. GIS-sDNA analytical result of the in-site characteristics of the settlements of the industry dominant type.

4.3. The In-Site Phenotype of the Settlement Space in History Dominant Type

4.3.1. Spatial Configuration Phenotype of the Settlement Space in History Dominant Type

The typical settlements in terms of the history dominant type include Xidiaotai, Zhouyangzhuang, Kuangergang-Gangbeixin-Nanxinzhuang, and Dalongzhuang (Figure 9 and Table 5). (1) The overall NQPDA showed that the overall mean values of the six rural settlements were in the sequence of Xidiaotai (of 1.68) > Kuangergan-Gangbeixin-Nanxinzhuang (of 1.04) > Zhouyangzhuang (of 0.81) > Dalongzhuang (of 0.51). The Xidiaotai core zone is in a grid distribution, Zhouyangzhuang in an axis-shaped distribution and the Kuangergang-Gangbeixin-Nanxinzhuang core zone with Nanxinzhuang as the core is in a grid-shaped distribution. Dalongzhuang has small-scale features a low proximity and high-value structures are in a cross-shaped distribution. (2) The analytical result of the degree of local integration shows that most of the six core zones of the six rural settlements lie in the center of the settlements. (3) The overall TPBt indicates that the overall TPBts of Xidiaotai and Kuangergang-Gangbeixin-Nanxinzhuang are higher; both villages have bridges over the canal and higher space exploitation degrees. (4) In terms of local TPBt, the road network with a high local TPBt value in all six villages is approximate to the local integration degree and in the discrete distribution.

Table 5. SDNA syntactic parameters of the settlement space in the history dominant type.

| History Dominant Type | Basic Parameters | | | | | Parameters of Space Syntax | | | |
| | | | | | | NQPDA | | TPBt | |
	Core Scope of Settlement Space	Analytical Range of Modeling Space	Local Core Radius	Amount of Segments	Mean Length/m	Overall Mean Value	Local Mean Value	Overall Mean Value	Local Mean Value
Xidiaotai	50.80 ha	556.02 ha	402 m	965	77.97	1.68	0.33	26.28	6.28
Zhouyangzhuang	33.5 ha	210.11 ha	326 m	408	80.68	1.04	0.16	18.28	4.30
Kuangergang/ Gangbeixinzhuang/ Nanxinzhuang	51.40 ha	738.05 ha	404 m	581	86.80	1.04	0.25	21.62	5.52
Dalongzhuang	16.93 ha	105.09 ha	232 m	93	176.11	0.51	0.11	12.00	3.13

4.3.2. In-Site Characteristics of the Settlement Space in History Dominant Type

The analysis of the in-site characteristics of a settlement space of the history dominant type shows (Figure 10) that: (1) Xidiaotai boasts a long history, with connections to Dongpingshu Historical Town of the Western Han Dynasty and the legend of Jiang Taigong Fishing. From the perspective of the in-site phenotype of proximity, the high-value center lay to the northeast of the canal and only a small number of commercial facilities were found there. The high trafficability network mainly covers the Dongdiaotai Bridge and Xidiaotai Bridge and the roads in the northern and western parts of the settlements have a high relevance for industrial resources. (2) Zhouyangzhuang is reputed as the "No.1 Village of Ethnic Groups in Tianjin" with Hexiwu Mosque rebuilt on ruins that were originally constructed in the Ming Dynasty. As to the in-site phenotype of proximity, an enormous number of public-service facilities were distributed in the area of high centrality, including the majority of commercial facilities. The area of high trafficability was distributed as arterial traffic in the east and south and has high relevance for Hexiwu Mosque and some commercial service facilities. (3) Kuangergang, Gangbeixin, and Nanxinzhuang are close to the Kuangergang Hub of the north canal. The imperial monument "Dao Liu Ji Yun" erected by Emperor Kangxi is found there. The central area of high value is in the settlements to the north, and have relevance for some public-service facilities and a low relevance for tourist resources of farm yards. Road sections with high trafficability characteristics are of a high relevance to arterial traffic and tourist resources. (4) Dalongzhuang used to be the mansion of the marquis of the Yang family. Later, the tenants of the Yang family scattered around Dalongzhuang, which became a natural village. The area with a high centrality value in the settlements is at the former sites of the mansion of the marquis and the Phoenix Platform, and is now primarily a residential space. An area of high TPBts comprised the roads in the center and south of the settlement, and was of high relevance to public-service facilities.

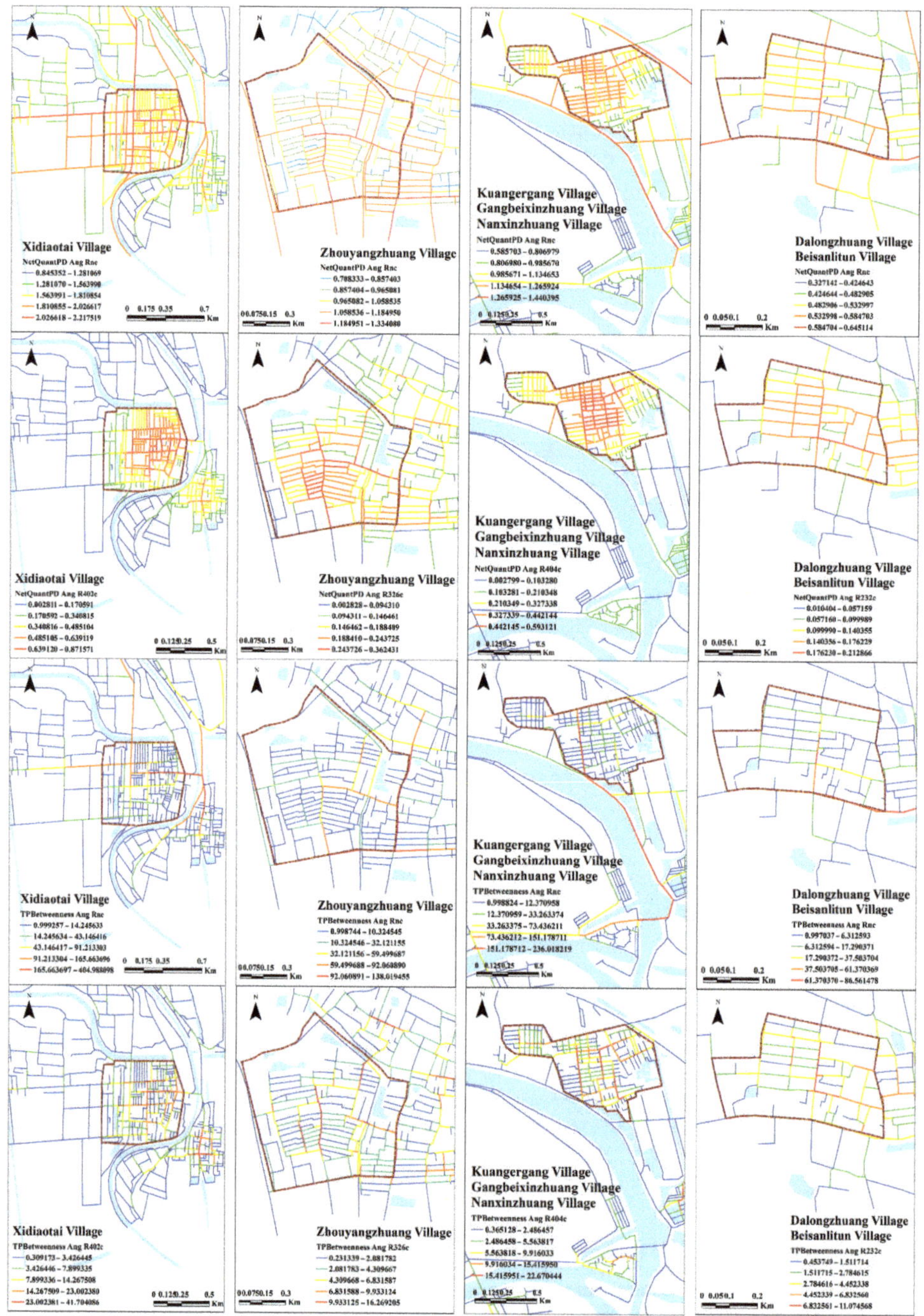

Figure 9. Spatial configuration analytical result of the settlement space in the history dominant type.

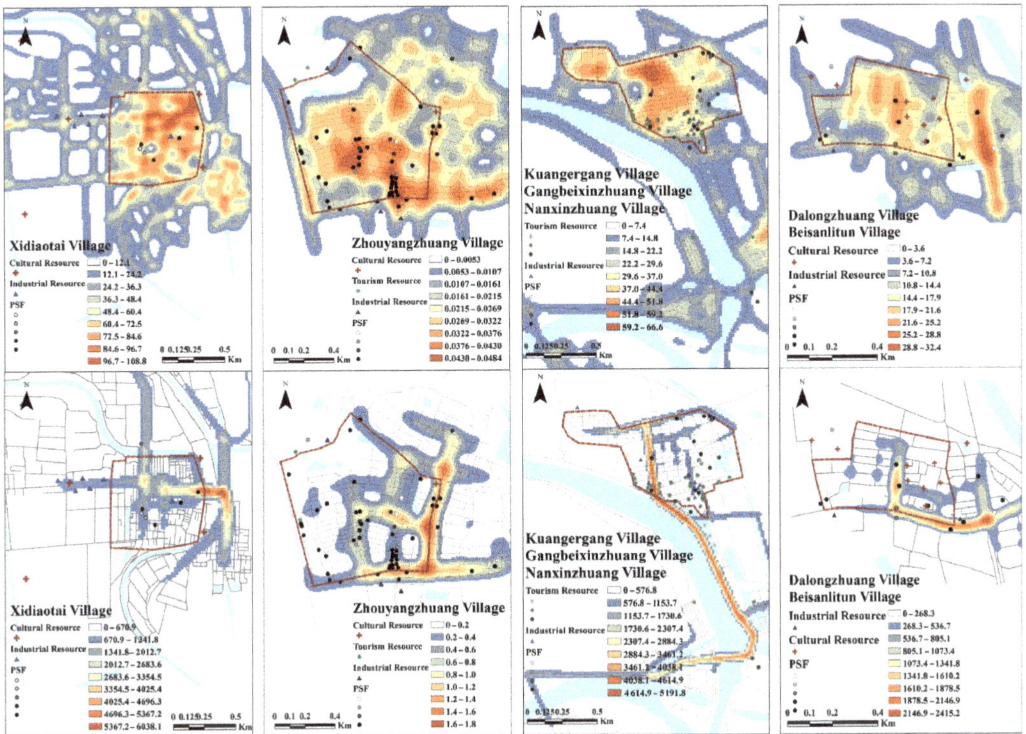

Figure 10. GIS-SDNA analytical result of the in-site characteristics of the settlement space in the history dominant type.

4.4. In-Site Phenotype of the Settlement Space in Folk Customs Dominant Type

4.4.1. Spatial Configuration Phenotype of the Settlement Space in Folk Customs Dominant Type

The settlements for the folk customs dominant type consist of Sigezhuang, Wangqinzhuang, Yangliuqing Historical Town, and Beisanlitun (Figure 11 and Table 6). (1) As to the overall integration degree, the overall mean values of the three rural settlements are in the sequence of Yangliuqing Historical Town (of 1.50) > Sigezhuang (of 1.16) > Beisanlitun (of 0.51). The main rural roads are evenly distributed in the overall core zone of Sigezhuang, and are in a square grid layout, covering the whole settlement space and featuring excellent accessibility. Yangliuqing Historical Town comprises tourist streets and traditional settlements with a high-value road network in a circular distribution. Beisanlitun features the smallest scale, only two high-value trunk roads and connections with the main rural space. (2) The local integration degree indicates that the local core zones of the three rural settlements are of a cohesive uni-core structure. (3) The overall TPBt manifests in that the TPBt of Yangliuqing Historical Town is the highest and the roads with trafficability characteristics are connected to historical blocks, traditional settlements, roads to the canal used by emperors, and the trunk roads in the historical town; the structures or roads with trafficability characteristics in Sigezhuang are have an overall integration degree, mainly distributed around the settlement with a road running through the core zone of the village; the overall TPBt of Beisanlitun is low, indicating that the rural settlements are close. (4) The local TPBt manifests in that the roads with a high local trafficability value for the three villages are similar to the local structures with some integration degree.

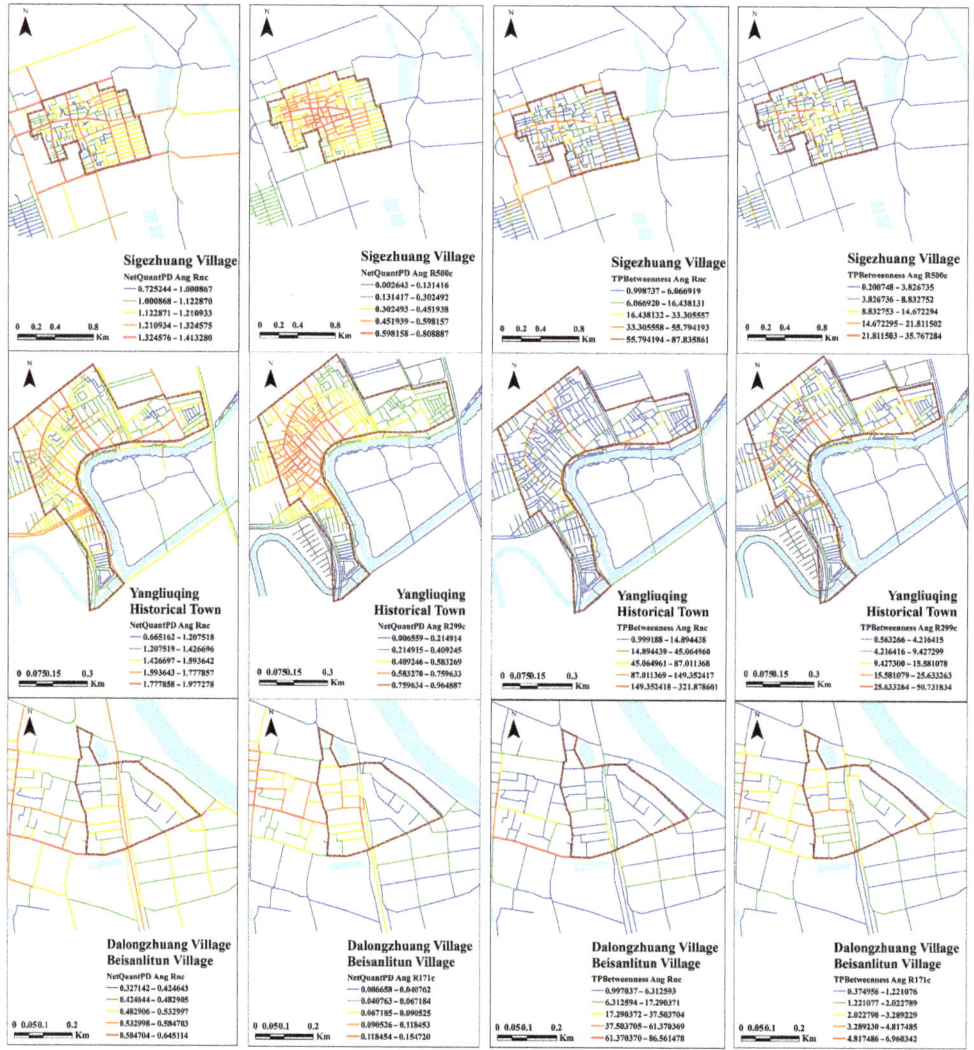

Figure 11. SDNA analytical result of the settlement space in the folk customs dominant type.

Table 6. SDNA syntactic parameters of the settlement space in the folk customs dominant type.

Folk Customs Dominant Type	Basic Parameters					Parameters of Space Syntax			
						NQPDA		TPBt	
	Core Scope of Settlement Space	Analytical Range of Modeling Space	Local Core Radius	Amount of Segments	Mean Length/m	Overall Mean Value	Local Mean Value	Overall Mean Value	Local Mean Value
Sigezhuang	78.72 ha	660.05 ha	500 m	543	84.09	1.16	0.41	16.35	7.16
Yangliuqing	28.11 ha	87.09 ha	299 m	866	36.70	1.50	0.49	23.45	9.00
Beisanlitun	9.20 ha	105.09 ha	171 m	93	176.11	0.51	0.07	12.00	2.29

4.4.2. In-Site Characteristics of the Settlement Space in Folk Customs Dominant Type

The analysis of the in-site characteristics of the settlement space in the folk customs dominant type shows that (Figure 12): (1) The in-site characteristics of Sigezhuang shows that the high-value areas lie in the northwest and have a high relevance tp public-service facilities and the intangible Bamboo Hobbyhorse Society cultural heritage site, as well as such cultural heritage sites as the Temple of Local God of the Land and the Temple of Master. The high-value road network comprises arterial traffic to enter the settlements and has a high relevance to public-service facilities. (2) Yangliuqing Historical Town was established in the Jin Dynasty and was the hub of grain transport and a gathering place for merchants in the Ming and Qing Dynasties, featuring great cultural diversity. From the point of view of in-site characteristics, high-value settlements have a high relevance to available public-service facilities, cultural heritage resources, and tourist and landscape resources. Tourist attractions boast abundant resources despite weak topological integration. (3) Beisanlitun was the garrison of the "2nd Garrison Troop" in the Yuan Dynasty and was established by Yuan troops to open up land for waste and to grow food grain. "Male and Female Brothers of Liu Xiaoguan", a storytelling script adapted from a strange tale in the village from famous storytelling script writer Feng Menglong of the Ming Dynasty, is still popular today. As to the in-site characteristics, the central area has a high value and primarily covers roads leading to the village and some residential space in the west of the village. There are a few public-service facilities in the village. From the perspective of the in-site phenotype of the TPBt, the eastern and western roads linking Beisanlitun and Dalongzhuang are high-value areas with a relevance for public-service facilities and industrial resources.

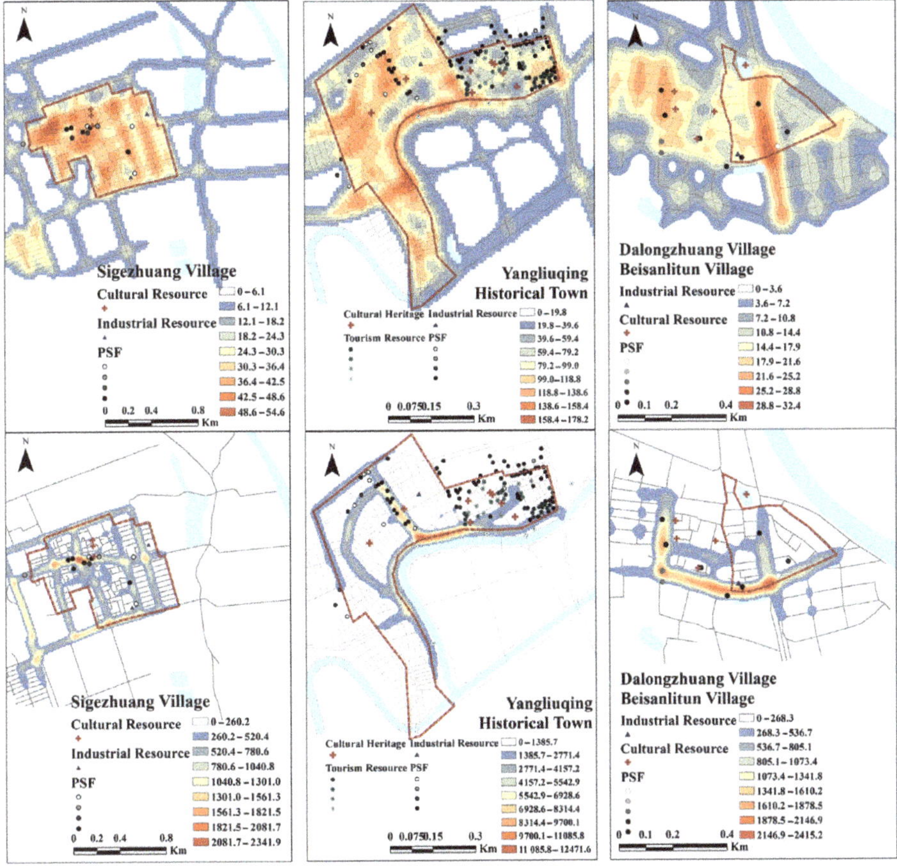

Figure 12. GIS-sDNA analytical result of the in-site phenotype results of the settlement space in the folk customs dominant type.

5. Discussion

5.1. In-Site Characteristics of the Settlements along Tianjin Section

Comparative analysis of the phenotype values of the overall proximity configuration shows (Table 7) that the value of the spatial configurations of settlements in the industry dominant type was 2.05, followed by 1.07 in the history dominant type, 1.06 for that in the folk customs dominant type and 0.92 for that in the grain transporting dominant type. Comparative analysis of the phenotype value of the overall TPBt configuration showed that the value of the spatial configurations of all kinds of settlements were, from high to low, 30.22 for the industry dominant type, 19.55 for the history dominant type, 17.27 for the folk customs dominant type, and 17.22 for the grain transporting dominant type. The research shows that the comparison results of the overall through-movement and the overall proximity were similar. The dynamics of the settlements in the industry dominant type were higher, the topological integration capacity and attraction of the central area of the settlements were stronger, and the data of the areas with high trafficability characteristics were greater. On the one hand, as important posts of commerce and trade and water transport terminals, the patterns from Erdaojie of Duliu Town and Wangqinzhuang were complete and the spatial configurations could reproduce the advanced in-site characteristics of industries. On the other hand, with the decline in the functions of grain transport, the disappearance of large quantities of native heritage and the influence of urbanization, settlements, such as the former ferries, wharves, irrigation and dredging channels, warehouses, gentry posts, garrison, and courier stations for grain transport, have accordingly lost their status.

The phenotype values of local proximity configurations show that the value of the spatial configurations of different types of settlements are, from high to low, 0.52 for the industry dominant type, 0.32 for the folk customs dominant type, 0.29 for the grain transporting dominant type, and 0.21 for the history dominant type. The comparative analytical results of the in-site phenotypes are similar. The dynamic scope of the settlements is highly related with the data of public-service facilities, somewhat related with the data of cultural heritage and tourist resources, and poorly related with that of industrial resources.

5.2. In-Site Genes of the Settlements along Tianjin Section

In-site genes represent the formation mechanism of the spatial characteristics particular to an area and is the fundamental reason for regional differences and diversities. The above analysis shows that settlements along the canal boast multiple, causal, and blended in-site phenotypes. (1) Multiplicity—the origin of the rural settlements—had direct relations with the status of grain transporting; the development of grain transporting expedited the industrial prosperity of rural settlements. Therefore, the in-site attributes of the rural settlements under the effects of multiple in-site rules boast the inherent law of multiple genes. (2) Reciprocal causation—rural settlements prospered with the prosperity of the canal and became ordinary with its decline. The water system along the Tianjin Section is large in the north and small in the south, showing a mild terrain difference. The water environment stood as a prominent problem for long water shortages in the south canal in Jinghai District. With the decline of grain transporting and the rise of rail transportation, some settlements along the canal gradually declined. (3) Blending: the settlements were a material representation of the politics, economy, society, and culture over the course of historical development. The prosperity of grain transportation was the direct reason for the development of settlements along the canal and gave rise to the diverse and characteristic culture. The culture of grain transporting was the dominant culture of the settlements along the Tianjin Section and the diverse culture of Tianjin and the immigrant culture, marine culture, Mazu culture, commercial and trade culture, as well as folk culture, brought by grain transportation composed the blended in-site characteristics of the settlements in Tianjin.

Table 7. Summary and comparison of the spatial configuration of the rural settlements along China's Grand Canal Tianjin Section.

Summary and Comparison of the Phenotype Value of the Overall Proximity (NQPDAn)	Type of Settlement	Mean Value
Pangzui village NQPDAnc; Jinguantun village NQPDAn; Wangqinzhuang village NQPDAnc; Xidiaotai village NQPDAnc; Dalongzhuang village NQPDAnc; Beisanlitun village NQPDAnc; Laomidian village NQPDAnc; Dongxicang village NQPDAnc; Xiaoshawo village NQPDAnc; Zhouyangzhuang village NQPDAnc; Sigezhuang village NQPDAnc; Majiakou village NQPDAn; Duliuerdaojie NQPDAnc; Jizhuangzi village NQPDAnc; KEG&GBXZ&NXZ village NQPDAnc; Yangliuqing Historical Town NQPDAnc — *NQPDAn value vs. Amount of the first 80 segments of the sDNA model of the spatial configuration of the settlements*	Grain transport-ing domi-nant type	0.92
	Industry dominant type	2.05
	History dominant type	1.07
	Folk customs dominant type	1.06
Summary and Comparison of the Phenotype Values of the Local Proximity (NQPDAn171c-500c)	Type of Settlement	Mean Value
Pangzui village NQPDA384c; Jinguangtun village NQPDA412c; Wangqinzhuang village NQPDA360c; Xidiaotai village NQPDA402c; Dalongzhuang village NQPDA232c; Beisanlitun village NQPDA171c; Laomidian village NQPDA327c; Dongxicang village NQPDA324c; Xiaoshawo village NQPDA242c; Zhouyangzhuang village NQPDA326c; Sigezhuang village NQPDA500c; Majiakou village NQPDA247c; Erdaojie of Duliu Town NQPDA401c; Jizhuangzi village NQPDA423c; KEG&GBXZ&NXZ village NQPDA404c; Yangliuqing Historical Town NQPDA299c — *NQPDAn171c-500c value vs. Amount of the first 80 segments of the sDNA model of the spatial configuration of the settlements; NQPDAn200c value*	Grain transport-ing domi-nant	0.29
	Industry dominant type	0.52
	History dominant type	0.21
	Folk customs dominant type	0.32
Summary and Comparison of the Phenotype Values of the Overall Through-Movement Degree (TPBtAnc)	Type of Settlement	Mean Value
Pangzui village TPBtAnc; Jinguantun village TPBtAnc; Wangqinzhuang village TPBtAnc; Xidiaotai village TPBtAnc; Dalongzhuang village TPBtAnc; Beisanlitun village TPBtAnc; Laomidian village TPBtAnc; Dongxicang village TPBtAnc; Xiaoshawo village TPBtAnc; Zhouyangzhuang village TPBtAnc; Sigezhuang village TPBtAnc; Majiakou village TPBtAnc; Duliuerdaojie TPBtAnc; Jizhuangzi village TPBtAnc; KEG&GBXZ&NXZ village TPBtAnc; Yangliuqing Historical Town TPBtAnc — *TPBtAnc value vs. Amount of the first 80 segments of the sDNA model of the spatial configuration of the settlements*	Grain transport-ing domi-nant	17.22
	Industry dominant type	30.27
	History dominant type	19.55
	Folk customs dominant type	17.27
Summary and Comparison of the Phenotype Values of the Local Through-Movement Degree (TPBtAn171c-500c)	Type of Settlement	Mean Value
Pangzui village TPBtA384c; Jinguantun village TPBtA412c; Wangqinzhuang village TPBtA360c; Xidiaotai village TPBtA402c; Dalongzhuang village TPBtA232c; Beisanlitun village TPBtA171c; Laomidian village TPBtA327c; Dongxicang village TPBtA324c; Xiaoshawo village TPBtA242c; Zhouyangzhuang village TPBtA326c; Sigezhuang village TPBtA500c; Majiakou village TPBtA247c; Erdaojie of Duliu Town TPBtA401c; Jizhuangzi village TPBtA423c; KEG&GBXZ&NXZ village TPBtA404c; Yangliuqing Historical Town TPBtA299c — *TPBtAn171c-500c value vs. Amount of the first 80 segments of the sDNA model of the spatial configuration of the settlements*	Grain transport-ing domi-nant	6.36
	Industry dominant type	9.49
	History dominant type	4.81
	Folk customs dominant type	6.15

6. Conclusions and Prospects

The paper applied GIS-sDNA software to quantitatively analyze the spatial configuration phenotypes and the in-site characteristics of 18 typical rural settlements along China's Grand Canal Tianjin Section and determined the different types of dynamic scope of the settlements. Overlapping analysis was conducted for data regarding cultural heritage, tourist resources, public-service facilities, spatial models and spatial configuration density of the settlements in the context of historical, synchronic and spatial dimensions so as to explore the phenotype and rules of the in-site genes of rural settlements along the Tianjin Section. Research found that the settlements there can be divided into 16 subtypes of four major categories. Among these, settlements in the industry dominant type, especially for commerce and trade exchange, featured higher dynamics and a stronger integration capacity of settlements; the core zones in the settlements showed high relevance to public-service facilities, some relevance for cultural heritage and tourist resources and the lowest relevancy with industrial resources. The settlements along the Tianjin Section feature multiple, causal, and blended in-site phenotypes. The research results provided new standards to categorize these settlements, new paths and methods to extract their in-site characteristics, new cognitive perspectives for conservation, and new theoretical support to conserve and renew these settlements. The present research results can enrich the cultural heritage resources of China's Grand Canal and it is suggested to use in research on settlements of canals in other areas to help extract the in-site characteristics of those settlements, clarifying the advantages of cultural resources, help with conservation, development, and management of the cultural heritage of those settlements and further prove the significance and value of this research in practice.

The research analyzed the macroscopic characteristics of settlements along the Tianjin Section. In the future, microscopic quantitative analysis of the in-site characteristics of the streets, lanes, and courtyards in these settlements can be made to explore the in-site language and spatial genes. In the future, the in-site development of these settlements can be researched according to the following aspects: (1) To conserve, inherit, and develop the settlements along the Tianjin Section, efforts should be made to give scope to in-site collaboration of the local management departments of all sections and strengthen the top-level management mechanisms. In-site evaluation should be made of settlements to clarify the in-site attributes of the settlements, influenced by multiple in-site laws, and establish an inheritance system of in-site phenotypes. (2) On the basis of repairing the ecological corridor of China's Grand Canal, efforts should be made to make use of the functions of the settlements in the context of the cavernous body, integrating the green corridor, parks, and water systems of the canal into the settlements, and setting up an ecological security pattern for the settlements. (3) Extract the in-site genes from the perspectives of regional spaces, settlements, street and lane spaces, building spaces and environmental spaces, and form the technological mapping of an in-site creation of settlements along the Tianjin Section. (4) Strengthen thought of in-site development of settlements featuring "integration of culture and tourism", and designing high-quality routes of different themes and categories, implementing the development of characteristic towns and villages and invigorate the living, ecological, cultural and industrial spaces of towns and villages.

Author Contributions: Conceptualization, Y.Z., J.-W.Y. and G.-M.B.; methodology, Y.Z. and Y.L.; software, Y.Z.; validation, Y.L., G.-M.B. and Y.L.; formal analysis, Y.L. and G.-M.B.; investigation, Y.Z., Y.-Z.D. and G.-M.B.; resources, Y.Z. and Y.-Z.D.; data curation, G.-M.B.; writing—original draft preparation, Y.Z.; writing—review and editing, G.-M.B.; visualization, Y.Z. and G.-M.B.; supervision, J.-W.Y.; project administration, J.-W.Y.; funding acquisition, Y.Z. All authors have read and agreed to the published version of the manuscript.

Funding: The paper is a phased achievement of the Tianjin Art Planning Project (E18007), National Undergraduate Training Programs for Innovation and Entrepreneurship (202114038004 and 202114038008) and Tianjin Ren'ai College Scientific Research Project (XX18004).

Institutional Review Board Statement: Not applicable.

Informed Consent Statement: Not applicable.

Data Availability Statement: Data derived from the current study can be provided to readers upon request.

Acknowledgments: This paper was financially supported by the Tianjin Art Planning Project (E18007), National Undergraduate Training Programs for Innovation and Entrepreneurship (202114038004 and 202114038008) and Tianjin Ren'ai College Scientific Research Project (XX18004). The authors are grateful to Hengyu Gu for the support on the spatial syntax technology he provided.

Conflicts of Interest: The authors declare no conflict of interest.

References

1. Malaterre, P.; Rogers, D.C.; Schuurmans, J. Classification of Canal Control Algorithms. *J. Irrig. Drain. Eng.* **1998**, *124*, 3–10. [CrossRef]
2. Boughriet, A.; Proix, N.; Billon, G. Environmental Impacts of Heavy Metal Discharges from a Smelter in Deûle-canal Sediments (Northern France): Concentration Levels and Chemical Fractionation. *Water Air Soil Pollut.* **2007**, *180*, 83–95. [CrossRef]
3. Ingersoll, C.G.; MacDonald, D.D.; Brumbaugh, W.G. Toxicity assessment of sediments from the Grand Calumet River and Indiana Harbor Canal in Northwestern Indiana, USA. *Arch. Environ. Contam. Toxicol.* **2002**, *43*, 156–167. [CrossRef]
4. Mekonnen, D.; Siddiqi, A.; Ringler, C. Drivers of groundwater use and technical efficiency of groundwater, canal water, and conjunctive use in Pakistan's Indus Basin Irrigation System. *Int. J. Water Resour. Dev.* **2016**, *32*, 459–476. [CrossRef]
5. Müller, J.C.; Dennis, H.; Alfred, S. Canal construction destroys the barrier between major European invasion lineages of the zebra mussel. *Proc. R. Soc. B Biol. Sci.* **2002**, *269*, 1139–1142. [CrossRef]
6. Biscaya, S.; Elkadi, H.A. Smart ecological urban corridor for the Manchester Ship Canal. *Cities* **2021**, *110*, 103042. [CrossRef]
7. Flemsæter, F.; Stokowski, P.; Frisvoll, S. The rhythms of canal tourism: Synchronizing the host-visitor interface. *J. Rural Stud.* **2020**, *78*, 199–210. [CrossRef]
8. Wang, X.; Han, J.; Xu, L.; Zhang, Q. Spatial and seasonal variations of the contamination within water body of the Grand Canal, China. *Environ. Pollut.* **2010**, *158*, 1513–1520.
9. Wang, X.; Han, J.; Xu, L.; Zhang, Q. Effects of anthropogenic activities on chemical contamination within the Grand Canal, China. *Environ. Monit. Assess.* **2011**, *177*, 127–139. [CrossRef]
10. Fu, C.L. The History of Canal Cities in China. In *Development History of Canal Cities in China*; Sichuan People's Press: Chengdu, China, 1985; Volume 11, pp. 58–68.
11. Lin, C.L. Grain transporting and the rise of Tianjin commercial city in Ming Dynasty. *Tianjin Soc. Sci.* **1984**, *10*, 85–91.
12. Yang, J.J.; Guo, L.X. Study on urban redevelopment planning along the canal in Hangzhou. *City Plan. Rev.* **2001**, *25*, 77–80.
13. Li, C.B.; Zhu, Q. Research on the Width of Canal Heritage Corridor Based on Heritage Distribution-Taking Tianjin canal as an example. *Urban Probl.* **2007**, *146*, 12–15.
14. Chen, W. *Walking on the Canal Line, a Study of the Historical Cities and Architecture along the Great Canal*; China Architecture & Building Press: Beijing, China, 2014.
15. Zhao, X. The Comprehensive Protection Approach for the Grand Canal Settlements in North Zhejiang Province based on the Concept of Historic Urban Landscape: A Case Study of Jiaxing. *Urban Dev. Stud.* **2014**, *21*, 37–43.
16. Cheng, Z.F.; Tang, S.Y.; Hua, H.L. A Study of Cognition and Identification of the Grand Canal Culture from the Perspective of Traditional Village Residents in Beijing-The Cases of Three Traditional Villages in Tongzhou. *J. Beijing Union Univ.* **2018**, *16*, 36–46.
17. Duan, J.; Yin, M.; Tao, A.J. "In-site" Protection: Cognitive Turn of Characteristic Towns and Villages Protection and Conservation, Implementation Approaches and Institutional Recommendations. *Urban Plan. Forum* **2021**, *2*, 25–32.
18. Hillier, B.; Hanson, J. The Logic of Space. In *The Social Logic of Space*; Cambridge University Press: Cambridge, UK, 1984; pp. 52–66.
19. Ding, C.B.; Gu, H.Y.; Tao, W. Progress in the Application of Space Syntax in Human Geography Research in China. *Trop. Geogr.* **2015**, *35*, 515–521.
20. Hillier, B.; Hanson, J.; Graham, H. Ideas are in things: An application of the space syntax method to discovering house genotypes. *Environ. Plan. B Plan. Des.* **1987**, *14*, 363–385. [CrossRef]
21. Hanson, J. Tradition and Change in the English House. In *Decoding Homes and Houses*; Cambridge University Press: Cambridge, UK, 1998; pp. 56–78.
22. Shokouhi, M. Legible Cities: The Role of Visual Clues and Pathway Configuration in Legibility of Cities. In Proceedings of the 4th International Symposium on Space Syntax, London, UK, 1 June 2003.
23. Hillier, B.; Penn, A.; Hanson, J.; Grajewski, T.; Xu, J. Natural Movement: Or, Configuration and Attraction in Urban Pedestrian Movement. *Environ. Plan. B Plan. Des.* **1993**, *20*, 29–66. [CrossRef]
24. Hillier, B.; Hanson, J.; Peponis, J. Syntactic Analysis of Settlements. *Archit. Behav.* **1987**, *3*, 217–231.
25. Hillier, B. A theory of the city as object: Or, how spatial laws mediate the social construction of urban space. *Urban Des. Int.* **2002**, *7*, 153–179. [CrossRef]
26. Hillier, B. The Knowledge that Shapes the City: The Human City Beneath the Social City. In Proceedings of the 4th International Space Syntax Symposium, London, UK, 1 June 2003.
27. Read, S. Space syntax and the Dutch city. *Environ. Plan. B Plan. Des.* **1999**, *26*, 251–264. [CrossRef]
28. Barkat, I.; Ayad, H.; Elcherif, I. Detecting the physical aspects of local identity using a hybrid qualitative and quantitative approach: The case of Souk Al-Khawajat district. *Alex. Eng. J.* **2019**, *58*, 1339–1352. [CrossRef]
29. Wernke, S.A. Spatial network analysis of a terminal prehispanic and early colonial settlement in highland Peru. *J. Archaeol. Sci.* **2012**, *39*, 1111–1122. [CrossRef]
30. Kim, Y.O.; Penn, A. Linking the spatial syntax of cognitive map to the spatial syntax of the environment. *Environ. Behav.* **2004**, *36*, 483–504. [CrossRef]
31. Wang, H.F. The dynamic relationship between social function and space and the morphological evolution of Huizhou traditional villages. *Architect* **2008**, *2*, 23–30.
32. Tao, W.; Lin, K.F.; Gu, H.Y.; Liao, C.M.; Liu, S.Y.; Ou, Q.Y. Spatio-Temporal Evolution of Shawan Ancient Town in Guangzhou from the Perspective of Spatial Syntax. *Trop. Geogr.* **2020**, *40*, 970–980.

33. Cooper, C.H.V.; Chiaradia, A.J.F. SDNA: 3-d spatial network analysis for GIS, CAD, Command Line & Python. *SoftwareX* **2020**, *12*, 100525.
34. Cooper, C.H.V. Predictive spatial network analysis for high-resolution transport modeling, applied to cyclist flows, mode choice, and targeting investment. *Int. J. Sustain. Transp.* **2018**, *12*, 714–724. [CrossRef]
35. Kang, C. Measuring the effects of street network configurations on walking in Seoul, Korea. *Cities* **2017**, *71*, 30–40. [CrossRef]
36. Ye, Y.; Li, D.; Liu, X. How block density and typology affect urban vitality: An exploratory analysis in Shenzhen, China. *Urban Geogr.* **2018**, *39*, 631–652. [CrossRef]
37. Ye, Y.; Yeh, A.; Zhuang, Y.; van Nes, A.; Liu, J. "Form Syntax" as a contribution to geodesign: A morphological tool for urbanity-making in urban design. *Urban Des. Int.* **2017**, *22*, 73–90. [CrossRef]
38. Gu, H.Y.; Huang, D.; Shen, T.Y. Space Syntax Model Verification and Application in Urban Design with Multi-source Data. *Planners* **2019**, *35*, 67–73.
39. Shao, R.Q.; Duan, J.; Qian, Y. Spatial Genes: New Approach of Planning and Design for Local Memory—A Case Study of the Former Capital Airport of the Republic of China in Nanjing. *Planners* **2020**, *36*, 40–46.

Article

Re-Examining Urban Vitality through Jane Jacobs' Criteria Using GIS-sDNA: The Case of Qingdao, China

Siyu Wang [1], Qingtan Deng [1,*], Shuai Jin [2] and Guangbin Wang [1]

1   School of Architecture and Urban Planning, Shandong Jianzhu University, Jinan 250101, China
2   Zhejiang Urban and Rural Planning Design Institute Co., Ltd., Hangzhou 310030, China
*   Correspondence: wangsiyu_chn@163.com

Abstract: This study focuses on the assessment of historic city vitality to address increasingly fragmented urban patterns and to prevent the decline of livability in older urban areas. In 1961, Jane Jacobs theorized urban vitality and found the main conditions that were required for the promotion of life in cities: diversity of land use, small block sizes, diversity of buildings with varied characteristics and ages, density of people and buildings, accessibility for all people without depending on private transport, and distance to border elements. Jacobs' criteria for urban vitality has had an indisputable influence on urban researchers and planners especially in the Anglo-American context. This perspective has influenced the development of New Urbanism and similar planning policies, such as neo-traditional communities and transit oriented development, yet her theories have to be more substantiated in Asia's developing cities, especially in China's historic cities. In order to verify the significance of Jacobs' urban vitality theory in Chinese historic cities, we develop a composite measure of 16 variables of built environment, and we test it using GIS-sDNA in a historic city with an aging population and low-income in Qingdao. A systematic approach to urban spatial analysis allows us to provide a detailed spatial interpretation of a historic city form. The results emphasize that historic cities vitality, far from being homogeneous, followed a multi-centered distribution pattern, which is related to the previous European planning of the region, where a grid-type pattern was more likely to disperse urban vitality. The results can serve as a useful framework for studying the livability and vitality of different areas of the city in different geographical contexts.

Keywords: urban vitality; Jane Jacobs; built environment; GIS sDNA; historic city

Citation: Wang, S.; Deng, Q.; Jin, S.; Wang, G. Re-Examining Urban Vitality through Jane Jacobs' Criteria Using GIS-sDNA: The Case of Qingdao, China. *Buildings* **2022**, *12*, 1586. https://doi.org/10.3390/buildings12101586

Academic Editors: Michael J Ostwald and Ju Hyun Lee

Received: 15 August 2022
Accepted: 25 September 2022
Published: 1 October 2022

Publisher's Note: MDPI stays neutral with regard to jurisdictional claims in published maps and institutional affiliations.

1. Introduction

The world is experiencing rapid urbanization, especially in developing countries, and while urbanization has significantly improved living standards, it has also brought about unprecedented urban problems, such as land Fiscal-driven urbanization, hollowing out of old urban areas, and "ghost cities" [1]. In this context, historic cities generally face double decay of social vitality and physical morphology, such as over-tourism, gentrification, aging and impoverishment of aborigines, aging infrastructure, and a deteriorating living environment [2–4]. Historic city vitality is an isomorphism of the urban morphology and the social activities behind it [5], determined by the spatial distribution pattern of the elements of the historic city form [6,7]. The unique urban scenes of the historic city have often undergone centuries of shaping and are significantly different from what is commonly thought of as vibrant urban areas [8,9]. With rapid urbanization, the original functional types and spatial patterns of the historic city can no longer meet the demands of modern life and require necessary adjustments to meet the needs of residents' lives and the needs of regional development. Numerous studies have shown that the distribution of urban vitality can be significantly influenced by the planning of urban form elements [10–14]. Therefore, in different urban contexts, how to evaluate the vitality of historical cities more rational and how to propose targeted strategies to enhance them has become a major research problem.

"The Death and Life of Great American Cities" (Jacobs, 1961) provided a landmark for urban planning theory with abiding influences [15]. This perspective is considered the benchmark for understanding how neighborhood vitality works today [16]. The mix of building functions and densities that meet the needs of citizens is fundamental to the urban vitality [17]. Her criteria have been recurrently rediscovered and revisited to respond to different needs and from different contexts and are still relevant in the present day. The outbreak of the COVID-19 has once again led to a consideration of Jacobs' theory and the revival of urban vitality. Observing the changes of the epidemic, it can be seen that traditional unit communities within historical cities have better epidemic prevention effects due to their relatively open boundaries and the diversity of functions within the community, carrying compounded functions within the residential units, than, for example, new commercial housing communities on the outskirts of large cities. Cities and other public institutions around the world are also explicitly or implicitly using her principles as part of planning strategies for the revitalization of local historic cities [18]. As a result, her theories have also received increasing attention from urban researchers around the world, who use her general view of urban vitality as a framework for reflecting on the success or failure of urban planning in ensuring urban quality [19]. The current urban vitality study aims to validate her ideas by assessing the degree of mixing and spatial distribution of buildings and facilities in the city [20–24].

In terms of the analytical approach to urban vitality, Jacobs argues that the city is an ever-expanding network of mobility, with population expansion, size, and traffic congestion problems expanding outward as the city itself becomes saturated [15]. Urban space is a collection of information flows and networks, and a growing number of studies prove that many urban phenomena are essentially network phenomena, such as the hierarchical scale, resilience, agglomeration, or scale efficiency of cities [25,26]. In network analysis methods, spatial information plays a leading role, containing location data, allowing us to study the physical underpinnings that govern and shape the flow of human information in urban spaces. [27]. More recently, the centrality assessment model and space syntax analysis have been used to evaluate the structural properties of street networks in an urban system. Street centrality indices representing closeness, betweenness, and straightness capture the skeleton of the urban system; these factors shape economic activities and land use intensity [28,29]. A research team from Cardiff University, UK, developed the spatial design network analysis (sDNA) tool [30], which proposes a line segment model that is based on the spatial syntax axis model, while the topological relationships between streets can be weighted with information such as street length and geometric angular distance to reflect real geospatial information, while compensating for the failure of spatial syntax to capture physical isolation and network efficiency, especially the navigation difficulties and psychological barriers of pedestrians [31]. Jacobs' vitality criteria are directly mirrored in sDNA, which incorporates four important indexes (density, connectivity, closeness, and betweenness) that are hypothesized to affect urban vitality in an urban system. In China, driven by both industrialization and urbanization, numerous satellite towns, industrial zones, commercial centers, and residential areas have emerged at the historic city fringe and pose significant challenges to sustainable urban development [32]. Until recently, empirical evidence in developing countries has indicated that the dynamic process and determinants of urban vitality can be different from the case studies in the United States or Europe due to their dissimilar historic urban morphology and territorial spatial planning policies. A lot of historic cities launched large-scale street planning projects and replicated the modern style of broad and grid roads from the US, thereby reducing their urban vitality [33].

The following questions are posed in this study: (1) What are the implications of Jacobs' criteria for urban vitality in the context of historic cities? (2) How to construct the evaluation index system of "JANE Index" of Asian historical cities based on Jacobs' criteria? (3) How to visualize and analyze the vitality of Qingdao's historical city through GIS-sDNA and other related methods, and propose targeted vitality regeneration strategies? Therefore, this study aims to enrich the existing empirical studies by assessing the level of vitality that is exhibited by each neighborhood of a historic city and formulate strategies to enhance urban vitality.

2. Jacobs' Criteria of Urban Vitality

Jacobs defines urban vitality as "the production of a diverse urban life consisting of human activities and living places" [15]. She proposed four main conditions for the creation of urban vitality: diversity of land use, small block sizes, diversity of buildings with varied characteristics and ages, and density of people and buildings, as well as two secondary conditions: accessibility without reliance on private transportation and distance from boundary elements [20].

Land use diversity, similar to architectural diversity of different characteristics and ages, is seen as paramount to maintaining urban vitality. She believes that the mix of functions creates spatial differentiation and increases the efficiency of the use of urban facilities, such as reducing commuter traffic and increasing the potential for public transportation use in a high-density environment [34,35]. On the other hand, building diversity also reflects the diverse mix of socio-economic groups [36]. Jacobs paid close attention to the "old building needs" of urban neighborhoods and argued that the healthiest areas of the city need not only old buildings, but also new buildings that are interspersed with old buildings. She noted that old buildings and new buildings require different levels of economic gain, i.e., the economic return that is generated in the building, and that new businesses tend to naturally emerge in buildings with low economic overhead. Th diversity index measures the degree of concentration, mix, and proportional relationships by measuring the number of different building types at a given scale [37]. The diversity index makes a morphological description, which calculates the number of buildings, number of functional types, and evenness in a site unit, quantifying the degree of spatial diversity [38]. When urban facilities show a high number and type with even distribution in space, this pattern is positively correlated with urban vitality.

Small-scale street blocks are based on the need for social spaces, which helps to create connections between people in specific areas of the city. This index is used to measure the connectivity characteristics of the street network, such as the number of intersections, the average distance between intersections, and other variables [39]. Studies have shown that potential traffic routes can be increased if the distance between intersections in a street network is shorter and the number of intersections is higher at the same scale. However, too short intersection distances with too many intersections can lead to a larger road network footprint and adversely affect the layout of other facilities [40]. A reasonable proportional and quantitative distribution of small-scale streets is positively correlated with urban vitality. Connectivity is only possible when streets vary in size and each size has a reasonable proportion of streets.

The influence of building and population density on vitality builds on the above indexes and emphasizes the degree of aggregation of population and activity places at a certain spatial scale [41,42]. The higher the building density, the higher the pedestrian traffic on the street. However, dense land occupation should not be confused with the construction of large apartment buildings, and areas with high housing density are often seen as a disadvantage. The presence of public services, commercial facilities, etc., is related to Jacobs' concept of the "eye on the street," which means that the constant presence of people creates a "natural surveillance" system that has also been shown to improve safety [43]. Compared to large public spaces, small infrastructures can be considered as potential social places. The POI kernel density calculation method has proven to be a valid method for vibrancy evaluation, for example for recreational facilities, small restaurant facilities, green infrastructures, and other types of calculations [7,44,45].

In terms of accessibility as one of the two supplementary conditions for urban vitality, in contrast to automobile-driven urban planning, studies have generally referred either to the supply of public transportation on the one hand, or to walkability conditions on the other [46]. In terms of walkability, sDNA's closeness index measures walkability within a specified radius. Boundary vacuum refers to the segregated forms of urban existence, in the form of barrier surfaces or lines, including functionally homogeneous urban spaces (i.e., large parks), natural elements that could act as barriers (i.e., rivers), and ground-level heavy transportation infrastructures (i.e., railways). Jacobs argued that these elements adversely affect urban vitality through man-made impermeable boundaries [22,46].

These six criteria that are necessary for the generation of urban vitality, as proposed by Jacobs, form the basis of the theoretical approach that was analyzed in this paper. Jacobs' influence on urban theory is most evident in Anglo-Saxon contexts. Her work has also inspired European, South American, and Asian urban studies on how to respond to urban development issues, such as the case studies of Barcelona, Cyprus, Santiago, and Seoul [20,21,46,47]. Revisiting Jacobs' theory of urban vitality is necessary because her focus on the incremental revitalization of already urbanized areas, rather than the one-time redevelopment of master plans [36] is consistent with the focus of current considerations in developing countries to address the low vitality of historic cities that are brought about by rapid urbanization. Although most applied studies that are based on Jacobs' original evaluation indexes have been conducted for various purposes, the means of handling the evaluation indexes are basically the same, usually transforming them into multiple morphological variables, adaptively weighting the set of values, and overlaying them into a comprehensive urban vitality index.

3. Methods

3.1. Study Area

This study focuses on the historic city of Qingdao, China (Figure 1). Qingdao city is located in the southeastern of Shandong peninsula (latitude 35°35′–37°09′ N, longitude 119°30′–121°00′ E) in eastern China, with a total area of about 28 km². The historic city area includes 14 historic conservation areas with a total area of 13.041 km² and a core protection area of 6.394 km², including 549 cultural relics protection units at all levels, 309 historical buildings, 1694 traditional style buildings. and 39 industrial heritage sites within the city area. The average population density is 8059 people/km². The distribution of landscape elements in each historic conservation area varies greatly, which is related to the adjustment of urban form planning and industrial structure in different historical stages. Until the end of the 19th century, Qingdao was a rural area with agriculture and fishing as the mainstay. In 1898, German colonialists forcibly leased Jiaozhou Bay, and Qingdao then entered colonial city stage. Urban planning of Qingdao was dominated by German colonialists, and the layout of city conformed to the topography and climate of the seaside hills. The urban residential area was divided into European and Chinese areas. The European area was spread along the southern coastal flat area, and the nature of the land was mainly for European residence, leisure, and entertainment, with outstanding landscape conditions. The Chinese area was located in the northern area of the city, including Dabaodao district, Taidong town, and Taisi town. Based on the unequal colonial zoning pattern of German-occupied Qingdao, there are huge differences between the European and Chinese districts in terms of urban location, plot scale, road width, building density, green landscape, and other aspects of the living environment. In 1902, the population density of the European areas was only 19.1 people/ha, and building density was only 20–25%; the population density of the Chinese areas was 417 people/ha, and building density was over 75%. In terms of green space, the European area accounted for 72.6% of the city's greenery, while there was no public green space within the Dabaodao district. Urban planning and construction in Qingdao during the German occupation period (1897–1914) established the spatial morphological diversity of Qingdao's historical urban area, resulting in six typical morphological types (Figure 2): (a) south of Guanhai Mountain (former European area), (b) south of Zhongshan Road (former European area), (c) Guanhai Mountain (former European area), (d) Dabaodao (former Chinese merchant settlement, (e) Taidong Town (former Chinese laborers settlement), and (f) Taisi Town (former Chinese poor settlement).

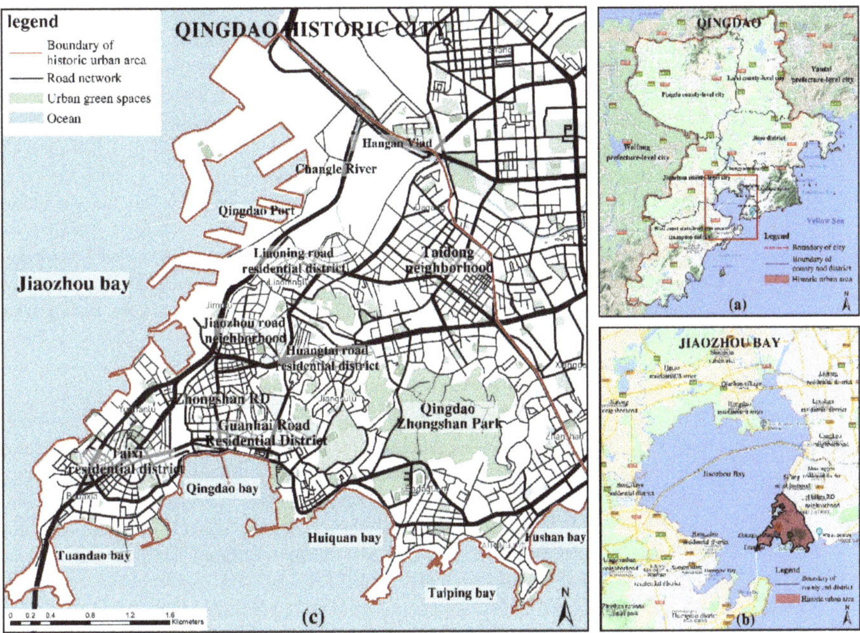

Figure 1. Study area. (**a**) Qingdao city area. (**b**) Location of Qingdao Historic City in Jiaozhou Bay. (**c**) Scope of Qingdao historic city.

Figure 2. Urban morphology types in Qingdao historic city.

We choose Qingdao as a case study of the application of Jacobs' theory to Asian cities because Asia, as the origin of human civilization and the most populous continent in the world, has accumulated a profound urban culture during the course of history, and its urban morphological development has the following characteristics: 1. High-density concentration of population and buildings, accompanied by a long-standing phenomenon of large differences in scale, class, and form in the central area. 2. Since 1897, Qingdao, as a modern colonial city, has experienced colonial urban planning led by Germany and Japan, and its urban development has both a technologically advanced industrial and trade base and a unique urban regional culture, with distinctive power in terms of street texture, neighborhood scale, green space ratio, and it has distinctive power and hierarchical differences in terms of street texture, neighborhood scale, green space ratio, building density, etc. It conforms to the general characteristics of Asian historical urban form while maintaining a unique regional culture.

3.2. Variables and Data Sources

This paper is based on Jacobs' theory of urban vitality, focusing on built environment characteristics, i.e., street form, land use, and human activity factors, adapted to the urban context of Qingdao, China, and translates Jacobs' theory into the general empirical model in Equation (1). Other variables were combined based on recent literature exploring vitality and its drivers in the Jacobs framework. The result is the adjusted JANE index, which includes 16 indicators that were compiled from different data sources. The model is based on the JANE index of urban vitality that was first constructed by Xavier Delclòs-Alió (2018). Table 1 details the variables that were used to construct these six dimensions and their respective data sources. Most of the data were obtained from official sources, while the rest were collected from spatial analysis by the research team using a geographic data system.

$$JANE\ Index = f(Div, Sca, Div_A, Con, Acc, Bou) \tag{1}$$

Div = Function diversity, (land use mixture, residential and non-residential ratio, commercial and facility ratio)

Sca = Road scale, (block size, road length density, connectivity and betweenness)

Div_A = Building age diversity, (average building age, building age diversity)

Con = Concentricity, (building and floor area density, public facility density, people density)

Acc = Accessibility, (road closeness, distance to public transportation)

Bou = Distance to large single-use buildings, large parks, surface large roadways, parking areas, and empty lots.

Building function diversity ($Div.$) is defined as the presence of land use mixture (D_1), residential and non-residential ratio (D_2) and commercial and facility ratio (D_3). In this study, based on the calculation method of information entropy, the Shannon–Weaver diversity index was used to calculate the land use diversity of spatial units using the main 11 land use function POIs (residential, commercial, work-related, recreational, and others) that were extracted from Amap, where n represents richness and p_i with proportional richness. When all building types in the dataset are the same, all p_i values are $1/n$. The larger the difference in building type richness, the larger the weighted geometric mean $\ln p_i$, and the smaller the corresponding D_1 value. If practically all the richness values are concentrated in one building type and the other types are very rare (even if they are numerous), the D_1 value is close to 0. When there is only one type in space unit, the D_1 value is equal to 0 [48,49]. The D_2 index was created with the expression as in the table, where Res_i refers to residential uses and $NonRes_i$ to non-residential uses. Both indices take values from 0 to 1.

Table 1. Conditions and indicators that were used in the analysis.

Jacobs Criteria	Indicators	Calculation Method	Formula	Number
1. Building function diversity (*Div*)	01. Land use mixture (D_1)	Shannon–Wiener index of building function.	$D_1 = - \sum\limits_{i=1}^{n} (p_i \cdot \ln p_i)$	(2)
	02. Residential and non-residential ratio (D_2)	Residential and non-residential balance (0–1).	$D_2 = 1 - \left| \frac{Res_i - NonRes_i}{Res_i + NonRes_i} \right|$	(3)
	03. Commercial and facility ratio (D_3)	Commercial facility and public facility mix (0–1).	$D_3 = 1 - \left| \frac{ComFac_i - PubFac_i}{ComFac_i + PubFac_i} \right|$	(4)
2. Road scale (*Sca*)	04. Block size (S_1)	Area of the traffic analysis zone	$S_1 = S_i$	(5)
	05. Road centerline density (S_2)	Total length of road centerline in TAZ unit.	$S_2 = \frac{L_i}{S_i}$	(6)
	06. Road centerline connectivity (S_3)	Number of interconnections per road in TAZ unit.	$S_3 = K_i$	(7)
	07. Road centerline betweenness (S_4)	Sum of geodesics that pass through a street in TAZ unit.	$S_4 = \sum\limits_{y \in N} \sum\limits_{z \in R_y} OD(y, z, x) \frac{P(z)}{links(y)}$	(8)
3. Building age diversity (*Div$_A$*)	08. Building age of construction (DA_1)	Average building age of construction.		
	09. Diversity of building age of construction (DA_2)	Diversity of average building age of construction (0–1).		
4. Concentricity (*Con*)	10. Buildings density (C_1)	Building footprint in TAZ unit.	$C_1 = \frac{M_i}{S_N}$	(9)
	11. Floor area density (C_2)	Total building area in TAZ unit.	$C_2 = \frac{Q_i}{S_N}$	(10)
	12. People density (C_3)	Number of population in TAZ unit.	$C_3 = \frac{N_i}{S_i}$	(11)
	13. Public facility density (C_4)	Kernel density values for public service facilities.	$C_4 = \sum\limits_{i=1}^{n} \frac{1}{r^2} k\left(\frac{x - x_i}{r}\right)$	(12)
5. Accessibility (*Acc*)	14. Road centerline closeness (A_1)	Network quantity penalized for distance: for all streets in TAZ unit.	$A_1 = \sum\limits_{y \in Rx} \frac{W(y)P(y)}{d_M(x,y)}$	(13)
	15. Distance to public transportation (A_2)	Nearest distance of TAZ unit to adjacent public transportation facilities.	$A_2 = D_0 - Min\left(D_{ip}\right)$	(14)
6. Boundary vacuum (*Bou*)	16. Distance to boundary vacuum (*DB*)	Nearest distance of TAZ unit to adjacent boundary vacuum elements.	$DB = D_0 - Min(D_{ib})$	(15)

Source: compiled by the authors.

For road scale (Sca), the indexes included block size (S_1), road density (S_2), connectivity (S_3), and betweenness (S_4). The street block size was first calculated by GIS, and then the road density, connection value, and penetration were calculated using sDNA with a radius of 15 min walking distance (1500 m). Where density indicates the length (L_i) of a road within specified radius. The connectivity value indicates the sum of the street and all other connected streets (K_i), and the node with a higher connectivity value is the center of the street network. Usually the streets with more intersections are more connected and the ends are less connected. Betweenness is a measure of centrality that is based on the shortest path, which indicates the likelihood and frequency of a person passing a node in a street network movement [31,40]. Since people prefer to follow straight paths and angular distances in daily life because they are easier to remember and faster on average, this paper emphasizes the importance of angular distance (the distance that is measured as angular change) and prefers to use the shortest angular path rather than Euclidean distance. Then, we use the angle-weighted TPBtA metric from the sDNA theoretical framework for the measurement with the expression as in the table where N is the set of streets, R_y is the set of streets within the proposed radius from street y, and OD denotes the ratio of beginning to end in the penetration, between the interval (0, 1). z and y denote the two endpoints of the penetration path, and x denotes the measurement point of the street penetration degree. TPBtA measures the "through" potential of a street, which is positively correlated with house prices and rents, traffic flow, population density, and commuting flows in the street network [50].

As for building age diversity (Div_A), Jacobs highlighted that cities need to ensure that buildings with different characteristics are present in order to guarantee a certain degree of socioeconomic diversity. "Plain, ordinary, low-value old buildings" are critically needed environments for small, entrepreneurial businesses and healthy districts and cities" [36,51]. With this intention, we first included building age of construction (DA_1) and diversity of building age of construction (DA_2), which correspond to the original concept of Jacobs. In Qingdao historic city, strict restrictions on the preservation of historic conservation areas, as well as phenomena such as tourism and gentrification, can clearly affect building age diversity. Too many old buildings can reduce the diversity of old and new buildings, thus affecting the self-organization of historic urban renewal and discouraging the generation of urban vitality.

In order to study the concentration criteria (Con), we have considered building density (C_1), floor area ratio (C_2), population density (C_3), and public service facilities density (C_4). Building density is the ratio of building projection area (M_i) to building land area (S_N), reflecting the open space rate per unit space and the density of building coverage. The higher building density reflects the higher degree of intensive use of urban space. The floor area ratio is a concentrated reflection of the development intensity of urban space, and a higher development intensity means a higher degree of land use. The population density was calculated from Landsat Enhanced Thematic Mapper (ETM) satellite images with a 100 m spatial resolution provided by Worldpop. Public service facilities density is calculated by kernel density tool of ArcGIS. The principle is to define a threshold range (a circle of radius r) that is centered on the location of selected elements, and the kernel density value reaches a maximum at each core element n_i and decreases as it moves away from n_i until it drops to 0 when the distance from n_i reaches the threshold r. Due to the properties of buildings that are distributed along roads, this method is suitable for analyzing the concentration characteristics of infrastructure points in urban networks [44,52]. In this study, kernel density is considered as a cluster characterizing the concentration, and the peak area indicates the area of vitality hotspots, and the expression is shown in the table, where k is kernel function, r is distance decay threshold, n is the number of elements in threshold range, and ($x-x_i$) is the distance from point x to sample x_i. The size of the bandwidth has a significant impact on the accuracy of the analysis results [53]. In order to accurately reflect the concentration characteristics and to consider the discrete degree of building distribution and its average influence range, the bandwidth is set to 1 km in this paper.

Accessibility (Acc) is defined by the closeness of the road network (A_1) and pedestrian accessibility of transportation facilities (A_2). Closeness calculates the accessibility of the spatial structure of the street network through sDNA, which measures the difficulty of directing each street within a given radius to all possible destinations. A path with high closeness usually has high accessibility and it is easier to reach a further place, so that people can more easily reach the space from the surrounding areas for activities and social interactions, thus stimulating space vitality [40]. In this paper, the angle-weighted NQPDA model in the sDNA theoretical framework is used as the expression of closeness, as shown in the table, where d_M (x, y) denotes the shortest angular distance from street x to y based on the metric system. Rx denotes the set of polylines starting from street x within the proposed radius. W(y) denotes the weight of polyline y. P(y) denotes the ratio of any polyline y within the radius, when in discrete space, P(y) = 1 if the point is within the search radius, otherwise

P(y) = 0; in continuous space, it is determined according to the ratio of radius to the length of segment, $0 \leq P(y) \leq 1$. NQPDA can reflect street accessibility and mobility potential and is closely related to diverse land uses [54]. Pedestrian accessibility of transportation facilities measures the convenience of pedestrian access to public transportation services and is mainly related to the distribution of public transportation facilities (e.g., bus stops, subway stations, etc.) on a spatial scale. If the public transportation stations are distributed within the scale that is comfortable and reachable by walking, the higher people's willingness to choose public transportation; otherwise, the willingness will decrease. In this paper, when discussing pedestrian accessibility, we mainly choose rail transit. Considering the quickness of rail transit station transfer, this paper takes the straight-line distance between each spatial unit and its nearest rail station as the A_2 index and uses the ArcGIS nearest neighbor analysis tool for spatial analysis, and the expression is shown in the table, where D_{ip} is the spatial linear distance between spatial unit i and rail transit station, and then the minimum value Min (D_{ip}) is taken among all the linear distances. D_0 is the minimum distance constant, due to the different spatial scales of the whole study area, the range of values is different, this paper takes the value of 1500. If the distance between the spatial unit and the rail station is smaller, the larger A_2 is.

Lastly, we incorporate in the analysis distance from border vacuums (Bou), taking into account single-use buildings (5000 m^2 and above), large transportation infrastructures, single-use extensive service or administrative buildings, and also large parks (5000 m^2 and above).

3.3. Data Processing and Calculations

For the study area, we selected the conservation planning area (2020–2035) of Qingdao historic city given by Qingdao Natural Resources and Planning Bureau, and firstly corrected the projection coordinate system for the main road layers in the road network that were extracted by OSM, and then merged and physically interrupted them to generate a total of 5949 street links and checked all the nodes. In the second step, the spatial unit division of the traffic analysis zone (TAZ) is used to convert the road network into polygon elements and to delete, trim, and merge the irregular road network in order to partition the study area into irregular polygons (Figure 3a). In addition, the fragmented parcels with very small areas were merged with neighboring parcels for the convenience of statistics and calculation, and the study area was divided into 2281 study units after final processing. In the third step, this paper selects and reclassifies a total of 11 functional types including residential land, land for public administration and public service facilities, land for commercial service facilities, land for roads and transportation facilities, and land for green spaces and squares according to the Urban Land Classification and Planning and Construction Land Standard (GB50137-2011) that was issued by the Ministry of Housing and Urban-Rural Development of China, including 10,301 for shopping, 7451 for living services, 7384 for restaurants, 3100 for residential areas and services, 2633 for transportation facilities, 1891 for culture and education, 1860 for medical facilities, 1252 for government offices, 848 for financial services, 834 for recreational facilities, and 215 for natural landscapes, totaling 37,769 POI data for spatial analysis (Figure 3b).

The number of each type of POI in each study unit and within 10 m of the boundary was counted in diversity analysis, and the Shannon index was calculated for each unit. Subsequently, we standardized the data for extreme differences in view of the diversity of each variable unit, divided all the results of index quantification into 8 categories in ArcGIS using the natural discontinuity method [55], and finally summed up to obtain the graded results of diversity evaluation of each study unit (Figure 4).

In the calculation of the street scale and accessibility, the road data were vectorized and preprocessed through processes such as coordinate correction on the ArcGIS platform. By breaking all of the "link" intersections and checking all of the intersections, we obtained 5949 road network segments. To remove the errors in the network, we ran the prepare network tool on the transport vector layer in the sDNA toolbox and obtained a road network spatial database of 4863 road network segments. Finally, we performed "integrated analysis" on the output model, choose "Angle" weighting as the calculation type in the parameter settings, and set the analysis radius to 1500 m to calculate the local spatial scale, where the radius type is set to "continuous space" to obtain more accurate calculation results. The output of the integrated analysis is the result of the spatial syntax analysis of each index (Figure 5).

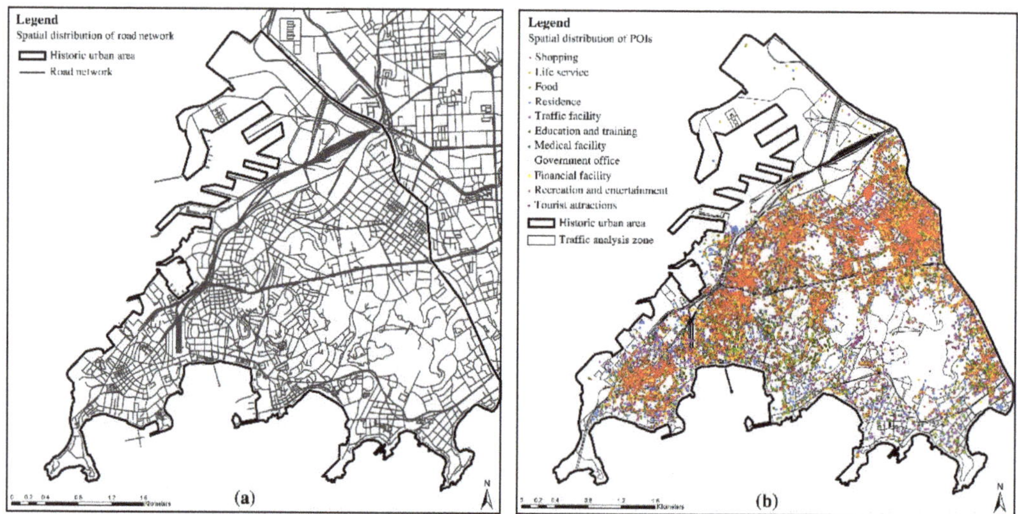

Figure 3. Required dataset type. (**a**) Road network distribution. (**b**) POIs distribution.

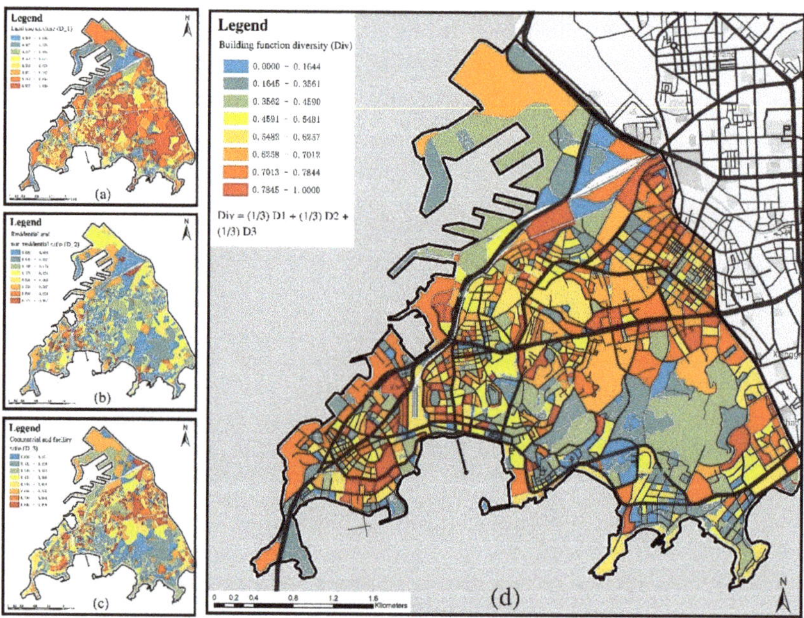

Figure 4. Building function diversity analysis. (**a**) Land use mix. (**b**) Residential and non-residential ratio. (**c**) Commercial and facility ratio. (**d**) Building function diversity composite index.

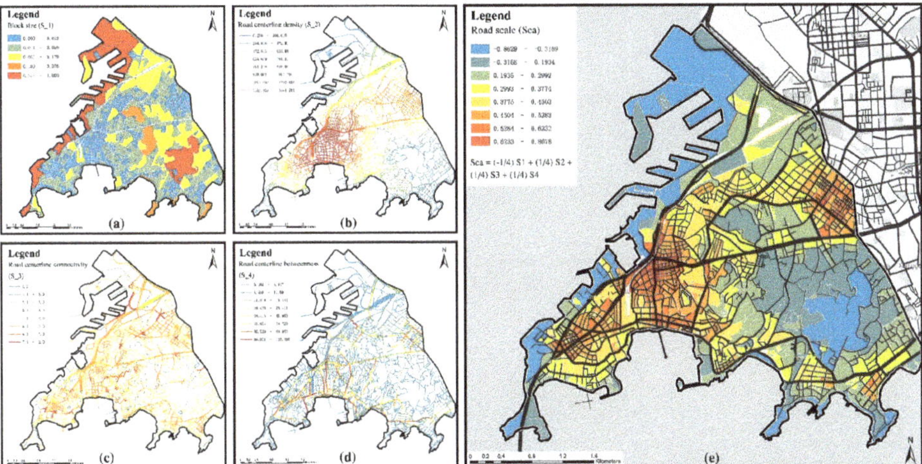

Figure 5. Road scale analysis. (**a**) Block size. (**b**) Road centerline density. (**c**) Road centerline connectivity. (**d**) Road centerline betweenness. (**e**) Road scale composite index.

In the building age diversity analysis, we used urban real estate statistics and connected them to TAZ units to obtain Figure 6. For the concentration analysis, this paper uses 100 m resolution open space population density statistics that were provided by WorldPop opensource dataset. There were four public service facilities of transportation, culture, education, and healthcare that were selected, including 2659 POIs of transportation service facilities, 215 cultural service facilities, 1919 educational service facilities, and 1866 medical service facilities, totaling 6659 POIs of basic service facilities. Then, we imported the obtained data into ArcGIS and carried out the kernel density calculation of point elements separately. Among the parameter settings, the image element size was set to 5.6 m, and the search radius was set to 600 m based on the 10-min walking radius. Finally, we used the entropy weight method to assign weights to the above four POI types, and assigned 0.0956, 0.4427, 0.1488, and 0.3128 weight ratios in turn, and overlapped with TAZ units to analyze the grading results of the comprehensive evaluation of concentration (Figure 7).

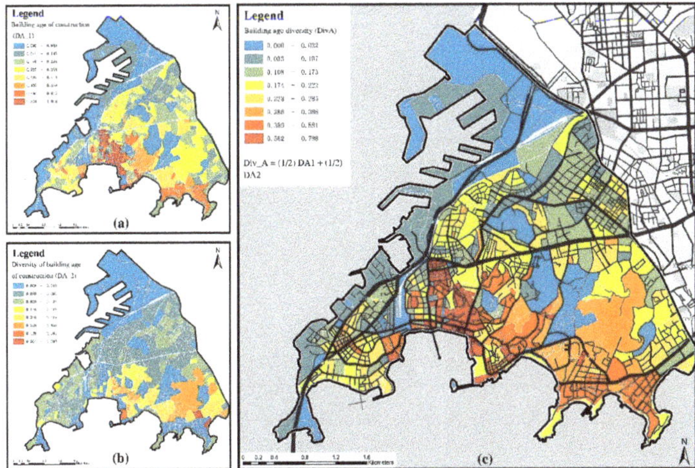

Figure 6. Building age diversity analysis. (**a**) Building construction age. (**b**) Building construction age diversity. (**c**) Building age diversity composite index.

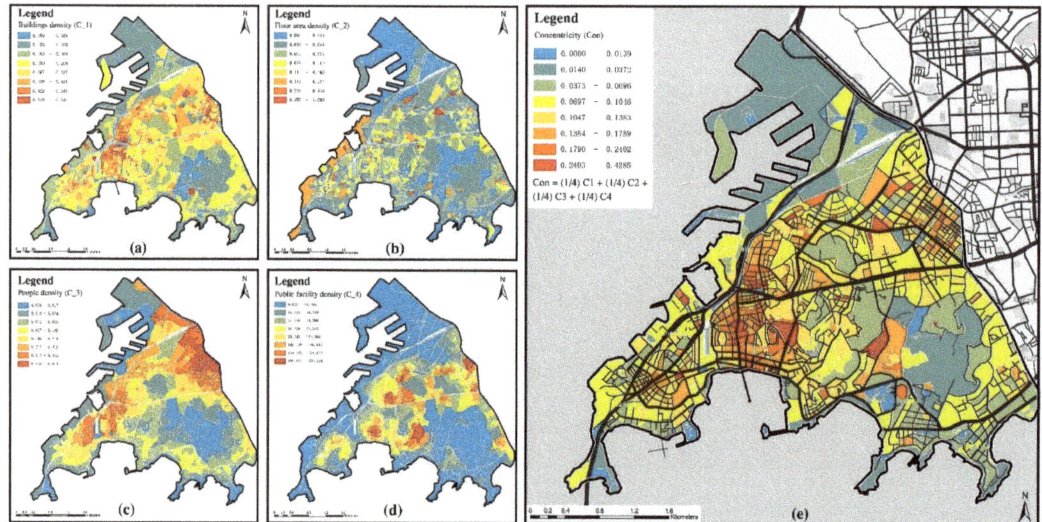

Figure 7. Connectivity analysis. (**a**) Buildings density. (**b**) Floor area density. (**c**) People density. (**d**) Public facility density. (**e**) Concentration composite index.

In order to combine the 16 constituent indexes with the corresponding urban vitality criteria, we weighted them by applying the raw z-values. Among the six dimension indicators, Jacobs regards the first four as the primary conditions for the generation of urban vitality. She believes that these four conditions must work together to generate urban vitality, and that the absence of any one of them will prevent the generation of vitality, so we use equal weights for the first 4 indicators. The last two conditions are considered as secondary conditions that are more independent of each other, but they also play an important role. In this paper, we adopt a more direct idea of weight assignment, i.e., 2:2:2:2:1:1, which is a simpler and more direct expression of Jacobs' criterion. The JANE index of urban vitality after overlay analysis is constructed as shown in (16).

$$JANE\ Index = \frac{1}{5}(Div) + \frac{1}{5}(Div_A) + \frac{1}{5}(Sca) + \frac{1}{5}(Con) + \frac{1}{10}(Acc) + \frac{1}{10}(Bou) \qquad (16)$$

4. Results

The result of mapping Jane Jacobs' urban vitality criteria in Qingdao historic city is presented in different maps in Figures 4–10. Comments on the side are made for each of the urban vitality conditions as well as for the final result in order to support their spatial interpretation.

4.1. Building Function Diversity

The spatial distribution of function diversity corresponding scores in Qingdao historic city is presented in Figure 4. Figure 4a–c show the spatial distribution of land use mixture (D1), residential and non-residential ratio (D2), and commercial and facility ratio (D3), respectively. The high values of D1 are mainly located in the central–eastern part of the country, with relatively low diversity in coastal land use, while D2 shows the opposite distribution pattern, with a relatively high ratio of residential to employment in the coastal zone. Figure 4d shows the weighted superposition of the three indicators using the entropy weighting method to obtain the composite index of Div.

Diversity was tested for spatial autocorrelation, given that the Global Moran's I was 0.29 and the z-score was 25.72, thus implying significant spatial correlation and clustering. The pattern of building function diversity in Qingdao is more complex: (1) The spatial pattern of the high-value area (Div $\geq$ 0.70) follows a banding pattern along Qingdao Jiaozhou road, and the further away from the main road, the more obvious the diversity decay, with a mean diversity index (Mdiv) of 0.77, including mean land use mix (MD$_1$) 0.75, mean residential-nonresidential ratio (MD$_2$) 0.31, and mean commercial-public facility ratio (MD$_3$) 0.82. Due to the limitation of the conservation plan of Qingdao's historic city, the diversity of Zhongshan Road and Taidong streets are low and unevenly distributed, compared to the higher diversity of land use in the surrounding urban areas. (2) Anselin Local Moran's I test shows that some areas show significant differences in land use mix within a relatively short distance, such as the intersection of Laiyang Road and Qingyu Road and Huiquan Commercial Square, where there are many high-low outlier areas, which are characterized by high values of local diversity in single-use building functional areas. This contrast is further revealed by the disaggregated nature of the raw data, and this micro-contrast is likely to be weakened if larger spatial scales (e.g., municipal scales) are used. (3) The area with low diversity value (Div $\leq$ 0.30) is mainly located in the eastern mountainous region with a mean diversity index (Mdiv) 0.09, including mean land use mix (MD$_1$) 0.15, mean residential-nonresidential ratio (MD$_2$) 0.02, and mean commercial-public facilities ratio (MD$_3$) 0.05, for instance, Taiping Mountain, Baguan Mountain, and Qingdao Mountain generally have lower land use diversity. Within the core area of the old city, the southern coastal neighborhoods such as Sifang Road, Guanhai Mountain, and Baguan Mountain reflect a certain degree of homogeneity in land use and a low diversity index.

In urban conservation planning of Qingdao that was formulated in 2020, Qingdao puts forward three categories of control requirements for construction height, the first of which requires the existing height to be maintained in the core protection area and no change is allowed. The second category requires that the construction height in the historic district of the main city shall not exceed 18 m, and in the towns of Taixi and Taidong shall not exceed 24 m, and the new buildings shall not exceed the adjacent historic buildings. The third category requires that the building heights in other areas shall not affect the overall mountain overlook view area and protect the overall view corridor. Combined with the results of the diversity distribution, the construction height restriction has a significant impact on the diversity of building functions, and the diversity of the main urban area is generally lower due to the 18-m building height limit, however, there are also individual phenomena, such as Parkson Shopping Center, Guangde Li 1898, and Yue Hik Lai Shopping Plaza in the Zhongshan Road neighborhood along the southern coast, which are much higher than the control height due to their early construction, and these buildings are affected by the conservation planning and development These buildings are affected by the conservation planning, development is reduced, and the customer flow is also reduced accordingly. The height limit in Taidong Town is higher than that of the main urban area, and thus has higher diversity characteristics, and it gathers more people traffic, however, because the central area of Taidong Street is stricter than the peripheral height limit, so the diversity distribution shows the characteristics of low center and high surrounding.

4.2. Road Scale

The spatial distribution of the road scale score is presented in Figure 5. Figure 5a to Figure 5d show the spatial distribution of block size (S1), road centerline density (S2), road centerline connectivity (S3), and road centerline betweenness (S4), respectively. The S1 high value area is mainly located in the port area along Jiaozhou Bay and the mountainous area in the southeast, and the street blocks are large in scale; the S1 low value area has a clear correlation with the S2 high value area, and the small scale street blocks increase the street density and have high centrality; S3 and S4 show the connectivity and passability of the local network. Figure 4d shows the Sca composite index that was obtained by weighted superposition of the four indicators using the entropy weighting method.

Road scale was tested for spatial autocorrelation, given that the Global Moran's I was 0.85 and the z-score was 71.70, thus implying significant spatial correlation and clustering. Since walking is more suitable for stimulating urban vitality than vehicular traffic, we focus on the spatial scale of people in the walking mode and set the search radius in the spatial syntax calculation to 1500 m, which is close to the 10 min walking circle range, covering central, daily, and active streets. According to Jacobs, smaller block sizes and shorter streets provide more favorable conditions for contact between people. The results of this index show two more significant spatial patterns: (1) High-value areas (Sca $\geq$ 0.52) identify areas that provide more public contact, which were planned during the German occupation period (1897–1914) in accordance with the European "garden city" concept of colonial planning, with a high-density road network, mainly in the central area of the old urban area, historic area of Taixi and Taidong, with a cluster layout. They are mainly located in the central area of the old city and in the historic areas of Taisi and Taitung and are laid out in clusters. These areas have a mean street-scale index (MSca) of 0.62, with a mean block area index (MS_1) of 0.22, a mean road density index (MS_2) of 0.73, a mean connectivity value (MS_3) of 0.58, and a mean betweenness value (MS_4) of 0.20. The newer urban development areas are considered to be areas with lower contact potential due to larger, more evenly spaced blocks planned and wider streets. (2) Low-value areas (Sca $\leq$ 0.30) that are located in the northern part of old urban center, which is bordered by Taidong historic area, and in southeast along Huiquan Bay and Taiping Bay, with a banding layout and mean street-scale index (MS_1) of 0.20, including a mean block area index (MS_2) of 0.67, a mean road density index (MS_3) of 0.25, a mean connection value of 0.38, and a mean penetration (MS_4) of 0.06. Various historic cities in the northern part of the city were planned during the massive land expansion program of the German occupation (1910–1914), with a low-density road network and lower contact opportunity values in these expansion areas compared to more compact road structures in old urban centers.

4.3. Building Age Diversity

The next criteria for urban vitality, building age diversity, allows us to distinguish the areas in which a larger presence of older buildings coexist with newer ones, from those that are newer and more uniform in terms of age. Building age diversity was tested for spatial autocorrelation, given that the Global Moran's I was 0.63 and the z-score was 52.97, thus implying significant spatial correlation and clustering. The results for this index reveal higher scores within the city's historic center, where a larger number of old buildings are intermingled with more recent structures that were built as part of the densification that has been occurring over the course of the past 123 years. The results show that: (1) High-value areas ($Div_A \geq 0.45$) that are distributed in addition to the historical centers, form important peripheral sub-centers in the southeast coastal area with a mean building age diversity index ($MDiv_A$) of 0.49, including a mean building age index (MDA_1) of 0.81, and a mean building age diversity (MDA_2) of 0.55, such as Yu Hill, Signal Hill historic conservation area, Huiquan Commercial Square, Badaguan, Taiping cape, due to its favorable climate and its historical use as a European retreat area, have more intact architectural types, mainly single-family houses, which are now mostly utilized as tourist resources such as resorts and sea baths. (2) Low-value areas ($Div_A \leq 0.18$), which are mainly located in Badaxia Street west of the railway station, and Dengzhou Road Street north of East—West Rapid Road, etc., show greater homogeneity in terms of building age, a feature that constitutes a negative factor in urban vitality. The mean building age diversity index ($MDiv_A$) 0.04, where the mean building age index (MDA_1) 0.07 and the mean building age diversity (MDA_2) 0.22 are identified in the Taixi and Taidong historic areas and Liaoning Road. These areas have lower housing and rental prices than higher value areas due to the government's Affordable Housing, Housing Provident Fund (HPF) and Low Rent Housing programs since 1994, which subsidize land allocation and reduce taxes on the one hand, and provide financial assistance to individuals on the other, allowing lower income families to enjoy lower rents.

4.4. Concentricity

Figure 7a–d show the spatial distribution of building density (C1), The floor area density (C2), people density (C3), and public facility density (C4), respectively, which all show more obvious spatial differences. There is an obvious spatial mismatch between building density and population density, and Taitung Town, which has a high concentration of population, but fails to provide housing density that meets the demand. Public facilities show a multi-core distribution, and the same problem of uneven distribution exists. Figure 7e shows the composite index after the weighted overlay of the four indicators.

Concentricity was tested for spatial autocorrelation, given that the Global Moran' s I was 0.65 and the z-score was 55.23, thus implying significant spatial correlation and clustering. Similar to the road scale evaluations, smaller street-scales, greater road density, connectivity, and betweenness also explained to some extent the distribution of concentration indexes. (1) High-value areas (Con $\geq$ 0.31) correspond first to high population and housing density neighborhoods in the central area of the old urban area and north of it, with a mean concentration index (MCon) of 0.28, including a mean building density index (MC_1) of 0.41, a mean floor area ratio index (MC_2) of 0.13, a population density index (MC_3) of 0.11, and a public facilities density index (MC_4) of 0.53. The local Moran's index test (Anselin Local Moran' s I test) shows that this area is mostly a high-high cluster area. (2) Although low-value areas (Con $\leq$ 0.14) show no clear trend in their spatial distribution, these areas are generally located in the urban fringe zone, corresponding to industrial sites or areas that are constrained by construction due to topography or the presence of large nature conservation parks, with a mean concentration index (MCon) of 0.05, where the mean building density index (MC_1) is 0.09, the mean floor area ratio index (MC_2) is 0.02, the mean population density index (MC_3) is 0.03, and the mean public facilities density index (MC_4) is 0.16. However, the presence of localized high-value areas can also be seen in the surrounding low-value areas. One explanation for this spatial pattern is that these areas are laid out with a high density of specific public service facilities, such as the historic district of Guanxiang Hill and its surrounding neighborhoods, which are dominated by medical service facilities.

In urban conservation planning of Qingdao, the requirements for building volume, layout, color, material, etc., are consistent with the historical features, that is, the environmental characteristics of small volumes of white walls, red tiles, and green trees, which attract Although the renovation of historical buildings for small entertainment facilities such as coffee shops and bars has limited the attraction of mixed functions to a certain extent, it has improved the attraction of tourists, and thus improved the characteristics of concentration. Such an agglomeration is most obvious in Zhongshan Road, Yushan, and Signal Hill neighborhoods.

4.5. Accessibility and Boundary Vacuum

The spatial distribution of the two complementary conditions for urban vitality are presented in Figure 8. Accessibility is mainly explained by the "center-periphery" structural logic of spatial syntax, i.e., the geometric law of decay from the center to the periphery, and its spatial distribution was tested for spatial autocorrelation, given that the Global Moran' s I was 0.92 and the z-score was 77.71, thus implying significant spatial correlation and clustering. Accessibility of Qingdao historic city shows a multi-polar layout, basically reflecting the public transportation network, with the shorter average distance from the geometric center of the city indicating more convenient transportation. The results show that (1) high-value areas (Acc $\geq$ 0.70) are concentrated in central areas of the old urban area with Zhongshan Road as the central axis, areas around the railway station and Taidong historic area, with a mean accessibility (MAcc) of 0.80, including a mean closeness (MA_1) of 0.69 and a mean traffic facility proximity (MA_2) of 0.05. The central area is characterized by a lower slope compared to the mountainous area, and its road network is planned to follow the topography and there are traffic calming areas, so the accessibility becomes higher. Another reason is the location of the railroad station at the end of the Jiaoji Railway, which connects the urban space. Under the influence of the east-west fast road network, closeness and public transportation accessibility appear to be multi-polar, and the Taitung urban area becomes the trend of urban structure in the 21st century. (2) The low-value area

(Acc ≤ 0.30) is concentrated in port area along Jiaozhou Bay, mountainous area such as Taiping Mountain, and the southeast coastal area, with a mean accessibility (MAcc) of 0.19, including a mean proximity (MA$_1$) of 0.16 and a mean transportation facilities proximity (MA$_2$) of 0.12. This area locates at the edge of southeast urban area, where the road network is less segmented and more interrupted; reducing the distribution of interrupted street network will help create more "reach" opportunities for people.

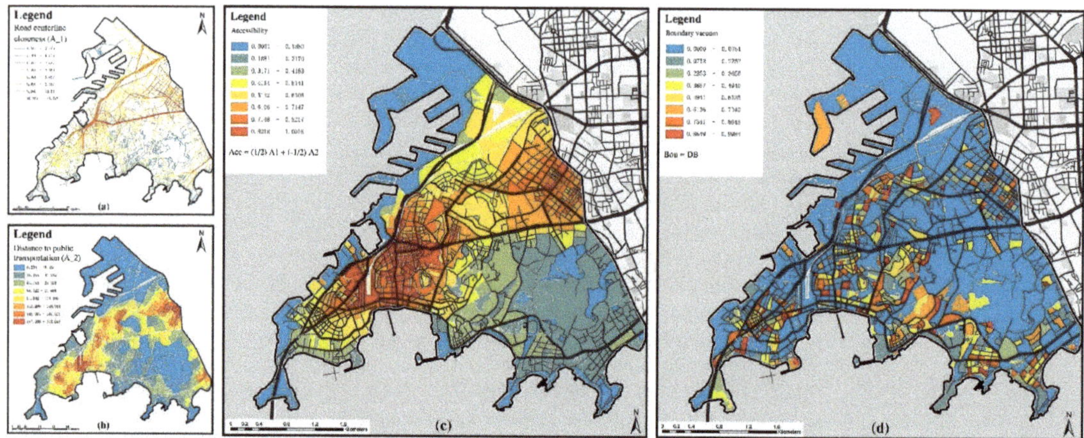

Figure 8. Accessibility and boundary vacuum analysis. (**a**) Road centerline closeness. (**b**) Distance to public transportation. (**c**) Accessibility composite index. (**d**) Distance to boundary vacuum.

In terms of distance from border vacuums, areas with high values are those that are distant to large infrastructures or large, single-used buildings that can potentially discourage street life. These boundary vacuum elements are mainly the Haibo River, Changle River, and Hangan Viaduct at the northern city boundary; Jiao Ning Viaduct extending eastward from the central old urban center; and the railway lines around the railway station. In the surrounding zone, the inner part of the old urban area is also covered with major parking lots in points and large parks such as nature reserves in polygon form. The Bou index measures the distance of each spatial unit from boundary vacuum elements, and its distribution pattern is tested for spatial autocorrelation, given that the Global Moran's I 0.33 and the z-score is 27.81, so there are significant spatial correlations and clustering. The results show that: (1) High-value areas (Bou ≥ 0.63) are mainly located along the southeast coastline, including Signal Hill, Yu Hill, and Baguan Hill, due to the superior natural environment and distance from facilities such as railway lines and large parking lots. Besides, there are points of high value areas that are distributed along Liaoning Road and Changle Road in the north. (2) Although the boundary vacuum elements are mainly located at urban boundaries, it is worth noticing that within the central area of Qingdao's historic city, there are also numerous streets that are close to the boundary vacuum and present low values. Since the planning of the German system at the end of the 19th century was motivated by the construction of Qingdao as a military base and industrial port, large infrastructures such as railroad stations, Jiaoji railroad, and ports were planned, and these infrastructural elements had a negative impact on human contact.

4.6. JANE Index in Qingdao Historic City

We then synthesized the six conditions of urban vitality in the JANE Index. Higher values of the JANE Index correspond to areas with a higher potential for urban vitality, while lower values indicate the lack of such conditions. The distribution of Qingdao historic city JANE index is shown in Figure 9.

The potential of urban vitality in Qingdao historic city presents a polycentric pattern, as we find high values of the JANE Index distributed in different sub-centers of potential vitality. In turn, different intensities of urban vitality potential are identified. As Jacobs' advanced, not to be picked up at the city scale, but is at the district and neighborhood levels that this is properly understood. In this sense, one center and four sub-centers of urban vitality were identified in Qingdao historic city. (1) The west end of Qingdao East–West Rapid Road, the south end of Jiaoji Railway, Taiping Road, and Jiangsu Road enclose the central area with the highest urban vitality, and the north end has part of the junction between the south and north districts of Qingdao, such as Zhongshan Road Street and Sifang Street, with a mean JANE index of 0.49 and a standard deviation 0.11 and more uniform spatial distribution. The area has a long planning history, significant street morphological differences, and more historic buildings, and thus has more intensive crowd activity (Con), more neighborhood contact opportunities (Sca), higher building age diversity (Div_A), and street accessibility (Acc). The general pattern of spatial patterns of urban vitality is again validated by the center–periphery logic: closer to the center means that higher urban vitality is achieved, while those areas with lower or no potential vitality are incorporated into peripheral contours of city. (2) The streets of Taidong, Liaoning Road, Badaxia and Jiangsu Road constitute the four sub-centers of urban vitality, with mean JANE scores of 0.44, 0.45, 0.43, and 0.39 and standard deviations of 0.08, 0.07, 0.07, and 0.08, respectively, and a more uniform spatial distribution. Compared with the traditional 100 m × 100 m grid in the historic center, these areas have a narrower street pattern, smaller block scale, and high housing and population density, and higher JANE indexes are mainly explained by two indicators of concentration (Con) and street-scale (Sca). (3) Similarly, it is evident that some edge areas have high JANE scores, such as Badaguan Street near Wushengguan Road and Dagang Street near Dagang Weisi Road, which are mainly explained by the street-scale (Sca) index. These areas with high-low cluster values confirm the multicenter trend of urban vitality. 4) The area with a low JANE index (JANE $\leq$ 0.30) is mainly located in northern Dagang Street, Yunnan Road Street, and the southern end of Badaxia Street, with a mean JANE index 0.24, 0.29, and 0.31 and a standard deviation 0.03, 0.05, and 0.07, respectively, and more uniform spatial distribution. The lower spatial vitality of this area is related to its location in the urban edge area, relative isolation, large street block area, low density, and single land use.

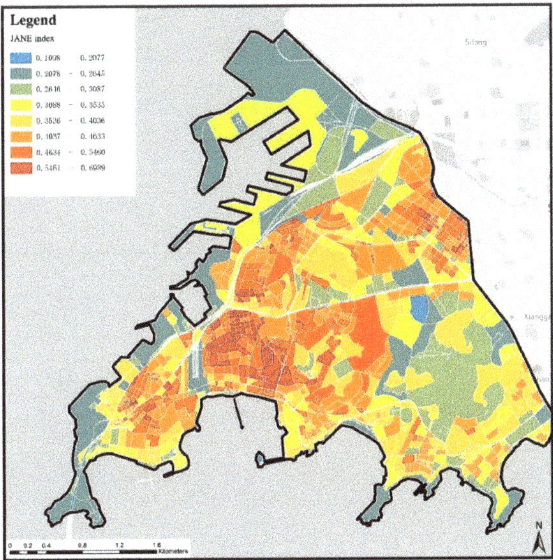

Figure 9. JANE Index of urban vitality.

Figure 10 shows the grading results of the Qingdao historic city JANE index, and Table 2 shows the statistics of TAZ units at each level. There were four urban vitality classes of Qingdao's historic city that were identified by JANE cluster analysis, labeled as high vitality areas, moderate vitality areas, low vitality areas, and non-vitality areas. The analysis results show that high vitality areas account for about 6.73% of the total area of the study area, with a mean JANE index of 0.568, standard deviation of 0.048, and extreme values 0.507 and 0.699, mostly located in historic central areas, and a few areas in the above four JANE index sub-centers. These areas are characterized by a small-scale street network, a mixed distribution of old and new buildings, and a mix of commercial and public facilities, producing a high accessibility and diversity value, a feature that is further enhanced by the increased distance from the boundary vacuum. The main axis of urban vitality in this area is from Zhongshan Road southward to the Trestle Bridge, where traditional Qingdao neighborhoods, shopping malls, and Trestle Park converge, reflecting a high level of diversity and neighborhood access. On the eastern side of the old city, Guanhai Hill and the area to its south are highly accessible due to the radial road network, creating high vitality points. Finally, in addition to the historic center, the neighborhoods of Weihai Road Pedestrian Street in Taidong Street are characterized by a higher concentration of urban vitality due to higher population density and commercial facility density.

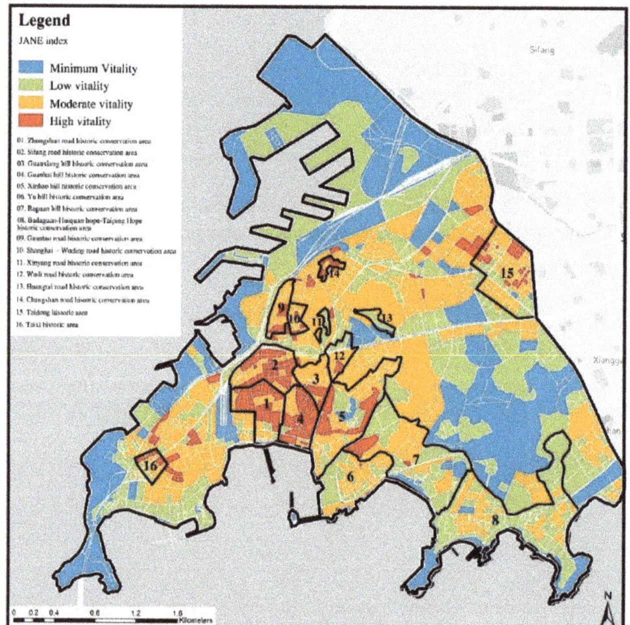

Figure 10. JANE index classification analysis.

Table 2. JANE index group statistics results.

Group	Count	Mean	Std. Deviation	Min.	Max.	Proportion
High Vitality	196	0.568	0.048	0.507	0.699	6.73%
Moderate Vitality	572	0.449	0.032	0.392	0.505	26.98%
Low Vitality	779	0.338	0.028	0.293	0.391	29.21%
Minimum Vitality	732	0.248	0.033	0.110	0.292	37.08%
Total	2279	0.356	0.104	0.110	0.699	100.00%

Source: compiled by the authors.

Secondary areas of moderate vitality zones account for approximately 26.98% of the study area, with a mean JANE index of 0.449, a standard deviation of 0.032, and extreme values of 0.392 and 0.505. In terms of the planned street patterns, the area is expected to exhibit a high potential level of street life, but the combination with other indexes reflects that central areas of the district are either categorized as moderate or even low vitality zones and tends to be a transitional buffer zone between high and low vitality zones. Compared with the neighboring high vitality zones, the approximate street scale and population density show that the mixed degree of building functions and density become the main reason for the formation of transition zones. For example, among Taidong streets, the layout of Weihai Road pedestrian street tends to have a positive impact on local vitality, while neighborhoods that are farther away from pedestrian streets have relatively lower vitality.

Finally, areas that are classified as low and no vitality account for 29.21% and 37.08% of Qingdao's historic city, respectively, with low vitality areas having a mean JANE index of 0.338, a standard deviation of 0.028, and extreme values of 0.293 and 0.391, and no vitality areas having a mean JANE index of 0.248, a standard deviation of 0.033, and extreme values of 0.110 and 0.292, which are often are located in urban edge zones with low population and building densities and in close proximity to border vacuums. These two vibrant environments correspond to being located on the periphery of a moderately vibrant zone, presenting very low accessibility values. The analysis reveals that places that are in contact with agricultural, natural, or industrial areas tend to present significantly low values of urban vitality; on the one hand they usually are near extensive mountainous areas with significant slopes, such as Qingdao Mountain, Taiping Mountain, and Zhushui Mountain, and on the other hand they are located in functionally single areas with larger block scales, such as Dagang and Xiaogang in the north, and Taiping Cape and Huizhuan Cape in the southeast coast.

5. Discussion and Conclusions

This study aims to recontextualize Jacobs' urban vitality principle through urban spatial analysis and analyzes the livability and urban vitality of Chinese historic cities in the context of rapid urbanization. The newly established JANE index consists of six criteria with a total of 16 indexes, and since urban vitality is considered a key measure of urban residents' happiness and quality of life, this study contributes to a discussion of the nature of the drivers of urban vitality and the way that they are distributed in space. To this end, the study creates a framework of indexes based on the Delclòs-Alió establishment and integrates the theoretical and methodological proposals that were made by recent literature on the subject, and performs a spatial analysis based on GIS-sDNA, the results of which are applicable to Qingdao historic city with its different muscular morphology. Jacobs has a rather important role in the history of the field of modern urbanism. From a practical point of view, this paper will help planners to combine morphological conditions that are conducive to vibrant streets, urban environments, and neighborhoods, thus promoting community, local spatial practices, and neighborhoods.

The results of this study show that conditions of urban vitality in today's cities are not necessarily related to centrality, a specific urban texture, or a certain income level, but may be a result of different combinations of certain urban characteristics. In this sense, this study validates the applicability of Jacobs' theory to cities in developing Asia, where more street activity and vitality is not singularly concentrated in central areas, although these areas are often characterized by higher population densities and mixed uses. Qingdao's historic urban vitality follows a multi-centered spatial pattern, with a more fragmented distribution of public services and coexists with a dense network of local businesses. Different combinations of indexes of urban vitality are distributed throughout the area, showing a multi-centered pattern covering a diverse range of urban texture with two characteristics: (1) The main areas with high potential values of urban vitality in Qingdao's historic city correspond to the oldest historic city center, which has the original dynamics of all vitality indexes. The square grid street pattern that was planned along the topography creates the conditions for urban vitality. As Jacobs argues, the mix of land uses and the distribution of more intersections produce a more compact and complex form that provides constant access to citizens and a diversity of facilities and services. (2) Medium and high potential vitality values are found equally in parts of the city that are away from the Old City Center, such as the Taidong and Taixi historic areas. It has long been marginalized due to topographical factors and the racial segregation motive in historical planning. In recent years, the area has gradually formed a sub-center of urban vitality in Qingdao with the combination of different vitality conditions such as increasing population density and diverse land use development, which attracts and directs not only short-distance mobility but also people that are traveling long distances.

After China established the historic and cultural city protection system in 1982, protection concepts gradually developed from single buildings to overall historic environmental protection in historic districts. Qingdao historic city was incorporated into the national historic and cultural city protection list in 1994, as Figure 8 delineates 14 historic conservation areas and 2 historic areas. Qingdao government strictly protects the physical environment of the historic city but implements relatively few policies and practices to improve the social life of residents and promote the renewal and maintenance of old public service facilities. Qingdao's historic conservation areas originated from the segregated division of European and Chinese districts in the urban planning during the German rule. Historic conservation areas, mainly in the former European district, are mainly villa residences, with large street block scale and high green space rate, but there is a problem of insufficient vitality, so in the renovation and reuse of such streets, commercial and public service facilities should be increased to improve the diversity of architectural functions. The former Chinese district is the main historical district that was originally for residential-commercial mixed function, with smaller street block scale, low green space rate, and poor livability. In the renovation and reuse of these kinds of streets, green space and public activity space should be increased. Based on the relationship between the Jacobs index system and the quantitative results, we established the following planning strategies to enhance the vitality of Qingdao's historic city:

1. Based on the results of diversity analysis, for single-function blocks, such as Badaguan and other blocks on the southeast coast, the mix of land use, residential-non-residential, and commercial-public facilities should be improved, and the interaction of various types of facilities should be exerted.
2. Based on the results of diversity and old building analysis, we will take measures to preserve the old buildings in Zhongshan Road and Sifang Road and revitalize them to meet functional needs.
3. Based on the results of small blocks analysis, some of the blocks along Jiaozhou Bay in the northwest maintain the historical block texture of short streets and small plots from the 1920s Japanese planning period, and the spatial structure of small blocks should be continued for new development sites.
4. Based on the results of concentration and old building analysis, measures are taken to maintain a reasonable high density of buildings, roads, and population in the neighborhoods of Yushan and Signal Hill, and to consider the use of old buildings as subsidized housing, so as to avoid excessive tourism and gentrification of the historic districts.
5. Based on the results of concentration, small blocks, and accessibility analysis, the density of bus route coverage and shared transportation facilities is reasonably increased for Zhongshan Road and other neighborhoods.
6. Based on the results of boundary vacuum analysis, large scale infrastructure construction should be avoided for the central city, and the crossing of boundary vacuum elements such as viaducts and railroads should be avoided, because these elements cut the integrity of the historical urban form.

There are certain limitations in this study. Firstly, urban vitality is a broad and complex concept involving social, economic, cultural, and other multi-dimensional contents.

Buildings **2022**, *12*, 1586

Although this study characterizes urban vitality by various parameters of LBS data with objectivity, it cannot accurately express the subjective perception of the public. For example, the satisfaction and perception of an active population on the space of historical neighborhoods mean the research results may be biased due to the data characteristics. Follow-up studies should calculate more accurate variable data that are based on streetscape images and explore human perception characteristics in detail. Second, POI data cannot present time information, a limitation that prevents the unique seasonal characteristics of historic urban vibrancy from being effectively measured. Follow-up studies can combine multi-source spatio-temporal data such as social media punch card data and cell phone signaling data to establish a more accurate urban vitality index system.

Author Contributions: Conceptualization, S.W. and Q.D.; methodology, S.W. and Q.D.; software, S.W. and S.J.; validation, Q.D. and G.W.; formal analysis, S.W. and Q.D.; investigation, Q.D. and G.W., resources, Q.D. and S.J.; data curation, S.W. and S.J.; writing—original draft preparation, S.W. and Q.D.; writing—review and editing, Q.D., S.W. and G.W., visualization, S.W., S.J. and G.W.; supervision, Q.D.; project administration, S.W. and Q.D.; funding acquisition, Q.D. All authors have read and agreed to the published version of the manuscript.

Funding: This research received no external funding.

Data Availability Statement: Data derived from the current study can be provided to readers upon request.

Conflicts of Interest: The authors declare no conflict of interest.

References

1. Jin, X.; Long, Y.; Sun, W.; Lu, Y.; Yang, X.; Tang, J. Evaluating cities' vitality and identifying ghost cities in China with emerging geographical data. *Cities* **2017**, *63*, 98–109. [CrossRef]
2. Wang, J. Problems and solutions in the protection of historical urban areas. *Front. Archit. Res.* **2012**, *1*, 40–43. [CrossRef]
3. Rosenthal, S.S. Old homes, externalities, and poor neighborhoods. A model of urban decline and renewal. *J. Urban Econ.* **2008**, *63*, 816–840. [CrossRef]
4. Mehanna, W.A.E.H.; Mehanna, W.A.E.H. Urban renewal for traditional commercial streets at the historical centers of cities. *Alex. Eng. J.* **2019**, *58*, 1127–1143. [CrossRef]
5. Gehl, J. *Life between Buildings*; Van Nostrand Reinhold: New York, NY, USA, 1987; Volume 23.
6. UNESCO, W. Recommendation on the historic urban landscape. In Proceedings of the Records of the General Conference 36th Session, Paris, France, 25 October–10 November 2011.
7. Wu, J.; Lu, Y.; Gao, H.; Wang, M. Cultivating historical heritage area vitality using urban morphology approach based on big data and machine learning. *Comput. Environ. Urban Syst.* **2022**, *91*, 101716. [CrossRef]
8. Whitehand, J.; Gu, K. Urban conservation in China: Historical development, current practice and morphological approach. *Town Plan. Rev.* **2007**, *78*, 643–671. [CrossRef]
9. Whitehand, J.W.; Gu, K.; Whitehand, S.M.; Zhang, J. Urban morphology and conservation in China. *Cities* **2011**, *28*, 171–185. [CrossRef]
10. Oliveira, V.; Medeiros, V. Morpho: Combining morphological measures. *Environ. Plan. B Plan. Des.* **2016**, *43*, 805–825. [CrossRef]
11. Kim, S. Urban vitality, urban form, and land use: Their relations within a geographical boundary for walkers. *Sustainability* **2020**, *12*, 10633. [CrossRef]
12. Martino, N.; Girling, C.; Lu, Y. Urban form and livability: Socioeconomic and built environment indicators. *Build. Cities* **2021**, *2*, 12–15. [CrossRef]
13. Xu, Y.; Jin, S.; Chen, Z.; Xie, X.; Hu, S.; Xie, Z. Application of a graph convolutional network with visual and semantic features to classify urban scenes. *Int. J. Geogr. Inf. Sci.* **2022**, *37*, 1–26. [CrossRef]
14. Huang, B.; Zhou, Y.; Li, Z.; Song, Y.; Cai, J.; Tu, W. Evaluating and characterizing urban vibrancy using spatial big data: Shanghai as a case study. *Environ. Plan. B Urban Anal. City Sci.* **2020**, *47*, 1543–1559. [CrossRef]
15. Jacobs, J. *The Death and Life of Great American Cities*; Random House: New York, NY, USA, 1961.
16. Laurence, P.L. The death and life of urban design: Jane Jacobs, The Rockefeller Foundation and the new research in urbanism, 1955–1965. *J. Urban Design.* **2006**, *09*, 145–172. [CrossRef]
17. Hall, P.; Pfeiffer, U. *Urban Future 21: A Global Agenda for Twenty-First Century Cities*; Routledge: London, UK, 2013. [CrossRef]
18. Klemek, C. Placing Jane Jacobs within the transatlantic urban conversation. American Planning Association. *J. Am. Plan. Assoc.* **2007**, *73*, 49. [CrossRef]
19. Lopes, M.N.; Camanho, A.S. Public green space use and consequences on urban vitality: An assessment of European cities. *Soc. Indic. Res.* **2013**, *113*, 751–767. [CrossRef]

20. Delclòs-Alió, X.; Miralles-Guasch, C. Looking at Barcelona through Jane Jacobs's eyes: Mapping the basic conditions for urban vitality in a Mediterranean conurbation. *Land Use Policy* **2018**, *75*, 505–517. [CrossRef]
21. Fuentes, L.; Miralles-Guasch, C.; Truffello, R.; Delclòs-Alió, X.; Flores, M.; Rodríguez, S. Santiago de Chile through the eyes of Jane Jacobs. Analysis of the conditions for urban vitality in a Latin American metropolis. *Land* **2020**, *9*, 498. [CrossRef]
22. Gómez-Varo, I.; Delclòs-Alió, X.; Miralles-Guasch, C. Jane Jacobs reloaded: A contemporary operationalization of urban vitality in a district in Barcelona. *Cities* **2022**, *123*, 103565. [CrossRef]
23. Paköz, M.Z.; Yaratgan, D.; Şahin, A. Re-mapping urban vitality through Jane Jacobs' criteria: The case of Kayseri, Turkey. *Land Use Policy* **2022**, *114*, 105985. [CrossRef]
24. Fujita, M.; Thisse, J.F. Economics of agglomeration. *J. Jpn. Int. Econ.* **1996**, *10*, 339–378. [CrossRef]
25. Derudder, B.; Witlox, F. An appraisal of the use of airline data in assessing the world city network: A research note on data. *Urban Stud.* **2005**, *42*, 2371–2388. [CrossRef]
26. Schläpfer, M.; Bettencourt, L.M.; Grauwin, S.; Raschke, M.; Claxton, R.; Smoreda, Z.; West, G.B.; Ratti, C. The scaling of human interactions with city size. *J. R. Soc. Interface* **2014**, *11*, 20130789. [CrossRef] [PubMed]
27. Batty, M. Defining urban science. In *Urban Informatics*; Springer: Singapore, 2021; pp. 15–28. [CrossRef]
28. Wang, F.; Antipova, A.; Porta, S. Street centrality and land use intensity in Baton Rouge, Louisiana. *J. Transp. Geogr.* **2011**, *19*, 285–293. [CrossRef]
29. Porta, S.; Latora, V.; Wang, F.; Rueda, S.; Strano, E.; Scellato, S.; Cardillo, A.; Belli, E.; Cardenas, F.; Cormenzana, B.; et al. Street centrality and the location of economic activities in Barcelona. *Urban Stud.* **2012**, *49*, 1471–1488. [CrossRef]
30. Cooper, C.H.; Fone, D.L.; Chiaradia, J.A. Measuring the impact of spatial network layout on community social cohesion: A cross-sectional study. *Int. J. Health Geogr.* **2014**, *13*, 11. [CrossRef] [PubMed]
31. Cooper, C.H. Spatial localization of closeness and betweenness measures: A self-contradictory but useful form of network analysis. *Int. J. Geogr. Inf. Sci.* **2015**, *29*, 1293–1309. [CrossRef]
32. Yang, D.F.; Yin, C.Z.; Long, Y. Urbanization and sustainability in China: An analysis based on the urbanization Kuznets-curve. *Plan. Theory* **2013**, *12*, 391–405. [CrossRef]
33. Yue, W.; Chen, Y.; Thy, P.T.; Fan, P.; Liu, Y.; Zhang, W. Identifying urban vitality in metropolitan areas of developing countries from a comparative perspective: Ho Chi Minh City versus Shanghai. *Sustain. Cities Soc.* **2021**, *65*, 102609. [CrossRef]
34. Maria Kockelman, K. Travel behavior as function of accessibility, land use mixing, and land use balance: Evidence from San Francisco Bay Area. *Transp. Res. Rec.* **1997**, *1607*, 116–125. [CrossRef]
35. Van Den Hoek, J.W. *Towards a Mixed-Use Index (MXI) as a Tool for Urban Planning and Analysis*; Urbanism: PhD Research; IOS Press: Amsterdam, The Netherlands, 2009; pp. 64–85.
36. King, K. Jane Jacobs and 'the need for aged buildings': Neighbourhood historical development pace and community social relations. *Urban Stud.* **2013**, *50*, 2407–2424. [CrossRef]
37. Kumakoshi, Y.; Koizumi, H.; Yoshimura, Y. Diversity and density of urban functions in station areas. *Comput. Environ. Urban Syst.* **2021**, *89*, 101679. [CrossRef]
38. Tucker, C.M.; Cadotte, M.W.; Carvalho, S.B.; Davies, T.J.; Ferrier, S.; Fritz, S.A.; Grenyer, R.; Helmus, M.R.; Jin, L.S.; Mooers, A.O.; et al. A guide to phylogenetic metrics for conservation, community ecology and macroecology. *Biol. Rev.* **2017**, *92*, 698–715. [CrossRef] [PubMed]
39. Salat, S.; Labbé, F.; Nowacki, C. *Cities and Forms on Sustainable Urbanism*; CSTB: Paris, France, 2011.
40. Fang, C.; He, S.; Wang, L. Spatial characterization of urban vitality and the association with various street network metrics from the multi-scalar perspective. *Front. Public Health.* **2021**, *9*, 677910. [CrossRef] [PubMed]
41. Ye, Y.; Li, D.; Liu, X. How block density and typology affect urban vitality: An exploratory analysis in Shenzhen, China. *Urban Geogr.* **2018**, *39*, 631–652. [CrossRef]
42. Xia, C.; Yeh, A.G.O.; Zhang, A. Analyzing spatial relationships between urban land use intensity and urban vitality at street block level: A case study of five Chinese megacities. *Landsc. Urban Plan.* **2020**, *193*, 103669. [CrossRef]
43. Anderson, J.M.; MacDonald, J.M.; Bluthenthal, R.; Ashwood, J.S. *Reducing Crime by Shaping the Built Environment with Zoning: An Empirical Study of Los Angeles*; University of Pennsylvania Law Review: Philadelphia, USA, 2013; Volume 161, pp. 699–756.
44. He, S.; Yu, S.; Wei, P.; Fang, C. A spatial design network analysis of street networks and the locations of leisure entertainment activities: A case study of Wuhan, China. *Sustain. Cities Soc.* **2019**, *44*, 880–887. [CrossRef]
45. Ma, F. Spatial equity analysis of urban green space based on spatial design network analysis (sDNA): A case study of central Jinan, China. *Sustain. Cities Soc.* **2020**, *60*, 102256. [CrossRef]
46. Sung, H.; Lee, S.; Cheon, S.H. Operationalizing jane jacobs's urban design theory: Empirical verification from the great city of seoul, korea. *J. Plan. Educ. Res.* **2015**, *35*, 117–130. [CrossRef]
47. Jalaladdini, S.; Oktay, D. Urban public spaces and vitality: A socio-spatial analysis in the streets of Cypriot towns. *Proc. Soc. Behav. Sci.* **2012**, *35*, 664–674. [CrossRef]
48. Hill, M.O. Diversity and evenness: A unifying notation and its consequences. *Ecology* **1973**, *54*, 427–432. [CrossRef]
49. Jost, L. Entropy and diversity. *Oikos* **2006**, *113*, 363–375. [CrossRef]
50. Powe, M.; Mabry, J.; Talen, E.; Mahmoudi, D. Jane Jacobs and the value of older, smaller buildings. *J. Am. Plan. Assoc.* **2016**, *82*, 167–180. [CrossRef]

51. Gao, S.; Wang, Y.; Gao, Y.; Liu, Y. Understanding urban traffic-flow characteristics: A rethinking of betweenness centrality. *Environ. Plan. B Plan. Des.* **2013**, *40*, 135–153. [CrossRef]
52. Zeng, C.; Song, Y.; He, Q.; Shen, F. Spatially explicit assessment on urban vitality: Case studies in Chicago and Wuhan. *Sustain. Cities Soc.* **2018**, *40*, 296–306. [CrossRef]
53. Anderson, T.K. Kernel density estimation and K-means clustering to profile road accident hotspots. *Accid. Anal. Prev.* **2009**, *41*, 359–364. [CrossRef] [PubMed]
54. Xiao, Y.; Sarkar, C.; Webster, C.; Chiaradia, A.; Lu, Y. Street network accessibility-based methodology for appraisal of land use master plans: An empirical case study of Wuhan, China. *Land Use Policy* **2017**, *69*, 193–203. [CrossRef]
55. Jenks, G.F.; Caspall, F.C. Error on choroplethic maps: Definition, measurement, reduction. *Ann. Assoc. Am. Geogr.* **1971**, *61*, 217–244. [CrossRef]

Article

Reshaping Publicness: Research on Correlation between Public Participation and Spatial Form in Urban Space Based on Space Syntax—A Case Study on Nanjing Xinjiekou

Mengyao Pan [1], Yangfan Shen [2], Qiaochu Jiang [1], Qi Zhou [1,*] and Yinghan Li [3,*]

[1] School of Architecture, Architectural History and Theory, Southeast University, Nanjing 210096, China
[2] Architectural Heritage Conservation Department, Architects & Engineers Co., Ltd. of Southeast University, Nanjing 210096, China
[3] Department of Architecture, School of Fine Arts, South-Central Minzu University, Wuhan 430074, China
* Correspondence: 101001781@seu.edu.cn (Q.Z.); 2022100@mail.scuec.edu.cn (Y.L.)

Abstract: This paper focuses on urban regeneration practices in central urban areas, aiming to find key points for reshaping the publicness of urban spaces by exploring the morphological features of public spaces and the spatial distribution patterns of public activities. Now that China's urbanization process has stabilized, large-scale regeneration is no longer applicable to the current urban environment, and urban morphology has proved to be significantly useful in understanding and designing the built environment. However, current research lacks quantitative studies on morphology and public activities, and thus is hardly instructive for the cognition and design of spatial morphology in specific locations. Therefore, this paper attempts to subdivide spatial morphology at the level of "micro-renewal" or "micro-renovation" in order to explore the impact of spatial morphology on public participation in cities. The site chosen for this study is Xinjiekou in Nanjing. As a key area of two important arteries in the center of Nanjing, Xinjiekou has been a gathering place for a variety of commercial forms, such as finance, retail department stores, restaurants, and entertainment, and has been the commercial and financial center of Nanjing since the 1940s. In an on-site observation of urban development and pedestrian flow in the Xinjiekou area, the study found that despite its status as the area with the highest degree of spatial accessibility and public participation, its public space has gradually lost its attractiveness to residents, who lack a sense of participation and place identity. Based on the study of urban public spaces, both accessibility and choice play an important role in increasing public participation. Therefore, this study combined observation and quantitative analysis of Space Syntax to obtain the distribution of accessibility, choice, and public activity. Based on the results of the analysis, this research uses GWR as the statistical method to clarify the correlation between different variables. The final conclusion is that when the space type is a path with high choice value and the paths are connected, the enhancement of accessibility and choice plays an important role in promoting public participation. This statistically based empirical study of testable correlations is very helpful for the perception of location-specific spaces with high levels of interpretability and confidence. Thus, it further guides the design and has a high reference value for future spatial planning.

Keywords: urban renewal; accessibility; place-making; typology; Space Syntax

Citation: Pan, M.; Shen, Y.; Jiang, Q.; Zhou, Q.; Li, Y. Reshaping Publicness: Research on Correlation between Public Participation and Spatial Form in Urban Space Based on Space Syntax—A Case Study on Nanjing Xinjiekou. *Buildings* **2022**, *12*, 1492. https://doi.org/10.3390/buildings12091492

Academic Editors: Michael J. Ostwald and Ju Hyun Lee

Received: 2 August 2022
Accepted: 16 September 2022
Published: 19 September 2022

Publisher's Note: MDPI stays neutral with regard to jurisdictional claims in published maps and institutional affiliations.

1. Introduction

In recent decades, the public nature of public spaces has been diminishing globally, suggesting that public space is losing its potential to provide opportunities for social interaction. Some superficial explanations suggest that the possible reasons for this stem from the suppression of the creative sphere in the context of homogenization constructed by globalization. Individuality and uniqueness have been destroyed and urban spaces tend to be similar. As described in the Beijing Charter, "a thousand cities all look alike" in the eyes of people today [1]. This is the dilemma facing today's cities, as the public memory carried by distinctive buildings and neighborhoods in urban public spaces is being destroyed with the rapid development. This in turn further promotes the loss of the meaning of public space, and gradually makes the public space in modern cities a meaningless transitional space where people find it difficult to develop a sense of belonging.

Many scholars have studied the publicness of urban public space in terms of its ownership, cognition, management, and public activities. A typical one is the Project of Public Space (2010). Based on a study of thousands of public spaces, this project concluded that a highly public space should have the characteristics of Social activities, Comfort and image, Accessibility, and Sociability to facilitate the interaction between the residents and the space. This view is not alone, as Arentze, Afonso, Ye et al. argue that the daily activities of individuals/families are driven by a set of universal needs, and thus travel time, comfort, and sociability are important factors that drive public engagement in public spaces [2–4]. At the same time this process of social–spatial interaction as a dynamic flow plays a considerable degree of permeability to its neighboring spaces, and thus on this level good public spaces will prove influential to neighborhoods beyond the limits of its own physical space. This idea is also similar to the one that was proposed by Jane Jacobs and William H. White in the 1960s in response to spatial deactivation—namely, the use of place-making, i.e., the use of a community's assets, values, and resources to create a place that promotes people's health, well-being, and happiness in order to enhance their sense of identity and the meaning of place [5–10].

The most prominent number of these related studies for spatial accessibility and morphology, Ye and Le Texier conclude that accessibility plays a crucial activity in public engagement, and the morphology of urban spaces will largely influence people's choices [4,11]. Although many studies have investigated public spaces in detail, these studies have not explored the association between public participation (publicness) and accessibility, choice and urban form, using a quantitative approach for specific sites from a design perspective. In addition, there are fewer studies on the quantitative analysis of spatial morphology, and the challenge of constructing correlations with actual social phenomena leads to difficulties in applying the results of the analysis to design.

At the same time, in the context of China's urban public space, the density, intensity, and height of the built environment have increased dramatically in the urbanization process since the implementation of the reform and opening-up policy 40 years ago [12]. The massive development and construction of urban centers has compensated for the lack of commercial space in the original areas. However, excessive commercialization has encroached on public space, while people are attracted to commercial services and neglect public space during their walks [13,14]. Commercial spaces pre-screen residents due to their ownership and the consumption or services they provide. As a result, there is a discontinuity and disintegration of spatial experience for some residents. This reduces the interaction between the individual and the environment, thus re-diminishing the resident's identification with place [15,16].

As Rowan Moore argues in his book *The Slow Born City*, cities should change gradually, with a combination of private investment, public interest, and legislative action, rather than destructive change. Therefore, this study will interpret the evolution of urban spatial forms in the process of urban development in the context of "micro-renewal" with a Space Syntax in order to obtain the basic characteristics of urban forms and spatial accessibility and choices. Secondly, we make long-term observations of the site and map the location of public activities. By calculating the correlation between these two types of spatial elements, the results are finally projected onto the different morphological types of urban public space to determine the role of urban morphology in promoting public participation. Based on this, this study attempts a combined qualitative–quantitative approach to public space, exploring the significance of spatial design in leading to positive human well-being, and the demonstration and testing of the feasibility of design potential. The lack of empirical analysis of spatial planning is remedied.

2. Literature Review

This chapter focuses on the definition of urban public space and the social significance it carries. On the basis of this chapter, the research objectives of this paper are combined with the research perspectives of choice, accessibility, and typology for the study of public space. Finally, it summarizes the Space Syntax and related research methods.

2.1. Publicness of Urban Space

Place and space have long been the focus of discussion in academic circles. In the context of urban design, the two can be used alternatively. Place is considered to be a space bearing social relations and identity [6–9]. Therefore, public engagement is particularly important in urban design. Lefebvre believes that active space (place with functions and meanings) is the result of people using space to implement daily necessities [17]. Urban design in turn endows place with meaning to reshape the daily life of the public.

As one of the first scholars to introduce the term "public space" into the field of urban study, Jacobs strongly criticized the functionalism-based urban reconstruction policy of American cities in the 1950s in *Life and Death of American Big Cities*. She believes that modernist planning has destroyed the vitality of urban streets and urban fabric in traditional cities, turning closely related urban blocks into isolated architectural individuals. The function-based division of urban space destroys the social value of space. The planner and designer must advocate the development of "public space" in "urban space", and take public space as a key element to promote the formation of good social exchanges and restore urban vitality in urban development and community construction [8]. In her view, the value of public space lies in that its existence can promote the communication of people from different social strata or groups in the city. Because of its nature of diversity and inclusiveness, public space plays a vital role in promoting social mutual understanding and integration as well as social stability and harmony, and serves as an important source of urban vitality.

The past 40 years have witnessed a greater engagement of factors such as capital, power, politics, and economy in the development and reform of contemporary Chinese cities. The philosophy of "place-making" echoes humans' basic needs for a high-quality living environment. According to Schultz, creating a place means expressing a sense of existence. This concept reflects the phenomenological essence of place-making. People realize their "settlement" in the world by creating place [18]. Therefore, providing people with a recognizable place that gives them a sense of belonging is the fundamental purpose of place-making. This purpose can be achieved, in part, through the physical construction of places and the organization of activities within them.

The most significant signal of the disappearance of the sense of place is the loss of appeal of cities to people's activities, the neglect of public space, and the lack of sense of identity and participation in the society. In developed countries, the mainstream understanding of urban design has become more and more inclined to the construction process as place identity [19]. Peter Buchanan believes that urban design is essentially about place-making, and place is not only a special space, but also the activities and events that make the city a place [20]. Among the five evaluation criteria of urban design, "vitality" is the first of the five criteria, which is an important goal of urban design [21]. Gehl believes that the vitality of the urban public space lies in the people in it and their activities. Only the users of space and the activities they participate in are the fundamental factors that determine the vitality of urban public space [10].

Therefore, the accessibility of public space refers not only to physical access, but also social inclusion, which is even more important. It should open to both "people" from different social classes, races, and groups and "different social activities" in accordance with space rights. This is similar to the concept of "publicness" in the field of political philosophy, which is reflected in space, that is, whether the meaning of spiritual and cultural levels expressed by space through the material environment is the most inclusive [22]. Human activity is the fundamental factor of vitality. A high-quality urban space can connect activities, images, and forms to foster a sense of place. The important goal of urban design is to restore or rebuild the harmonious relationship between people and place, and turn the public space into a public realm [13].

2.2. Typologies and Morphology of Public Space

According to Conzen, urban morphology is the study of the structural forms of cities, which aims to explain the spatial layout and composition of urban organizations and public spaces. He possed material characteristics and symbolic meanings, the power to create, expand, diversify, and change forms [23]. Scholars such as Sadeghi and Abarkan argue that a systematic interpretation of the forms and types of man-made urban fabric developed over time and combined with social change can provide a good understanding of the present society and the potential needs for future development [24,25]. Yang et al. argue that urban form explains to some extent the logic of urban functioning, and that spatial-typology interacts with social conditions to shape public life through relevant architectural and urban design theories [26]. These different typological combinations of public spaces are thus not only the emerging "urban public sphere" but also the "connecting parts" of a space [26]. Representing and describing urban organization according to its formative processes is the basis for understanding and planning the evolution of the urban realm [7].

The research exploring the association between public identity, participation, and the shape of urban public space is broadly divided into the following assessment perspectives. First is the individual types of public space itself. Lynch's monograph classifies spaces into perceptible landmark, node, path, distinct, and edge, based on psychological cognitive research [21]. Yılmaz Çakmak classifies public spaces into point-like space-activity and linear space-passage, based on spatial morphology and activity patterns [27]. Benjamin classifies public spaces into point-like space-activity and linear space-passage based on urban morphology. He discussed the historical development process by classifying spaces into food production areas, parks and gardens, recreational spaces, plazas, streets, transport facilities, and incidental spaces, based on their function of use [28]. In this way, he discusses the reproduction process of the daily life of the residents and the types of space during the historical development.

Next, the physical environment is characterized by a combination of public space forms, which are mainly focused on accessibility and permeability. The accessibility perspective, which can be said to be a common and valued perspective in recent years, was assessed by Zhou, Lotfi, and Le Texier et al. as the value of accessibility for residents' participation in public green spaces, which is driven by the intrinsic need for recreation and accessibility is strongly related to residents' need for participation [11,29,30]. A. Yavuz's study shows that permeability has positive implications for increasing public participation [31].

Individual-based cognition is also an important perspective in the study of public space. Öztürk, Thirumaran et al. showed that the visual features during movement can attract interest and thus generate memory and identity [32,33]. Mahendra showed that cultural features and symbols have a strong role in visual attraction and can convey the place identity through visuals by extracting elements from the urban forms of India [34]. The lack of visual and physical coherence may lead to a more utilitarian organization, with the concept of function gradually shifting from the external to the internal space. A building often becomes more like a separate part, separated from its environment.

2.3. Methods for Identify Publicness in Urban Space

In addition to social observation and agent-based simulation methods [35,36], the current quantitative analysis methods for urban research are divided into two main categories, one of which is spatial syntactic analysis based on network analysis [37,38]. The other is spatial clustering analysis, based on location [39,40]. Space Syntax is a typical method which was proposed by Bill Hillier based on social network theory and graph theory, where paths are considered as objects to be brought into the network for analysis [41]. Thus, the interpretation in social network theory converts the closeness centrality of the time required for information to be sequentially propagated from each node to all other nodes into paths that represent the time for each node that constitutes the path network to reach the location of the remaining nodes, i.e., path reachability is expressed as integration in Space Syntax [37,41]. The betweenness centrality, which is interpreted in social network theory as the number of times a node acts as a bridge to the shortest path between two other nodes, is represented in Space Syntax as choice, which refers to the location through which a person has a high probability of reaching other paths from the current path [37,41]. This network-based analysis has a very significant role in studying cities. Zeng et al. utilized Space Syntax to conclude that accessibility is a key element of connection types, which can connect above ground, underground, indoors, and outdoors, forming spatial sequences with different thematic and morphological characteristics [42,43]. Nilufar utilized Space Syntax to analyze the accessibility of Dhaka city during its evolution and explored the process of urban core change, while pointing out that the locational advantages of areas with high accessibility are significant [44].

Location-based spatial clustering analysis, mainly based on the distribution distance between locations as weights, generates individual clusters to explain their possible causes according to the phenomenon of heterogeneity in their distribution [39,40,45]. Telega used kernel density estimation for urban functional areas to obtain the potential value of different areas for walking in combination with the accessibility of paths [46]. Delso et al. combined network analysis with kernel density analysis to assess the possibility of pedestrian obstacle impacts to improve walking conditions and accessibility [39].

3. Background: The Evolution of Spatial Organization in Xinjiekou

The site of this research is located in the northeast area of Xinjiekou which is adjacent to Zhongshan Road to the west, Hongwu North Road to the east, Zhongshan East Road to the south, and Changjiang Road to the north (Figure 1). The formation of the northeast area spatial pattern of Xinjiekou is directly related to the opening of Zhongshan Road and Zhongshan East Road in 1928. Guofu road to the north (now Changjiang Road), Hongwu North Road to the east, and the important internal channels such as Tangfang bridge and Qingshi street have always been there since the beginning of development. By 1929, the fundamental layout of the northeast area was established, basically tallying with the time when the pattern of the Xinjiekou area was formed.

The name "Xinjiekou" can be traced back to the early Ming Dynasty (1368s). In 1928, Zhongshan Road, Zhongshan East Road, Zhongshan South Road, and Hanzhong Road were built before Dr. Sun Yat-sen's coffin was brought back to be buried at the mausoleum on Zijin Mountain. At the intersection of these roads, a ring square with a statue of Sun Yat-sen was set up. It was then that the basic pattern of Xinjiekou area took shape. In its 90 years of expansion and development, the area has also been affected by the structural changes of urban morphology (Figure 2).

After the outbreak of the World War II in 1937, urban development entered a state of chaos. Xinjiekou once became an important window for political promotion. From the 1950s to 1980s, the fragmented retail businesses in Xinjiekou were replaced by large department stores, so that a neat and uniform business pattern was initially formed. The evolution of the spatial structure of Xinjiekou at this stage was characterized by the outward expansion of the built-up area. Building density was enhanced from the street to the interior of the parcel, and the proportion of commercial business land grew significantly.

In the 1990s, China began rapid urbanization. The urban fabric of Xinjiekou area radiated to the surrounding area, which marked the subdivision of business office functions and the emergence of the central business district.

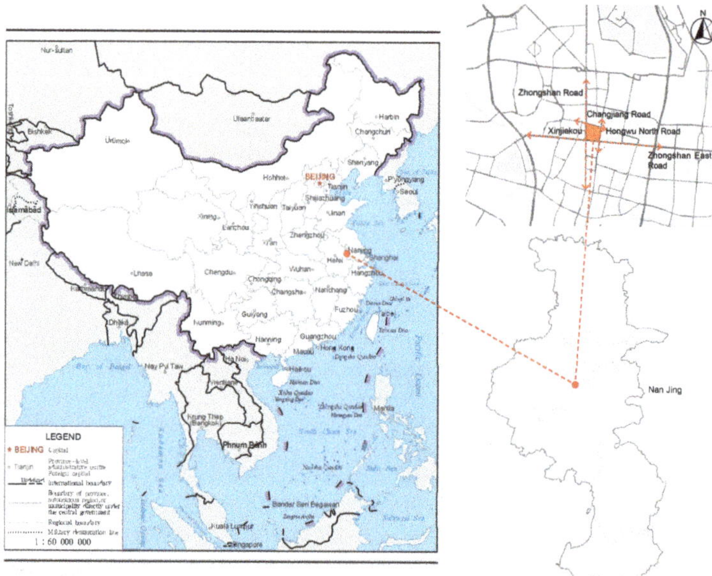

Figure 1. The site is located in north of Xinjiekou which is the central city of Nanjing, Jiangsu Province, China. (Left, China Standard Map from http://bzdt.ch.mnr.gov.cn/, accessed on 25 June 2022).

Figure 2. The spatial structure of Xinjiekou evolved and the public participation of the residents. In the 1920s–1946 it was mainly as a square for walking through, in 1964–1990s the public participation opportunities of the residents increased, and the participation mode was mainly political activities and leisure activities. In the 2000s onwards, motor vehicles occupied the main position and the public participation capacity decreased. (History photos from http://dfz.nanjing.gov.cn/, accessed on 18 April 2021.)

Figure 2 shows the historical changes in the spatial organization of Xinjiekou and its surroundings between 1920s and 2020s. It gradually shifted from a plaza with a high level of public participation to a plaza–pedestrian public space with mainly motorized traffic and indirect participation of the residents, and eventually to a public space for walking only. The residents' participation and identity gradually decreased. The space is gradually segmented and the sense of belonging is gradually fragmented.

4. Research Methodology

As previously described, the public nature of cities is characterized by public participation in everyday life, thus accessibility is theoretically beneficial in shaping the public nature of space and the typology of space is also seen as an effective way to increase the participation of residents [11,29,30]. Therefore, the objective of this study is to obtain correlations between the distribution of people's activities in a site and the organization of accessibility and space-typologies [28].

In order to understand the site more comprehensively, this study is divided into two steps (Figure 3). First, we will analyze the evolution of the spatial structure of the site in order to obtain the spatial accessibility, and this step aims to obtain the potential of public participation at a larger scale. If the sites selected for this study have a high accessibility in the urban evolution and a high and low distribution of task activities within the sites, then zoning the different spatial types and comparing them will be meaningful. The transition from the scale of the context in which the site is located to the site and the local scale of the site will allow us to better understand the association between spatial form and publicness (Table 1).

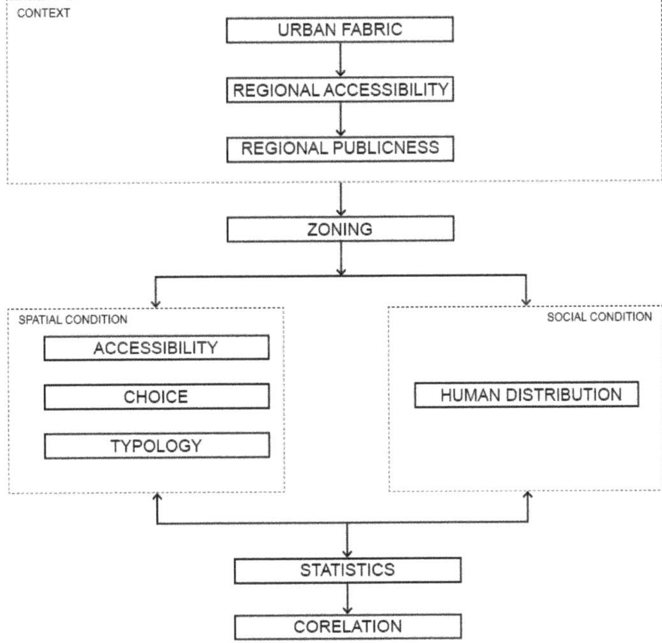

Figure 3. The publicness of public space in Xinjiekou area and the spatial morphology correlation research route.

Table 1. Indicators and definitions of spatial publicness research.

Variables	Indicators	Methods	Definition
Number of participants		Observation	The more people are active in a certain location, the more publicness exists there.
Choice	Betweenness centrality	Space Syntax	The number of shortest paths through a location is calculated, and the higher the number, the higher the betweenness centrality. This indicates the potential likelihood of residents choosing that path.
Integration	Accessibility	Space Syntax	The sum of the distances from one location to all other locations. The smaller the sum, the shorter the path from that location to all other locations. Indicates the accessibility of residents.

The second step is zoning into the site itself. The first step is to observe the scene and map the spatial distribution of people. In this study, observation points were identified and photographed and counted every 50 m along the paths within the site. The more people are active in a certain location, the more publicness is demonstrated. Secondly, we classify the public spaces in the site and analyze their intermediate centrality. The gentle transition of the intermediate centrality of the spatial form indicates good connectivity and a coherent spatial experience for the residents, which is conducive to the creation of publicness in the space. Secondly, the accessibility of different spaces is analyzed to show the potential of residents' participation. Finally, the correlation between the number of public participants and the mediated centrality and accessibility is determined statistically.

Since this project is based on a positive design, we compare accessibility and mediated centrality before and after the design to predict the effect of spatial design on reshaping the publicness of the space. The space and its adjacent spaces will become more coherent and accessible, thus shaping an urban space that has the potential to become a public realm.

In analyzing accessibility and mediated centrality metrics, this study uses Space Syntax. Space Syntax, proposed originally by Professor Hillier, aims to systematically depict the interaction between individuals and their society and environment. The heterogeneity of space and everyday life is described by quantifying the configurational characteristics and scales of urban space [37,41]. The statistical method uses Geographically Weighted Regression (GWR) to determine the degree of fitness and confidence of the variables. GWR, proposed by Professor Fotheringham, can reduce the influence of location distribution on the statistical results and show the correlation characteristics of variables in a local group compared to the traditional ordinary least squares [47,48].

5. Results and Discussion

5.1. Context of the Spatial Organization of Xinjiekou

As mentioned earlier, the spatial organization of Xinjiekou underwent a politicized and commercialized transformation during the historical development of the city's central district. The newly opened road in 1928 gradually became the center of the entire city's central district. The results of the analysis according to Space Syntax reflect the gradual strengthening of the INTEGRATION of the northeast part of Xinjiekou from 1898 to 1933 to 1945 (Figures 2 and 4). This indicates that the physical accessibility of the Xinjiekou area gradually improved, allowing more access for city residents. At the same time, the location of Xinjiekou has maintained a high accessibility from 1945 to 2012, and the historical images in Figure 4 show that Xinjiekou has had a high level of public participation for a long time. The square of Xinjiekou has also become an important public space for political demonstrations and rallies.

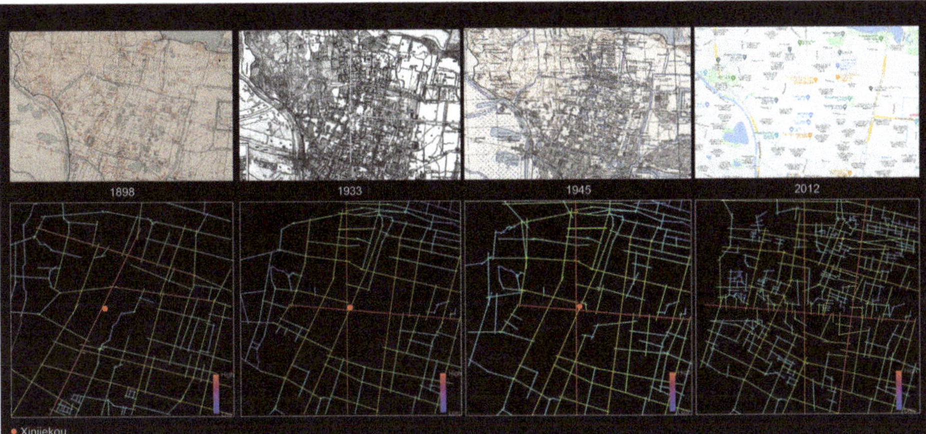

Figure 4. The evolution of the spatial structure of Xinjiekou on its spatial accessibility from 1898s to 2012s. (History map from http://dfz.nanjing.gov.cn/, accessed on 10 July 2022, digital map from https://www.google.com.hk/maps, accessed on 15 July 2022).

The accessibility of Xinjiekou and its context indicates the possibility of public participation on a regional scale, i.e., in a location that is easily accessible to the public and where daily activities are frequent. However, the scale of this macro-analysis is too large, ignoring the impact of local space shaping on the individual participants and on the cognition of human scale. It is meaningful to discuss the public participation in Xinjiekou base in this context, because the high accessibility and high participation attributes make this study representative of a post-zoning analysis of its spatial characteristics [4,11].

5.2. Correlation of Public Participation with Choice and Integration in the Northeast Area of Xinjiekou

As mentioned in the research methodology, we first made a comprehensive observation of the site after zoning to the Xinjiekou area. As shown in the Figures 5 and 6, a total of 146 observation points were selected for this study, and the location and number of people passing by were photographed to record their activities. The original space of Xinjiekou after the simple renovation in 2012 was a mixed-use pattern of commercial building space, residential buildings, and traditional residential houses.

In this study, we divided public spaces into two categories. The first category was path, a type of space in which the public participates by walking, with little stopping, and the width of the path is within 3–10 m and the length is between 50–500 m. The location is usually on one side of the building with a small amount of landscape. The second type of space is node, where the public participates in more diverse ways, socializing, stopping, resting, etc. Its size is 50 m × 50 m or more, which can create plazas, gardens, etc.

By observation, the spatial ratio is partially imbalanced, with commercial building spaces large and orderly, and residential buildings small and scattered, and the spatial sequence is in a fractured state. The commercial and residential spaces are self-contained and lack integrity. The public space connecting the different physical spaces is reduced to a non-architectural public domain, where residents rarely stop, walk, and move around, and have a low sense of belonging. From the diagram, we can see that the distribution of characters is mainly located in the central paths of the site and at the entrances and exits of some of the plots. At the same time the distribution of locations of higher public participation is disorderly and discontinuous.

At the same time, we analyzed the integration and choice of the site (Figure 6). According to the analysis results, we can see that the paths with higher choice are located around the base, and the overall choice of the site is low, while the center is special as a bridge between residential and commercial areas with higher choice. The accessibility analysis results show that, except for the residential paths with lower accessibility, the accessibility of the rest of the commercial locations is more homogeneous.

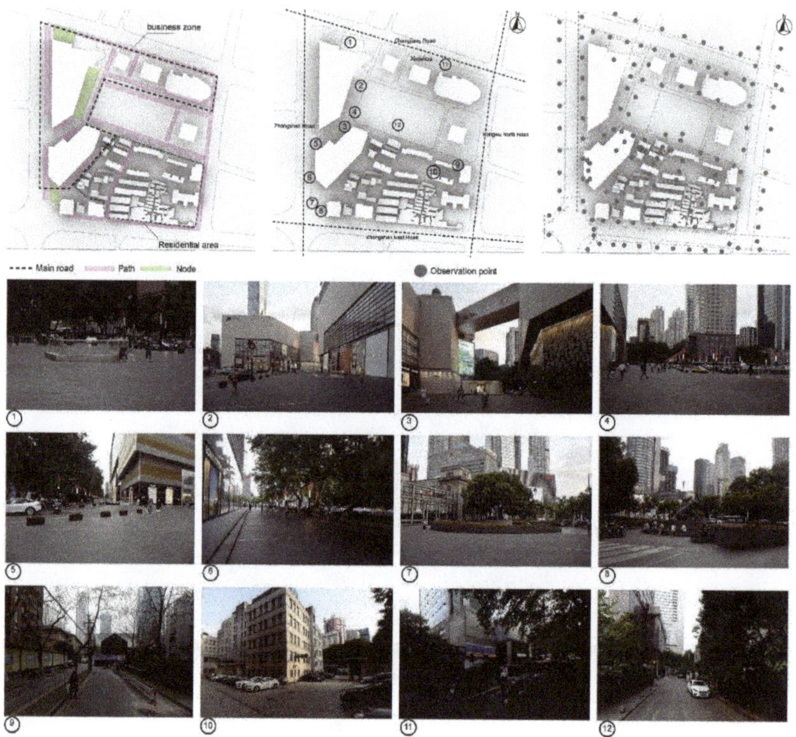

Figure 5. Top left, the distribution of public space types in the site with the distribution of building functions. Upper middle, index of actual photos of the site. Upper right, location of observation points for recording the number of public participants. Bottom, actual photos of the site.

Figure 6. On the left, the distribution of the number of residents' public participation. In the middle, the analysis of the spatial syntax on the choice value of urban morphology. On the right, the spatial syntax analysis of the integration value of urban form.

Based on this, we used GWR to analyze the correlation between the number of public participants, choice, and integration (Table 2 and Figure 7). At the same time, we can see that the R2 of the model fit is 0.392, and the fit of the model constructed by choice, integration, and the number of participants is relatively average. The number of effective parameters (ENP) is 13.801, which indicates that there are still variables not considered in the model except for choice and integration. And the p value of the variable is less than 0.01, indicating that the analysis result is very important. Meanwhile, the overall effect of choice on public participation is positively correlated, while the overall effect of integration on public participation is negatively correlated, which is different from what we usually think. This (Figures 6 and 7) suggests that in small-scale neighborhoods, the coherence of the paths will drive public participation, while the accessibility of the paths does not drive residents to participate in the activities on the site, thus making the public space lose its publicness.

Table 2. Diagnostic information of the statistical model of the number of public participants based on GWR and CHOICE, INTEGRATION.

Number of observations:		146
AICc:		374.831
R2:		0.392
Effective number of parameters (trace (S)):		13,801
Residual sum of squares:		88.739
p-value	Choice	0
	Integration	0.004
Parameter Estimates	Choice	0.374
	Integration	−0.257

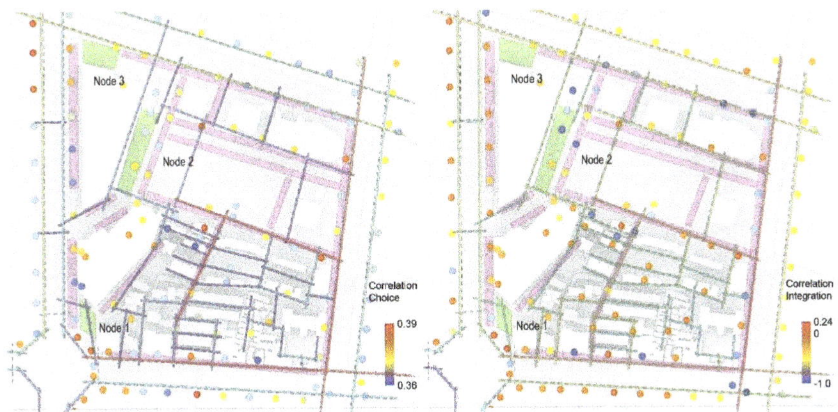

Figure 7. Distribution of the correlation between public participation, and choice and integration.

In addition, among the spatial types of paths (Figures 6 and 7), the overall performance is that high accessibility presents a positive correlation with public participation when the paths with higher choices are connected. Accessibility presents a negative correlation with public participation when the paths are not connected. In the spatial type of node, choice–integration correlation is consistent with node correlation, so the strategy to promote public participation in the spatial type of node should be adopted to increase choice–integration in parallel.

5.3. Pedestrian Flow Design Strategy for Xinjiekou Public Space

The design reconfigured the pedestrian flow based on the correlation between public participation and choice and integration as summarized by the site. We retraced the choice and integration of traffic on the site from 2005 to 2012, and the overall spatial organization remained unchanged except for the closure of pedestrian access around the road intersection (Figure 8). The choice of pedestrian flow in the site is locally enhanced and the paths of high choice are not coherent, and the accessibility of pedestrian flow in the site has been improved. Based on the previous analysis, we can understand that in this case the public participation of the residents will be weakened, thus weakening the public nature of the public space of the site.

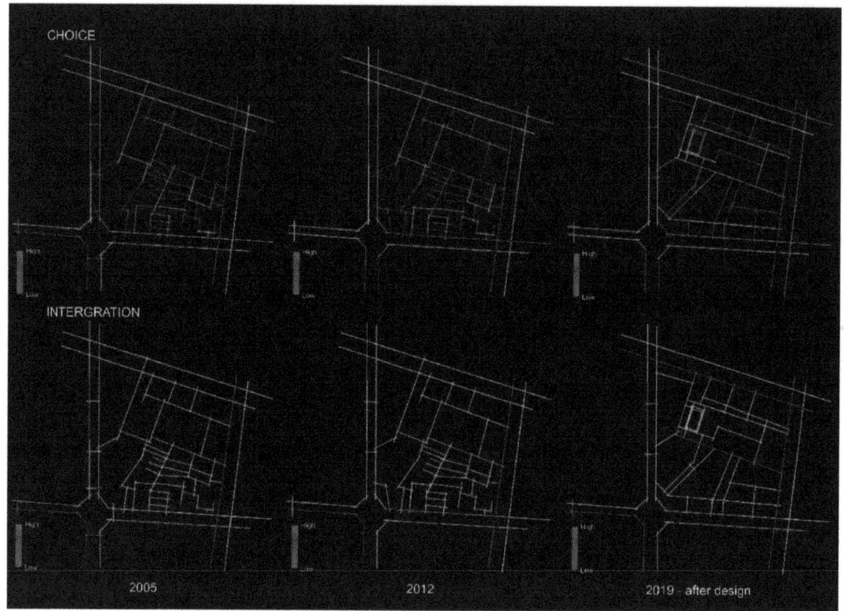

Figure 8. Spatial accessibility and choice analysis, from 2005 to 2012 to 2019.

Based on the above analysis, we can understand that the public participation of the residents will be weakened in this case, thus weakening the public space of the site. According to the analysis results, we can find that the choice-value of the pedestrian flow is higher and the paths with high choice-value are more coherent. At the same time, the integration of these paths has also improved significantly. Based on the previous analysis, we believe that the public space in this case has increased the potential for residents to participate in the site, thus increasing the public nature of the public space.

In conclusion, this study developed an urban design strategy based on field observations and quantitative analysis of Space Syntax. We created the publicness of public spaces in urban centers by developing situational strategies based on an understanding of the evolutionary characteristics of urban organization. In addition to satisfying functional use, it increases the possibility of interaction and communication between residents.

6. Conclusions

Many of China's urban design strategies tend to position themselves in terms of commercial value and neglect the creation of the human element of urban life. This has led to increasing social separation and a loss of the sense of urban belonging. This has had a negative impact on both the city and its residents. We believe that full awareness of the potential needs of society and future development is an important prerequisite for design. In this study, we also argue

Sadeghi, Abarkan, and Yang et al. that urban form explains to some extent the logic of urban function, that spatial-typology interacts with social conditions and can provide for the shaping of public life and publicness of space [24,25,41].

In this study, we first examined Xinjiekou in its contexts through the Space Syntax method to verify whether the smaller area selected for the study was representative of the study framework. Second, through the quantitative analysis of Space Syntax and social observations, it was determined that public participation tends to choose walking paths with a higher number of shortest paths through a location and that these locations are interconnected. Under this condition, accessibility has a catalytic effect on public participation. If these high-choice walking paths are not connected, residents are less likely to choose such areas for their activities. Later in the design, we organized the pedestrian circulation in the site based on the findings of the analysis, thus contextualizing, and enhancing, the possibilities and potential for future residents to participate in the daily life of the public space.

In this paper, the study provides a pre-feasibility hypothesis for the design that combines the correlation between the quantitative analysis of spatial morphology and the spatial distribution characteristics of social phenomena to construct a public participation-oriented cognitive framework of urban morphology. The method not only effectively delineates the interaction process of minute forms to the overall human–spatial, but also has a statistical-based verification process with high interpretability and confidence. The designs generated based on this framework are expected to facilitate the evolution of public space into the public realm upon completion. We hope that this combination of observation and quantitative analysis will validate and optimize design theory in anticipation of enhanced placemaking in urban public spaces to improve the lives of residents.

The current study has the following shortcomings that we hope to explore in future research.

1. This study only considered the analysis of integration and choice of public spaces while the number of potential variables exceeded the number of predetermined variables by a large margin according to the statistical diagnosis results, which indicates that other elements in the built environment still influence residents' choices. For example: different services and functions;

2. The fitness between the variables in this statistical model in the study is not excellent; based on experience, it is possible that the spatial distribution of distance has an impact on the analysis results, and later studies will consider optimizing the impact of location distance on the model fitness.

Author Contributions: Conceptualization, M.P.; methodology, M.P.; software, Y.S.; investigation, M.P., Y.S. and Q.J.; resources, Q.Z.; data curation, Q.Z. and Y.L.; writing—original draft preparation, M.P.; writing—review and editing, M.P.; visualization, M.P.; supervision, Q.Z. and Y.L.; project administration, Q.Z. and Y.L.; funding acquisition, Q.Z. and Y.L. All authors have read and agreed to the published version of the manuscript.

Funding: This research was supported by the Architects and Engineers Associates of Southeast University (No. 51778123).

Institutional Review Board Statement: Since this study did not involve personal privacy nor questionnaires, etc., and only the authors' own observations were used as data collection, ethical approval was not applicable.

Informed Consent Statement: Informed consent was obtained from all subjects involved in the study.

Data Availability Statement: Not applicable.

Conflicts of Interest: The authors declare no conflict of interest.

References

1. The UIA Beijing Charter. *J. Archit.* **2000**, *5*, 1–8. [CrossRef]
2. Arentze, T.A.; van der Waerden, P.J.H.J.; Bergen, J.W.; Timmermans, H.J.P. Measuring the Quality of Urban Environments: A Need-Based Micro-Simulation Approach. *Appl. Spat. Anal. Policy* **2009**, *2*, 195–209. [CrossRef]
3. Afonso, A.G.; Ergin, E.; Fatah gen. Schieck, A. Flowing Bodies: Exploring the Micro and Macro Scales of Bodily Interactions with Urban Media Installations. In Proceedings of the 2019 on Designing Interactive Systems Conference (DIS '19), San Diego, CA, USA, 23–28 June 2019; Association for Computing Machinery: New York, NY, USA, 2019; pp. 1183–1193. [CrossRef]
4. Ye, C.; Hu, L.; Li, M. Urban green space accessibility changes in a high-density city: A case study of Macau from 2010 to 2015. *J. Transp. Geogr.* **2018**, *66*, 106–115. [CrossRef]
5. Al-Kodmany, K. Placemaking with tall buildings. *Urban Des. Int.* **2011**, *16*, 252–269. [CrossRef]
6. Friedmann, J. Place and Place-Making in Cities: A Global Perspective. *Plan. Theory Pract.* **2010**, *11*, 149–165. [CrossRef]
7. Gu, K.; Li, Y.; Zheng, X. A typological approach to planning. *J. Urban. Int. Res. Placemaking Urban Sustain.* **2019**, *12*, 373–392. [CrossRef]
8. Jacobs, J. *Death and Life of Great American Cities*; Random House: New York, NY, USA, 1961; p. 474.
9. Najafi, P.; Mohammadi, M.; Le Blanc, P.M.; van Wesemael, P. Insights into placemaking, senior people, and digital technology: A systematic quantitative review. *J. Urban.* **2022**, 1–30. [CrossRef]
10. Gehl, J.; Matan, A. Two perspectives on public spaces. *Build. Res. Inf.* **2009**, *37*, 106–109. [CrossRef]
11. Le Texier, M.; Schiel, K.; Caruso, G. The provision of urban green space and its accessibility: Spatial data effects in Brussels. *PLoS ONE* **2018**, *13*, e0204684. [CrossRef]
12. Kropf, K. Urban tissue and the character of towns. *Urban Des. Int.* **1996**, *1*, 247–263. [CrossRef]
13. Banerjee, T. The Future of Public Space: Beyond Invented Streets and Reinvented Places. *J. Am. Plan. Assoc.* **2001**, *67*, 9–24. [CrossRef]
14. Knox, P.L. Creating Ordinary Places: Slow Cities in a Fast World. *J. Urban Des.* **2005**, *10*, 1–11. [CrossRef]
15. Miao, P. Brave New City: Three Problems in Chinese Urban Public Space since the 1980s. *J. Urban Des.* **2011**, *16*, 179–207. [CrossRef]
16. Wu, F. Planning centrality, market instruments: Governing Chinese urban transformation under state entrepreneurialism. *Urban Stud.* **2018**, *55*, 1383–1399. [CrossRef]
17. Lefebvre, H.; Nicholson-Smith, D.; Lefebvre, H. *The Production of Space*; Blackwell Publishing: Malden, MA, USA, 1991; ISBN 978-0-63118-177-4.
18. Resmini, A.; Rosati, L. Place-making. In *Pervasive Information Architecture*; Elsevier: Amsterdam, The Netherlands, 2011; pp. 63–87. ISBN 978-0-12382-094-5.
19. Stedman, R.C. Is It Really Just a Social Construction?: The Contribution of the Physical Environment to Sense of Place. *Soc. Nat. Resour.* **2003**, *16*, 671–685. [CrossRef]
20. Lang, J. The Place Shaping Continuum: A Theory of Urban Design Process. *J. Urban Des.* **2014**, *19*, 2–36. [CrossRef]
21. Lynch, K. *The Image of the City*; Publication of the Joint Center for Urban Studies; M.I.T. Press: Cambridge, UK, 1960; ISBN 978-0-26212-004-3.
22. Akkar, M. The changing 'publicness' of contemporary public spaces: A case study of the Grey's Monument Area, Newcastle upon Tyne. *Urban Des. Int.* **2005**, *10*, 95–113. [CrossRef]
23. Kulesza, M. Conzenian Tradition in Polish Urban Historical Morphology. *Eur. Spat. Res. Policy* **2015**, *21*, 133–153. [CrossRef]
24. Sadeghi, G.; Li, B. Urban Morphology: Comparative Study of Different Schools of Thought. *Curr. Urban Stud.* **2019**, *7*, 562–572. [CrossRef]
25. Abarkan, A. Typo-morfologi: Metoden och dess tillämpning på bebyggelsesmönster. *Nordic J. Archit. Res.* **2013**, *13*, 1–2.
26. Yang, Y.; Kossak, F. Building Typology as a Generator of Urban Form for Urban Design Project. *Cities Assem.* **2022**, *1*, 147–159. [CrossRef]
27. Yılmaz Çakmak, B.; Topçu, M. An evaluation of urban open spaces in Historical City Center of Konya in the context of pedestrian mobility. *J. Hum. Sci.* **2018**, *15*, 1827–1846. [CrossRef]
28. Stanley, B.W.; Stark, B.L.; Johnston, K.L.; Smith, M.E. Urban Open Spaces in Historical Perspective: A Transdisciplinary Typology and Analysis. *Urban Geogr.* **2012**, *33*, 1089–1117. [CrossRef]
29. Zhou, X.; Parves Rana, M. Social benefits of urban green space: A conceptual framework of valuation and accessibility measurements. *Manag. Environ. Qual. Int. J.* **2012**, *23*, 173–189. [CrossRef]
30. Lotfi, S.; Koohsari, M.J. Analyzing Accessibility Dimension of Urban Quality of Life: Where Urban Designers Face Duality Between Subjective and Objective Reading of Place. *Soc. Indic. Res.* **2009**, *94*, 417–435. [CrossRef]
31. Yavuz, A.; Kuloğlu, N. Permeability as an indicator of environmental quality: Physical, functional, perceptual components of the environment. *World J. Environ. Res.* **2014**, *4*, 13.
32. Öztürk, S.; Işinkaralar, Ö.; Ayan, E. Visibility Analysis in Historical Environments: The case of Kastamonu Castle and its Surrounding. *J. Curr. Res. Soc. Sci.* **2018**, *8*, 405–412.
33. Thirumaran, K.; Balaji, G.; Prasad, N.D. (Eds.) *Sustainable Urban Architecture: Select Proceedings of VALUE 2020*; Lecture Notes in Civil Engineering; Springer: Singapore, 2021; Volume 114, ISBN 978-9-81159-584-4.

34. Mahendra, I.M.A.; Paturusi, S.; Dwijendra, N.K.A. The meaning of local culture elements and urban elements as forming the identity of the klungkung urban area, Bali, Indonesia. *PalArch's J. Archaeol. Egypt/Egyptol.* **2020**, *17*, 11563–11580.
35. Batty, M.; Desyllas, J.; Duxbury, E. The discrete dynamics of small-scale spatial events: Agent-based models of mobility in carnivals and street parades. *Int. J. Geogr. Inf. Sci.* **2003**, *17*, 673–697. [CrossRef]
36. Liu, H.; Silva, E.A.; Wang, Q. *Creative Industries and Urban Spatial Structure: Agent-Based Modelling of the Dynamics in Nanjing*; Springer International Publishing AG: Berlin, Germany, 2015.
37. Hillier, B.; Tzortzi, K. Space Syntax. In *A Companion to Museum Studies*; Macdonald, S., Ed.; Blackwell Publishing Ltd.: Malden, MA, USA, 2006; pp. 282–301. ISBN 978-0-47099-683-6.
38. Netto, V.M. 'What is space syntax not?' Reflections on space syntax as sociospatial theory. *Urban Des. Int.* **2016**, *21*, 25–40. [CrossRef]
39. Delso, J.; Martín, B.; Ortega, E. A new procedure using network analysis and kernel density estimations to evaluate the effect of urban configurations on pedestrian mobility. The case study of Vitoria–Gasteiz. *J. Transp. Geogr.* **2018**, *67*, 61–72. [CrossRef]
40. Yang, J.; Zhu, J.; Sun, Y.; Zhao, J. Delimitating Urban Commercial Central Districts by Combining Kernel Density Estimation and Road Intersections: A Case Study in Nanjing City, China. *ISPRS Int. J. Geo-Inf.* **2019**, *8*, 93. [CrossRef]
41. Freeman, L.C. Centrality in social networks conceptual clarification. *Soc. Netw.* **1978**, *1*, 215–239. [CrossRef]
42. Zeng, R.; Shen, Z. Post-Occupancy Evaluation of the Urban Underground Complex: A Case Study of Chengdu Tianfu Square in China. *J. Asian Archit. Build. Eng.* **2022**, 1–16. [CrossRef]
43. Zhang, S.; Yu, P.; Chen, Y.; Jing, Y.; Zeng, F. Accessibility of Park Green Space in Wuhan, China: Implications for Spatial Equity in the Post-COVID-19 Era. *Int. J. Environ. Res. Public Health* **2022**, *19*, 5440. [CrossRef]
44. Nilufar, F. *Urban Morphology of Dhaka City: Spatial Dynamics of Growing City and the Urban Core*; Asiatic Society of Bangladesh: Dhaka, Bangladesh, 2010.
45. Long, Y.; Gu, Y.; Han, H. Spatiotemporal heterogeneity of urban planning implementation effectiveness: Evidence from five urban master plans of Beijing. *Landsc. Urban Plan.* **2012**, *108*, 103–111. [CrossRef]
46. Telega, A.; Telega, I.; Bieda, A. Measuring Walkability with GIS—Methods Overview and New Approach Proposal. *Sustainability* **2021**, *13*, 1883. [CrossRef]
47. Shabrina, Z.; Buyuklieva, B.; Ng, M.K.M. Short-Term Rental Platform in the Urban Tourism Context: A Geographically Weighted Regression (GWR) and a Multiscale GWR (MGWR) Approaches. *Geogr. Anal.* **2021**, *53*, 686–707. [CrossRef]
48. Oshan, T.; Li, Z.; Kang, W.; Wolf, L.; Fotheringham, A. mgwr: A Python Implementation of Multiscale Geographically Weighted Regression for Investigating Process Spatial Heterogeneity and Scale. *ISPRS Int. J. Geo-Inf.* **2019**, *8*, 269. [CrossRef]

Article

Structural Landmark Salience Computation in Compact Urban Districts with 3D Node-Landmark Grid Analysis Model: A Case Study on Two Sample Districts in Changsha, China

Yang Guo [1,2], Xijun Hu [1,2,*] and Jia Tang [1,2]

[1] Department of Landscape Architecture, Central South University of Forestry and Technology, Changsha 410004, China
[2] Hunan Big Data Engineering Technology Research Center of Natural Protected Areas Landscape Resources, Changsha 410004, China
* Correspondence: t20040191@csuft.edu.cn; Tel.: +86-137-5518-3379

Abstract: Mastering the relationship between urban landmarks and urban space morphology in urban planning, landscape planning, and architectural design helps maintain the intelligibility of compact urban districts. The objective of the present study was to numerically determine the structural salience of various landmarks in an urban environment and use it to interpret the intelligibility of the city. Combining the measurement method of 3D visibility and the related principles of space syntax, this study develops a new 3D Node–Landmark Grid Analysis Model (3D NL GAM) for structural salience computation of urban landmarks. In this study, a numerical approach is used to construct a 3D simulation model. Firstly, the visibility of each decision node to landmarks in an urban environment, using a 3D digital model, is measured using the 3D isovist component of Rhinoceros and Grasshopper software. Secondly, links among wayfinding decision nodes and landmarks are established to form a 3D NL GAM. The normalized angular integration of decision nodes and the normalized angular choice of landmarks are computed using the principle of space syntax. Thirdly, the structural salience of landmarks is determined with a function of landmark visibility, spatial properties of landmarks, and wayfinding decision nodes. Finally, a case study was carried out by using a 3D NL GAM to analyze three types of urban areas located in Changsha. The results indicated that large-scale natural landscapes have a higher structural salience among the types of landmarks. The structural salience of architectural landmarks in the combined spatial form of combining tall and low building groups has a clear advantage over the form dominated by high-rise building groups. Raising the height of landmark buildings can modify the structure of the grid analysis model and improve the people aggregation of urban space. The 3D NL GAM can quantify the spatial properties and landmark structural salience of a city and can effectively assist in the evaluation of the intelligibility of built or future urban environments.

Keywords: structural salience; landmark visibility; space syntax; 3D isovist; node–landmark grid analysis model

Citation: Guo, Y.; Hu, X.; Tang, J. Structural Landmark Salience Computation in Compact Urban Districts with 3D Node-Landmark Grid Analysis Model: A Case Study on Two Sample Districts in Changsha, China. *Buildings* **2023**, *13*, 1024. https://doi.org/10.3390/buildings13041024

Academic Editors: Michael J Ostwald and Ju Hyun Lee

Received: 24 February 2023
Revised: 4 April 2023
Accepted: 11 April 2023
Published: 13 April 2023

1. Introduction

Compact cities are an important means of sustainable development and are conducive to the intensification of urban resources, but they also present higher requirements for urban planning, landscape planning, and architectural design [1–3]. UN-Habitat advocates moderately compact and high-density cities, and suggests that the population density of future urban planning should be greater than 150 people per hectare [4]. This has led to higher population densities, mixed-function land use, more centralized public facilities, compact residential built communities within a given urban area, and the development of vertical space. "Intelligibility" plays a crucial role in the quality of urban space [5].

However, high density and high-rise clusters in compact cities tend to block views in urban spaces, which affects the permeability of urban spaces and undermines urban intelligibility. Therefore, intelligibility in urban environments has become an important topic in urban design for compact cities. In order to ensure the quality of urban space and enhance urban intelligibility, urban designers need to think about the following questions: (1) how to obtain comfortable perception density in urban street space [6–8], (2) ensuring the synergy of landmarks, (3) how to garner the necessary landscape resources, (4) the synergy between specific location and the overall environment [9,10], and (5) other urban image elements in relation to urban residents' activities. Among them, where clarifying the structural relationship of landmarks in compact urban areas, accurately describing the structural salience of landmarks in a compact urban environment is a key issue for enhancing urban intelligibility.

Urban intelligibility is the original goal of urban imagery in terms of how people orient themselves and navigate the city [5,11]. People's actions are governed by their visual perception and influenced by the characteristics of urban imagery. In essence, urban imagery describes the structural salience of the city's elements in relation to each other [9,12]. The integration of the five elements of urban imagery reflects the behavior of people moving in the city, i.e., for a person moving in an urban area, this is the process of perceiving and acting in this area, with boundaries determining the scope of analysis, paths as channels, nodes as decision points for action, and landmarks as references. Therefore, the structural salience of landmarks is based on the integration of nodes, paths, areas, and boundary forms, expressing the role of landmarks in the spatial structure of urban areas and their relationship with other elements [13].

The structural salience of landmarks has substantial implications for wayfinding and navigation in cities [13–16]. Caduff et al. generated landmark route navigation using a landmark saliency-weighted path algorithm and compared landmark routes with shortest-distance routes, which demonstrates the advantage of landmark routes for navigating route selection in most cases [17,18]. Simulation analysis agent-based modeling has shown [19] that the route selection model containing landmarks provides directional guidance and environmental characteristic variability, thus expressing more urban information. That is, the urban information conveyed by landmarks is related to the visibility of landmarks, and the analysis of landmarks in urban areas should consider not only neighboring but also distant landmarks. In terms of the location relationship between landmarks and pathfinders, landmarks are usually classified into global and local landmarks [15,20]. Global landmarks refer to landmarks that can be seen from a large distance, and a global reference system, such as the efficacy of a compass in landmark routes. Local landmarks can only be seen from a small distance; they are only visible in a limited area and only from a specific method [15,20]. Studies have shown that both local and global landmarks are used for decision making [15,20–22].

However, there has been a lack of quantitative characterization metrics and scientific analytical methods to describe the structural salience of landmarks in urban spaces [23]. In terms of landmarks and human wayfinding needs, landmarks are defined as "geographic objects that constitute human mental representations of space" [24], and landmark representations are the result of the local integration of visual, motor, and spatial information [25]. Klippel and Winter determined the structural salience of landmarks from the semantic expression of the positional relationship between landmarks and navigation routes. This is a kind of qualitative research method focused on navigation route arrangement [26], where it is difficult to form a difference comparison for the structural salience of landmarks. Winter also proposed an extension of the advance visibility assumption, in which landmarks can be identified early along the roadway with valid landmark structural saliency [27], but does not provide a measure of visibility in complex 3D urban environments and miscellaneous road grids. Therefore, the structural salience of landmarks in urban environments is a unique property of the cognitive and physiological trilateral relationship between the landmark feature itself, its surroundings, and the observer's perspective [11,14,28].

Space syntax theory provides methods for measuring spatial node accessibility and directional guidance in cities [29]. Claramunt and Winter adopted the method of mathematical graph theory and space syntax to analyze the saliency of regions, boundaries, streets, and nodes and formed a structural model of urban element recognition [12,30]. It has been shown that the principles of space syntax are suitable for the structural description of urban spatial forms, and quantitative analytical results can be obtained. However, this analytical approach does not deal with distant landmarks on the one hand, nor does it consider the impact of the visibility of nearby landmarks on the urban structure model. Therefore, based on the two-dimensional limitation of urban grids, space syntax cannot adequately describe landmarks [31]. Enhancements based on 3D isovist generation techniques have promoted the development of spatial syntactic analysis toward 3D spatial analysis. These have been realized in spatial analyses regarding the measurement of urban spatial openness [32], the spatial morphological characteristics [33], as well as visual permeability measurements [34]. Detailed representations of spatial visibilities have been implemented in these studies, but they have not been synchronously implemented for the visibility of a particular object.

The above literature gives an important insight that if a 2D grid analysis model for an urban network can be developed into a 3D grid analysis model, then the objective of accurately quantifying the structural saliency of landmarks is possible. Two conditions must be satisfied to achieve this objective. First, visibility measurements of urban landmarks are implemented for all the wayfinding decision nodes in a 3D environment. Second, the spatial properties of landmark locations and the spatial properties of wayfinding decision nodes are implemented for directional guidance and spatial accessibility. With the improvement of computer hardware and software technology, the current measurement technology for both 3D digital models of cities and visibility data of spatial places has been significantly improved [35,36]. Meanwhile, there is a mature research paradigm for analyzing the relationship between urban spaces using space syntax [37,38]. In order to achieve these two fundamental conditions, the next section is devoted to a literature review on the construction of a grid analysis model based on space syntax, and the presentation of a 3D isovist.

Based on the literature and analysis, this study develops a 3D Node–Landmark Grid Analysis Model (3D NL GAM) to incorporate the 3D visibility data of landmarks into the spatial syntactic analysis model. In the context of compact cities, this paper aimed to clarify the integrated relationship between landmarks and textural features in urban areas. In particular, the relationship between natural landscape resources, street grids, the combined forms of different architectural groups, and the landmark's own form was examined and used to guide the evaluation of urban intelligibility and the urban design.

2. Literature Review

2.1. Grid Analysis Model Based on the Principle of Space Syntax

Space syntax is a scientific reconstruction of how people interact with their environment. This helps to understand the relationship between humans and their environment by creating a model that connects real spaces with human cognitive structures through logical spaces (employing signs, symbols and representations to create schematized multidimensional spaces) [39]. After more than 50 years of development, space syntax has become a paradigm for spatial research, and is linked to other disciplines or research fields [40]. Using the principles of space syntax, the city is understood as a network formed by space, thus establishing a link between human behavior and network science in the urban environment [41,42]. The analytical method of space syntax can be used to analyze the urban street networks, functional layouts, data on urban points of interest, as well as the structure of urban centers and other mobile features of human agglomeration and dispersion in the urban environment [43–46].

Traditional grid analysis models for space syntax include convex map analysis [47], axial map analysis [48], and Visibility Graph Analysis (VGA) [49,50]. An urban network is used to abstractly express the location, connectivity relationships, and the structure of the city's nodal spaces in a region-wide system, as well as to form a corresponding structural form with various types of analytical models [51–53]. The approach to grid construction is consistent with the principles of mathematical graph theory, i.e., a grid graph is composed of nodes and lines, where nodes correspond to the vertices of the graph, in mathematical graph theory, and lines correspond to the edges of the graph [30]. Thus, there are two core tasks for constructing an analytical model:

1.   Investigating as to how the logical space expresses the real space of the city;
2.   Establishing links between spatial nodes.

Traditionally, there are three ways to construct a grid. The first way is to consider streets as spatial nodes, with axes acting as the logical space of streets and intersections between the streets acting as connecting edges, which is used to calculate the motion properties of street space in the form of an axial analysis map [29]. The second way treats street intersections and turning points as spatial nodes and street segments as connecting edges, which is then used to analyze the spatial movement properties of street intersections in the form of convex analysis maps [12,30]. The third way belongs to VGA. First, isovist is employed to express a field of view; an isovist is the set of all visible objects of a viewpoint formed according to set viewpoint parameters [54]. Essentially, an isovist is 3D, and only a cross-section containing the viewpoint is taken in the viewable analysis model construction, i.e., a 2D isovist [54,55]. The grid spacing as a raster is filled in the view plane of the urban space, and a cell of the raster represents a node, which is the logical space of the grid. The connectivity between nodes is determined by applying the visibility principle of isovists to determine the accessibility between nodes. All nodes contained in the 2D isovist generated by a node are neighbors of that node, and the domains are connected to each other so that they can be used to analyze the spatial mobility properties of all nodes in the urban planar space.

The essence of grid analysis model construction is to reflect the relationship between spaces in as simple and accurate way as possible [56]. Batty has compared the above three grid analysis models, where the setting of the axis and the convex is set manually with some subjectivity, while the raster can be automatically generated after setting the accuracy [56]. Based on the VGA analysis model, it can automatically generate more axial analysis maps consisting of axes, while the convex itself is a subset of the raster plane, and the removal of some unwanted cells becomes the convex analysis map [56]. Therefore, the second method describe can be understood as a simplified version of the traditional VGA graph, where the intersections are located on the same linear street and close elevations are all neighborhood nodes. That is, n intersections on the same linear road are neighbors of each other, and the grid is a graph with a combination of n vertices and C_n^2 bar edges.

Integration and choice express the motion properties of spatial nodes. The integration originates from the concept of node closeness centrality in graph theory, i.e., the smaller the cumulative value of the distance from the point to all other points, the more it indicates that the node is close to the center in the system [12,30]. The integration is expressed by the space syntax as the reciprocal of the depth of a spatial node relative to other spaces; therefore, the integration reflects the structural relationship of a node with others [57]. The higher the integration of a node, the better the accessibility, and thus the easier it is for a moving population to cluster or stay at that spatial node. This choice originates from the probability of a node intervening in the shortest path between two nodes in graph theory, and is referred to, in graphs, as betweenness centrality [12]. Space syntax theory uses choice to analyze the degree of movement through a node and is a powerful tool for predicting the potential of pedestrian and vehicular movement. Spaces with high global choice records lie on the shortest paths from all origins to all destinations [29]. The choice usually describes the movement through a place rather than possession or a stay.

In order to compare the spatial property metrics of different urban areas, a normalized approach is used to calculate the spatial properties of nodes [46,58,59]. Earlier space syntax theory was usually limited to the comparison of nodes within the same grid system. However, in urban design, it is often necessary to compare the differences between the spatial attributes of different areas of the city, or those between similar nodes of different cities; therefore, the space syntax method incorporates the total depth and total number of nodes of the grid system into the integration degree calculation, resulting in the Normalized Angular Integration (*NAIN*) [58], which is calculated as follows:

$$NAIN = \frac{(Nc)^{1.2}}{TD_i} \tag{1}$$

where *NAIN* is the normalized angular integration degree, N_c is the total number of nodes in the urban grid, and TD_i is the total topological distance from node i to all other nodes [58].

Similarly, the Normalized Angular Choice (*NACH*) is described using the relationship between the ratio of selectivity and total depth of that node in the grid system, as given by:

$$NACH = \frac{\log(Ch_i + 1)}{\log(TD_i + 3)} \tag{2}$$

where *NACH* is the normalized angular choice, Ch_i is the choice of node i in the grid, and TD_i is the total topological distance from node i to all other nodes [58].

The calculation of spatial properties can be done by the UCL Depthmap software, which was the first professional software developed by University College London for space syntax calculation [60,61]. Later, it was extended to Rhinoceros and Grasshopper software platforms as plug-ins for spatial syntactic operations, such as DeCodingSpaces Toolbox for Grasshopper [62], which integrates and extends existing methods for urban spatial analysis.

The previous section has already mentioned the problem of the breakthrough of the space syntax construction model at the level of 3D analysis. It is clear from the spatial logic of space syntax that real progress can only be achieved by accurately representing the problem of spatial nodes and their connections [47,56]. Essentially, the analysis of the structure of urban networks represents an attractive model for describing urban phenomena. Landmark structures are used as attractors into the urban grid, whose spatial properties are computed in the same way as those of the conventional 2D grid analysis model [41,63,64]. Obviously, it is not possible to use 2D isovists to check the visibility of all landmarks in a complex urban space by measuring each spatial node in the grid, but if a 3D isovist can achieve this goal, it is possible to construct a 3D NL GAM.

For 3D NL GAM, the computation of spatial properties should separately consider the *NAIN* of the wayfinding decision nodes and the *NACH* of the landmark location nodes. According to the multiplier effect of grids [29,65–67], the greater the integration of the nodes at the location of the path decision node, the greater the number of people gathered at this location. Therefore, the influence of the landmarks is greater. Additionally, choosing a higher number of of landmark location nodes indicates that landmarks are also more directive for going to the destination in the direction of the wayfinding route. Landmarks can be more involved in the navigation path [19,68,69]. Therefore, the structural salience of landmarks is positively correlated with the *NAIN* of wayfinding decision nodes and the *NACH* of landmark location nodes.

2.2. The Representation of 3D Isovist

The 3D isovist is an extension of the 2D isovist, and expresses visual data in a 3D field of view [54]; however, a 3D isovist is easy to understand but difficult to express [35]. In GIS, the 3D isovist is used to determine the visual extent, by the use of Boolean operations, of the line of sight emanating from the viewpoint with the terrain of the digital elevation model [70]. Morello and Ratti used the DEM as the basis for generating a 3D isovist and isovisi-matrix [31], and this approach further explains the influence of the five elements of urban imagery on the intelligibility of the urban environment. However, the 3D isovist formed by the DEM is still not a true 3D view representation. Since the topographic data only correspond to an "X-Y" planar reference system, it is difficult to express the visibility relationships of complex urban building forms, for example, building overhangs, bridges, and other cases with multiple "Z" values [32,71].

The increase in the computational power of 3D modeling software platforms has effectively facilitated the generation of 3D isovist methods. It is first necessary to build a 3D geometric model in a 3D software platform (e.g., AutoCAD, Sketchup, Rhinoceros, etc.), and then to determine the vantage points from which the field of view needs to be generated. The usual approach is to use the 3D isovist as a set of lines of sight emanating from the vantage point locations, and have the lines of sight fill the viewing space [72]. In this way, the 3D isovist itself becomes a geometric form consisting of lines of sight. Using parametric plugins in the 3D software platform makes the lines of sight Boolean in nature, with 3D geometry, and, eventually, the 3D isovist becomes a line structure with the viewpoint as the starting point of all lines of sight, and the endpoints of the lines of sight as the visual objects of the spatial place. Bhatia, Chalup and Ostwald generated a 3D isovist using Sketchup software and used it to analyze the structural salience of a single building space to simulate the experience of walking inside a building space [73]. Gewirtzman used Rhinoceros software to create 3D models and generated a collection of three-dimensional views of viewpoints connected to mesh surfaces based on the Grasshopper plug-in, which enabled the use of Boolean operations to analyze views and geometric entities, and was applied to the visibility measurement of real 3D scene models [72,74]. Bielik et al. developed a 3D isovist computing component based on the Rhinoceros and Grasshopper platforms [75], and the generated 3D isovists achieved highly accurate visibility measurements.

Essentially, the analytic model construction of spatial syntax is a perceptual description of visual perception in real space. Because all procedures, activities, and relationships in parametric design are explicitly defined [76], the parametric generation of 3Disovist also reflects the use of parametric design in spatial analytic models. In this paper, Rhinoceros and Grasshopper are used to construct a 3D isovist. First, the 3D isovist view structure is inherited from one or more parameters [77,78]. Meanwhile, the Grasshopper platform is able to transfer the geometric properties of the 3D isovist view structure to the subsequent spatial property operations, thus realizing the idea of a parametric model based on graphical metrics [62,79,80].

3. Materials and Methods

In this paper, a 3D grid analysis model was established for measuring the structural saliency of urban landmarks. We built the analytical model (Figure 1) as follows: first, a 3D geometric model of the neighborhood was built in the 3D modeling platform Rhino 7.0, and the 3D isovist component was run in Grasshopper to obtain accurate landmark visibility data. Then, the links between decision nodes and landmarks were established from the visibility data to form a 3D grid model that satisfied the spatial syntactic analysis, and the spatial motion properties of decision nodes and landmark locations were obtained. Finally, the expression function for structural landmark saliency was formed from the visible dominance index of landmarks and the motion properties of spatial nodes.

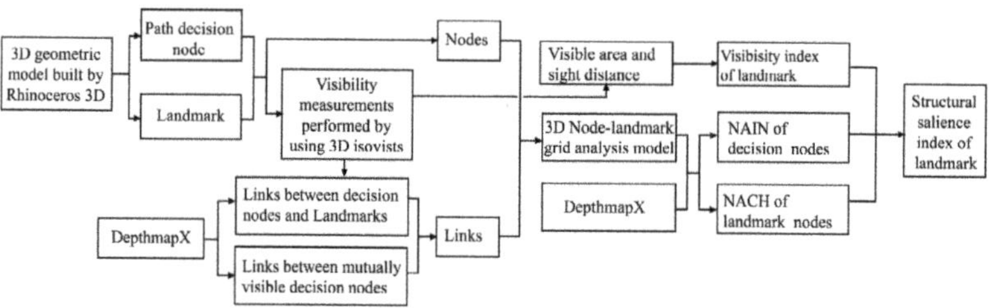

Figure 1. Research framework.

3.1. 3D Isovist-Based Landmark Visibility Measurement Method

Algorithms using 3D isovists achieve the visibility of landmarks measured from all decision nodes, as a fundamental condition for 3D NL GAM construction. In this study, the 3D isovist component on Rhino 7.0 and Grasshopper software platforms is used to generate 3D isovists located at the viewpoint of the decision node. In the software platform, a 3D isovist is represented as a line-of-sight structure originating from a viewpoint, and is generated according to the viewing direction, viewing range, and line-of-sight generated density, parameter settings, etc. In the process of wayfinding, people adjust the direction of their head's rotation, so that a horizontal viewing range of 180° can usually be reached [81]. When a decision point is reached, it is possible to turn around and identify the surroundings. Thus, the horizontal viewpoint can reach 360°, and this field of view is defined as the panoramic 3D isovist [39]. The vertical viewpoint includes the vertical downward viewpoint α and the vertical upward viewpoint β. In the vertical viewing, the color of the visible object is not easily perceived when the upward viewpoint β is greater than 30°, and people are more accustomed to the viewpoint centerline, which is generally directed downward by 10° when standing to ensure the safety of traveling [40]. Thus, $\beta = 30°$ was taken. Although the actual downward perspective is often greater than the upward perspective because the viewpoint is very close to the ground, the visible volume is smaller, thus, we set $\alpha = 15°$.

In an ideal state without occlusion, the panoramic viewshed of the viewpoint consists only of equidistant lines of sight at a set viewing range (Figure 2), forming a sphere with the radius of the length of the line-of-sight. Figure 2a,b represent a vertical sector view. Taking the angle between the vertical lines of sight as 1°, the number of lines of sight of a vertical sector is $\alpha + \beta$ roots. Rotating the vertical sector by 360° horizontally and copying one vertical sector every 1° forms the structure of a panoramic field of view (Figure 2c). Then the total number of lines-of-sight (T_{los}) of the ideal field of view is:

$$T_{los} = 360 \times (\alpha + \beta + 1) \tag{3}$$

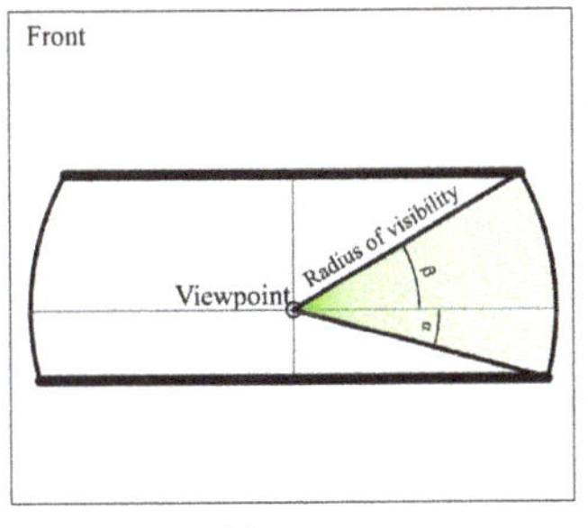

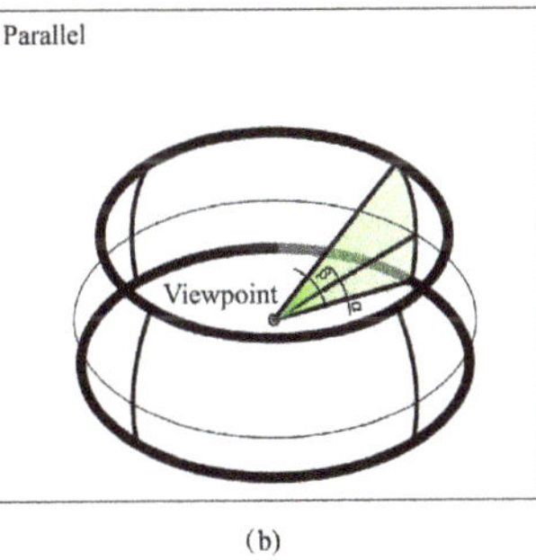

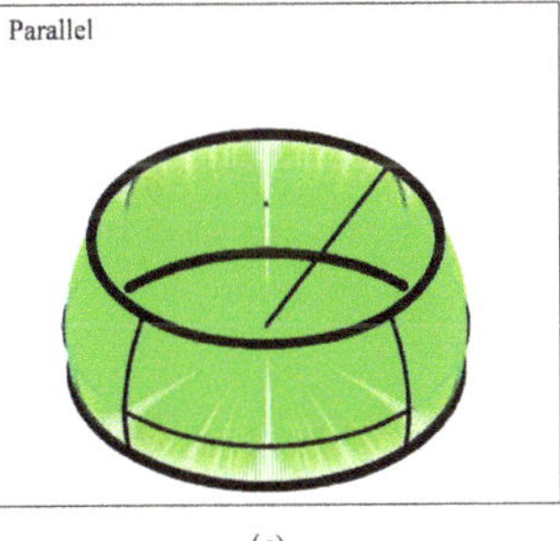

(a)　　　(b)　　　(c)

Figure 2. A 3D isovist line-of-sight composition structure, (**a**) generating a vertical 2D sectoral line-of-sight composition structure from the viewpoint, (**b**) a horizontal circular array centered on the viewpoint, (**c**) a 3D isovist originating from a viewpoint.

Subtracting the area of the upper and lower crowns from the area of the sphere, it follows that the visible area represented by the end of each line-of-sight is:

$$A_{los} = \frac{\pi R^2 (\sin \alpha + \sin \beta)}{180(\alpha + \beta + 1)} \tag{4}$$

where A_{los}: the visible area of each line-of-sight; R: the length of the line-of-sight.

In a line-of-sight obstacle environment formed by a 3D geometry model, the panoramic viewshed is divided into three line-of-sight types:

1. No obstacle line-of-sight: the initially set viewing range is maintained, representing the visibility of the sky;
2. Ground line-of-sight: line-of-sight projected onto the ground;
3. Obstructive line-of-sight: line-of-sight is obstructed by geometry. At the same time, the reference number and length of the line-of-sight, as well as the reference number of the geometry onto which the line-of-sight is projected, are recorded in the 3D isovist dataset.

Since the obstructed environment only changes the length of the initial line-of-sight distance and does not change the density of lines-of-sight in the 3D isovist, the visible area at the endpoint is only related to the length of the line-of-sight. In Equation (4), the visible area expressed by the endpoints of every line-of-sight projected onto a landmark is cumulatively the visual area of that landmark corresponding to the decision node.

3.2. Simulation Model of a Standard Urban Neighborhood

A standard residential neighborhood (10 min walkable range) model (Figure 3) was simulated in the Rhinoceros3D platform to express the 3D NL GAM construction process. The area of the simulated neighborhood was about 100 hm^2, the average number of floors of all residential buildings was 10, and the population size was 30,000, which comprised a typical compact urban neighborhood. The urban spatial form is the external appearance of the urban tissue, which consists of four categories of geometric models: natural landscape resources, residential buildings, public buildings, and urban public facilities, etc., which are subdivided as follows:

- Natural landscape resources: mountains, forests, rivers, lakes, etc., within the city;
- Neighborhood blocks: urban residential area units made up of residential building groups;
- Public buildings: civic buildings of various functional types, including office buildings, shopping malls, hotels, conference centers, etc.
- Urban public facilities: streets, sidewalks, street trees, parks, squares, parking lots, etc.

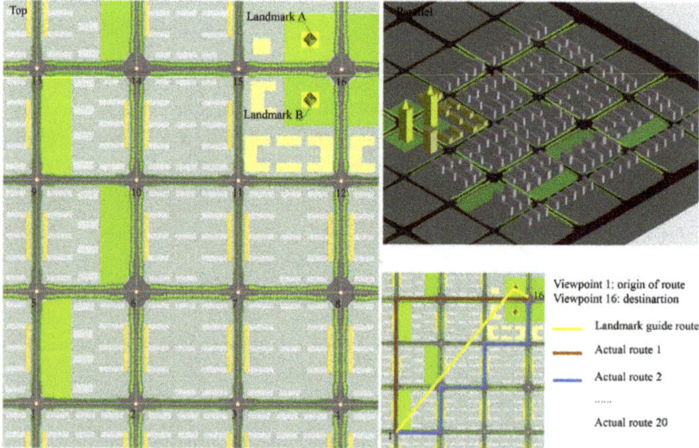

Figure 3. Simulated 3D geometric model of a standard urban residential neighborhood.

Analyzing the street space from a 2D grid alone makes it difficult to determine a way-finding path and can easily lead to getting lost. The block consisted of eight streets with sixteen street intersections. This arrangement made it difficult to distinguish the path orientations of the street intersections and resulted in too many undetermined paths. For example, if intersection 1 is the starting point and intersection 16 is the ending point, and each crossing is counted as 1 topological distance, then the shortest path distance is 6 topological depths, and 20 shortest paths can be arranged.

Landmark-based route selection methods can better reproduce human navigation [14]. It can be seen in Figure 3 that if Landmark A is visible from intersection 1, the shortest path guided by the landmark is V1-Landmark A-V16, i.e., the topological distance of the landmark-guided path is 2. Thus, with the intervention of landmarks, the choice of route intersection becomes less important, and the choice of any actual path does not lead to getting lost but only toward the landmark, and the destination can be reached by walking in the direction of the landmark. Therefore, to implement landmark intervention analysis, it was necessary to determine the setting of the path decision node and the landmark visibility at the decision node.

3.3. Determination of Path Decision Nodes in 3D Grid

Determining path decision nodes for wayfinding or navigation in urban space is the first task in constructing 3D NL GAM. Firstly, the visibility differences between road sections and intersections were dynamically simulated in a 3D environment (Figure 4). The method involved route selection and parameter setting for walking, cycling, and driving [74], and arranging motor vehicle lanes, bicycle lanes, street trees, and sidewalks in a street environment, as follows:

- The trees on either side of the trail were spaced 8 m apart and had a net height of more than 2.8 m below the branches of the canopy;
- The pavement trees were spaced 6 m apart, with a net height of more than 1.8 m below the crown branches;
- The line-of-sight triangle was set at the intersection.

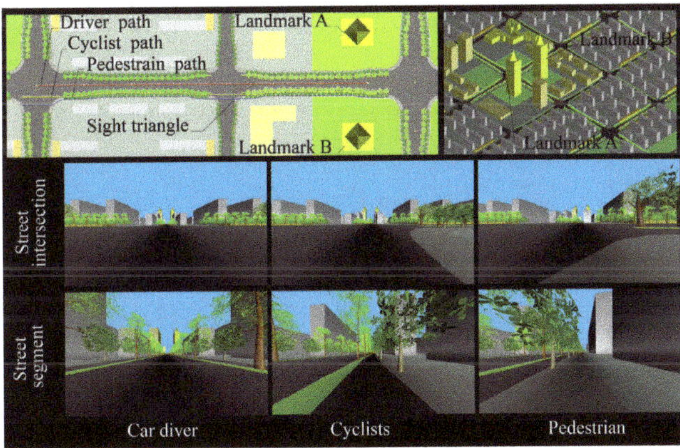

Figure 4. Dynamic snapshots of the visual perception of three types of routes, including pedestrian, bicycle, and driving, as simulated on the Rhino 3D software platform.

Then, using Rhino 7.0 3D model-making software, a dynamic snapshot was set up with a camera focal length of 12 mm and the shot direction facing the next viewpoint:

- The height of the walking viewpoint was 1.6 m from the ground, the distance between the viewpoints was 30 m, and the playing speed of the dynamic snapshot corresponded to the walking speed of the person, which was 5 km/h;
- The riding viewpoint was 1.75 m from the ground height, the viewpoint spacing was 30 m, and the playback speed of the dynamic snapshot corresponded to the riding speed, 10 km/h;
- The driving viewpoint height from the ground was 1.35m, the viewpoint spacing was 50 m, and the playback speed of the dynamic snapshot corresponded to the speed of the motor vehicle, 20 km/h.

Thirty-two architecture students were assigned to view dynamic snapshots to compare the differences in landmark visibility between the 3 path types (Figure 4). The results indicated that the intersection enabled more visible information to be obtained. At the intersection location, on the one hand, the road was visible in both directions, and on the other hand, the street trees were less obscured by the city buildings. When in the roadway area, only one road was visible, and street trees were highly obscured by buildings. Figure 4 shows that two landmark buildings can be seen in the distance at the intersection, but the visibility of the landmark buildings in the section area was significantly reduced. It follows that the intersection location should be identified as the decision node, and the intersection point of the road medians is taken to express the logical space of the intersection.

In addition to the intersections, additional decision nodes should be created in accordance with the axial graph analysis principle. This indicates that the urban grid should satisfy line-of-sight continuity, and the decision nodes should be set at the locations where line-of-sight transitions are generated. The decision nodes were required as follows:

1. Street intersections and street turns within the analysis area are decision nodes of the 3D grid;
2. For arc roads, consecutive decision nodes should be set and the maximum spacing requirement should be satisfied;
3. For roads with significant slopes, the decision node should be set at the top of the slope;
4. When two roads intersect, path continuity can only be formed by secondary roads.

3.4. Expression Functions for Landmark Visibility in 3D Grid

Landmark visibility measurement consists of two objectives; one is the visibility of the decision point to the landmark, and if it is visible, the decision point forms a connection with the landmark. The other objective is to measure the visible area of the landmark and the distance of the decision point from the landmark if the landmark is visible, and to quantify the dominance of expressing the visibility of the landmark.

Computation of the viewing data of the decision nodes in the standard urban neighborhood environment was performed by running the dynamic parameters program with the 3D isovist component on the Rhino 7.0 and the Grasshopper software platforms. The ideal viewing range was set to 5000 m, the horizontal view angle was 360°, the vertical view angle was 45°, and the viewpoints and obstacles were picked up. After the 3D isovist component calculation, the visible data of 16 decision nodes were obtained (Figure 5). The data consisted of various types of lines of sight, and each viewpoint accumulated a total of 16,560 lines-of-sight. To facilitate landmark visibility, only the obstructive sightlines are shown in Figure 5, which shows that the greater the number of projected sightlines of the geometry, the warmer the displayed color of the geometry and, with fewer sightlines, the cooler the displayed color. Based on the corresponding values of the length of the sightlines and Equation (4), the visible areas of the viewpoints to the two landmark buildings were obtained for all decision nodes (Table 1).

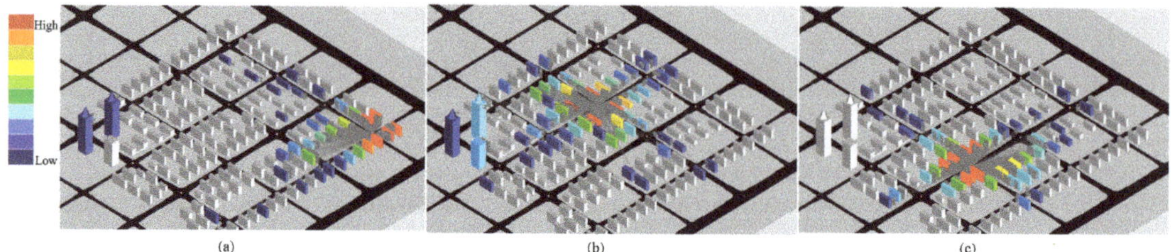

Figure 5. Measuring the visibility of landmarks using 3D isovists. (a) V_1, (b) V_7, (c) V_{10}.

Table 1. Visible area of landmarks from each viewpoint.

Landmark	V_1	V_2	V_3	V_4	V_5	V_6	V_7	V_8	V_9	V_{10}	V_{11}	V_{12}	V_{13}	V_{14}	V_{15}	V_{16}
A	1898.9	0	1587.1	1128.8	0	0	217.1	1438.5	479.3	0	0	4979.1	3477	3354.9	3963.4	2378.8
B	798.4	0	2273.7	3461.2	0	0	1752.8	4578.1	0	0	0 *	4880.9	3229.4	3533.5	4370.1	2343.9 *

* Unit: m^2; 0: invisible.

Landmark visibility dominance is an expression of the visual perception of landmark visibility at the decision node location. The main influencing factors of people's visual perceptions of landmarks include visual acuity, sight distance, and visible area [82]. The initial setting of the 3D isovist is a constant viewing range; that is, the volume of the panoramic field of view is consistent within the same urban area, representing a normalized field of view. Thus, the comparison of visual perception metrics for different landmarks at a decision node is achieved in the same normalized viewshed. That is, the landmark visibility metric is directly expressed as a function of the visibility area of the viewpoint to the landmark and the view distance:

$$VISIBILITY_i = \frac{A_{lm}}{D_{sight} + 1} \tag{5}$$

$VISIBILITY_i$: landmark visibility index; A_{lm}: landmark visible area; D_{sight}: average length of lines of sight.

4. Salience Computation of Structural Landmarks in Urban Neighborhood

4.1. Computation of Structural Landmark Salience

The decision node locations and landmark visibility measurements were determined to form a 3D NL GAM, which was used to compute the spatial mobility properties of node and landmark locations. The computation of the spatial properties in this paper is performed on the DepthmapX software, that is, after importing the spatial node data obtained in the Rhino and Grasshopper software platforms into the DepthmapX software platform, the grid analysis model is further constructed and computed in the DepthmapX software. To illustrate the differences before and after landmark intervention in the grid, the changes in the spatial movement attributes of nodes and landmarks in the GAM were compared for the unlink and link cases (Figure 6, Supplementary Materials Table S1). In the unlink case (Figure 6a), it expressed only a 2D node link. The VGA result indicated [61,83] that the *NAIN* value for all intersections was 1.16. Thus, it was difficult to determine if there was any difference in the spatial mobility properties of the decision nodes from the 2D grid alone. Based on the landmark visibility measurements of the 16 decision nodes determined with the 3D isovist method, the link between nodes and landmarks was established (Figure 6b). In this case, the spatial movement attributes of both landmarks and nodes reflected significant differences (Figure 6c).

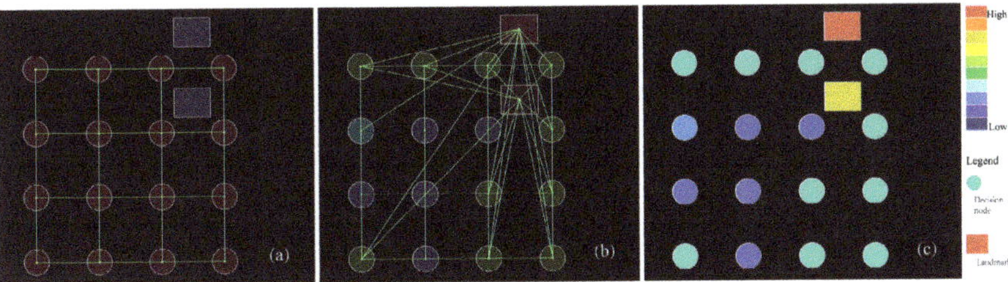

Figure 6. Space syntax computation. (**a**) Nodes are unlinked with landmarks, (**b**) landmarks are linked with nodes, (**c**) the *NAIN* values of each node in the 3D NL GAM.

While the decision node was concerned with spatial accessibility, the landmark location was concerned with spatial selectivity; that is, the *NAIN* of the decision node and the *NACH* of the landmark location were obtained for the 3D NL GAM using the space syntax principle. According to the multiplier effect of grid grouping [33,44–46], the greater the *NAIN* of the nodes of the path decision point location and the higher the number of people gathered at this place, the greater the influence of landmarks. At the same time, higher *NACH* values of landmark location nodes indicate that the landmarks are also more directional along the wayfinding route to the destination. However, there is a special case where the decision node, destination, and landmark are adjacent to each other (all decision points that have a view of the landmark are in the same straight-line street), and the *NACH* of the landmark is zero. From the grid effect, landmarks with a degree of zero are named as local landmarks. Thus, for a decision node, the structural salience of a landmark is computed as follows:

$$SSL_i = (NAIN_i \times VISIBILITY_i) \times (NACH_{lm} + a) \tag{6}$$

where SSL_i: structural salience of a landmark corresponding to path decision node i; $NAIN_i$: normalized angular integration of path decision point location nodes; $VISIBILITY_i$: landmark visibility index; $NACH_{lm}$: normalized angular choice of landmark location nodes; a: initial bootstrap effect constant of local landmarks.

Having determined the landmark salience values for each decision node (Figure 7), it was possible to compare the advantages of the locations of the landmarks in the grid with the locations of the decision nodes. The values of each structural landmark salience-related factor in the standard neighborhood NL GAM were substituted into Equation (6) to obtain the landmark salience for the 16 viewpoints (Figure 7a). The structural salience of a landmark in the grid was the average of the salience of the landmark with respect to all decision nodes in the grid for all decision nodes. The structural salience of Landmark A was 6.24 and that of Landmark B was 5.99 (Figure 7b).

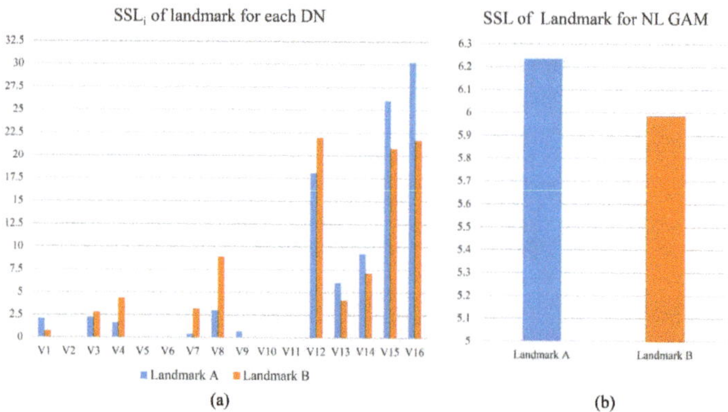

Figure 7. Structural salience of landmarks in the standard block. (**a**) Structural salience of landmarks corresponding to each decision node, (**b**) structural salience of landmarks in the grid.

4.2. Computation Procedure of Structual Landmark Salience for Two Typical Samples of Districts in Changsha

A typical urban sample area is a compact urban residential area with a 15 min walkable living radius, typically around 400 hectares. It has distinct natural and architectural landmarks within its viewing area. At the same time, the building groups in the area have the typical spatial form of a compact city. In this paper, two tracts located in Changsha, Hunan, China (Figure 8, Table 2) were selected, which are the Tianxin Pavilion District (TP-District) (Figure 9) and the Xinhe Delta District (XD-District) (Figure 10), which two urban areas in Changsha, China, and both are typical high-density compact urban neighborhoods along the riverfront, with significant spatial morphological differences between them (Table 2). Table 2 presents the basic situation of the two zones. The landmarks in the area were initially identified based on natural landscape resources, human resources, building volume, building height and architectural shape [11,28]. The final landmark analysis objects were identified based on relevant public results from the Changsha Planning Information Center combined with expert consultation recommendations. In this study, three types of 3D NL GAM were established based on the morphological features of each neighborhood:

1. TP-District, an area typically formed by a combination of high- and low-rise architectural groups;
2. XD-District, a typical area pattern formed mainly by high-rise building groups; the height of the proposed A2-Beichen Landmark was 268 m (Figure 11a), with $H_{a2} = 268$ m as the default in the later text;
3. For the XD-District, the height of the proposed A2-Beichen Landmark was set at 400 m (Figure 11b), with $H_{a2} = 400$ m as the default in the latter part of the text.

Computation of structural landmark salience for the above three morphologies of urban areas was needed to achieve three research objectives: First, we compared the structural salience differences between landmarks in the two types of neighborhoods with the combined architectural group morphologies. Second, quantitative results were used to compare the structural salience of large-scale natural landscapes, super-tall architectural landmarks, multistory buildings, and human landmarks. Finally, we analyzed the effect of changing the morphology of individual building landmarks (A2-Beichen Landmark) on the structural salience of landmarks.

Table 2. Comparison of the spatial morphological characteristics of the two sample urban districts.

City Tissue	TP-District	XD-District
Block size	435 hm^2	204 hm^2
Population density	275 persons/hm^2	350 persons/hm^2
Building Groups Form	Building groups with a combination of high- and low-rise buildings	High-rise dominated building groups
Land Use	High degree of mixed functions, with residential land accounting for 28% of the total construction land	Low degree of mixed functions, with residential land use accounting for 52% of total construction land
Large-scale natural landscape landmarks	Yuelu Mountain, Tianma Mountain, Xiangjiang River, Orange Island	Xiangjiang River
Architectural and cultural landscape landmarks	A total of 14, located adjacent to or within the area, including 9 high-rise buildings. IFS-West Tower is the tallest building in Changsha, with a height of 452 m	A total of 27, with 11 high-rise buildings adjacent to or located in the area; at present, China CITIC Bank is the tallest building in the area, the height of 268 m. Nine super high-rise buildings are distant landmarks facing the area across the river
Construction situation	Mature built environment, all landmark buildings have been constructed	Mature built environment, but the A2-Beichen Landmark is a proposed project. There are 2 publicized schemes; the design height of scheme 1 is 268 m, and scheme 2 is 400 m

We followed the steps below to build the 3D NL GAM for computing the structural salience of landmarks in the grid:

Step 1: Make a 3D model of each urban district. On the ArcGIS platform, geographic information data are formed based on raster images of terrain and building vector data. Then, these are converted to a DXF vector data format, and imported into the Rhino 7.0 3D software platform, in order to generate the 3D model (Figures 9b and 10b).

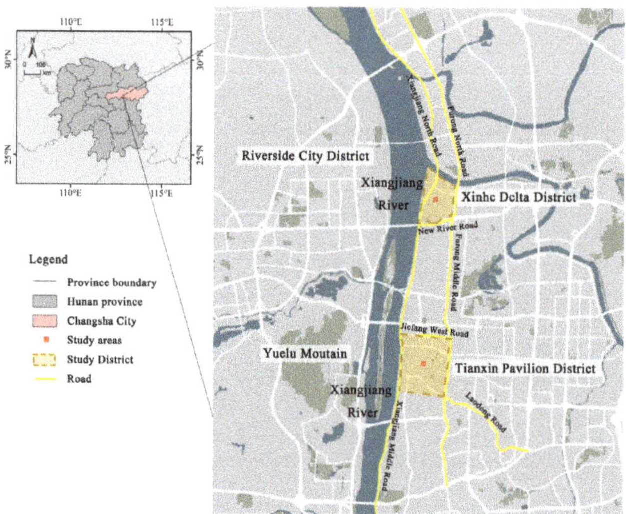

Figure 8. Regional location map of the sample districts.

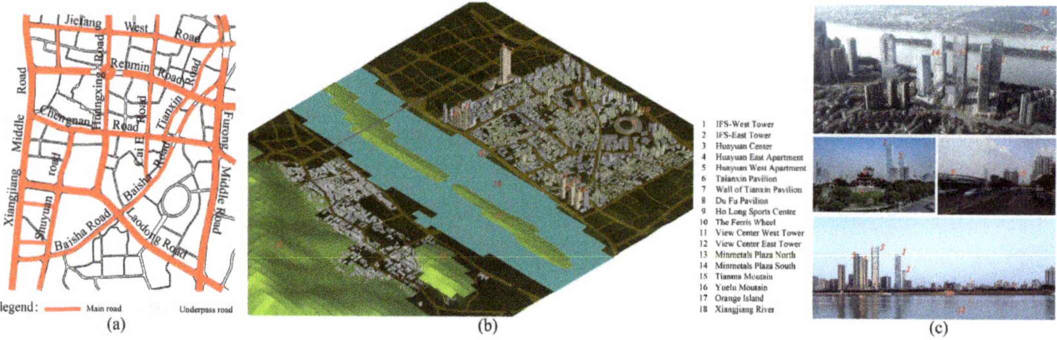

Figure 9. TP-District. (**a**) Regional road network, (**b**) 3D geometric model of the district, as well as the surrounding environment, (**c**) Photographs and partial renderings of landmarks.

Step 2: Determine the landmarks of the area. Landmarks of urban districts include both neighboring landmarks and distant landmarks.

Step 3: Determine the decision nodes. The decision node number is set according to the name of the main road where the decision node was located and classified as a riverfront road decision node (Riverside-RD), north–south road decision node (NS-RD), east–west road decision node (EW-RD), or north–south to east–west road decision node (NS-EW-RD). Cumulatively, there are 44 decision nodes in the TP-District (Figure 12a) and 43 decision nodes in the New River Delta area (Figure 12b).

Step 4: Measure landmark visibility for all path decision nodes (Figure 13). On the Grasshopper platform, the initial setup parameters of the 3D isovist are kept consistent with those of the standard urban neighborhood model, and a dynamic program is written to record and filter the sightline data projected to different landmarks from each viewpoint, as well as to calculate the visibility of landmarks according to Equation (5).

Step 5: Calculate the movement attribute of node space (Figure 14). The 3D NL GAM is formed by establishing relationships between route decision nodes and between decision nodes and landmarks. Perform the graph analysis operation on the DepthmapX platform, edit the node normalized angle integration degree attribute according to Formula (1), and edit the node-normalized angle choice degree attribute according to Formula (2).

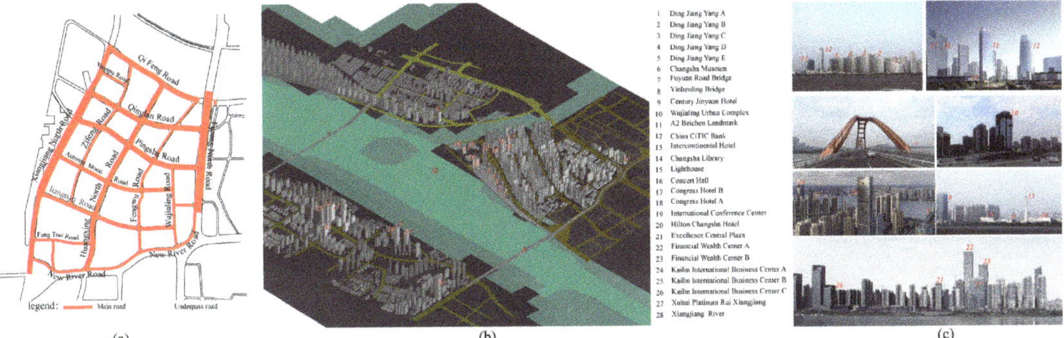

Figure 10. XD-District. (**a**) Regional road network, (**b**) 3D geometric model of the district, as well as the surrounding environment, (**c**) Photographs and partial renderings of landmarks.

Figure 11. The A2-Beichen landmark as a proposed project. (**a**) Height is 268 m; (**b**) height is 400 m.

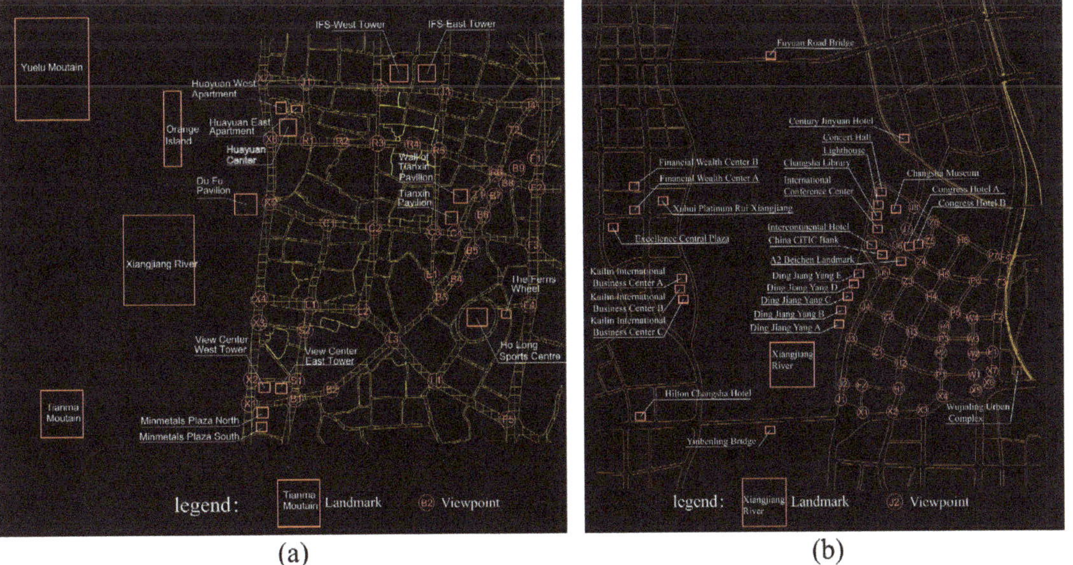

Figure 12. Route decision nodes and landmarks on area streets. (**a**) TP-District; (**b**) XD-District.

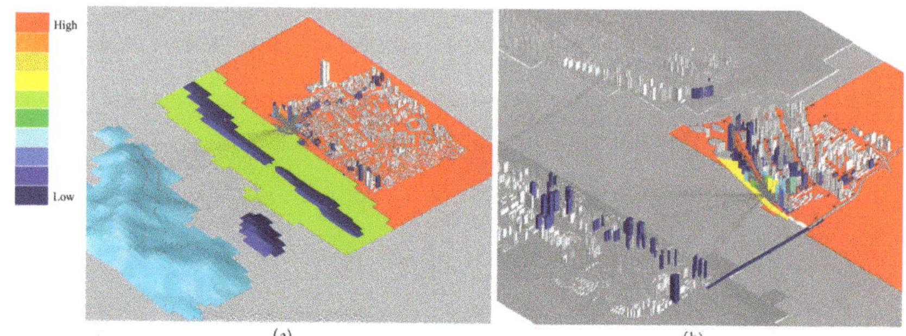

Figure 13. Measuring the visibility of landmarks using 3D isovists. (**a**) TP-District; (**b**) XD-District.

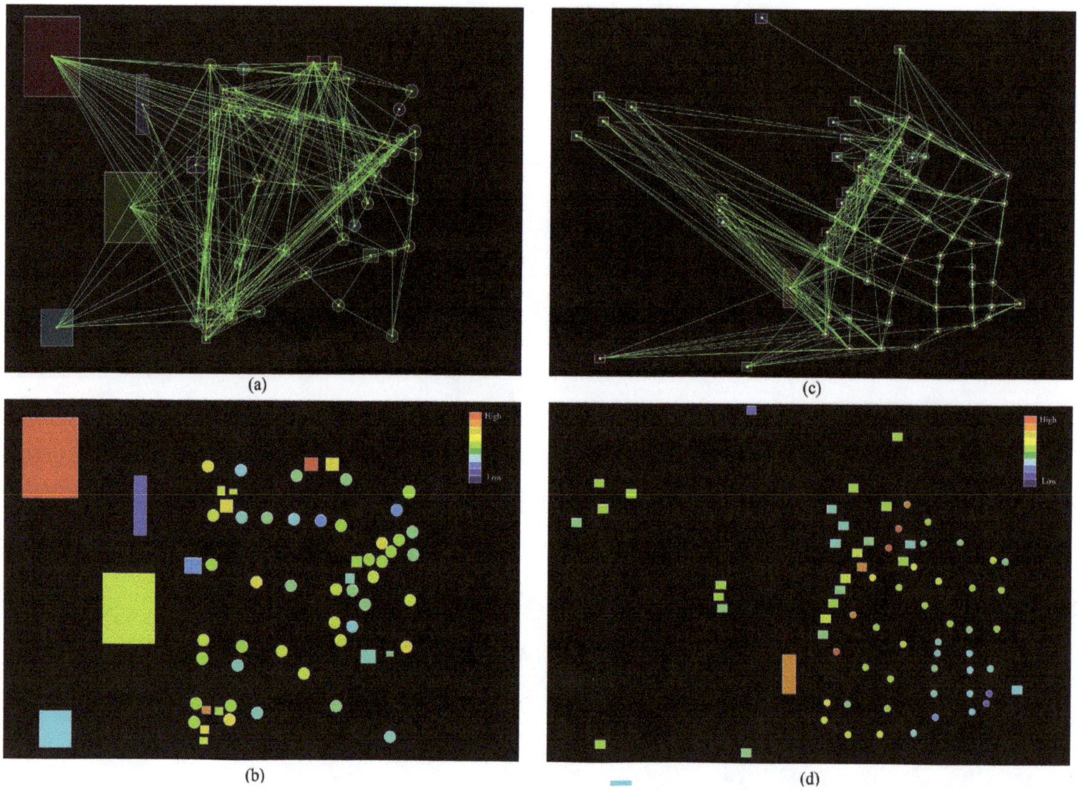

Figure 14. Space syntax computation. (**a**) 3D NL GAM of the TP-District, (**b**) *NAIN* of each node and landmark in the TP-District, (**c**) 3D NL GAM of the XD-District, (**d**) *NAIN* of each node and landmark in the XD-District.

Step 6: Calculate the structural salience of the landmarks. Calculate the structural salience of landmarks corresponding to each decision node according to Formula (6) and calculate the structural salience of landmarks in the whole grid system.

5. Results

5.1. Spatial Movement Properties of 3D NL GAM

When comparing the *NAIN* values of the three grid decision nodes (Figure 15, Tables S2–S4), the TP-District had obvious advantages. The average *NAIN* value in the TP-District was 1.128; when H_{a2} = 268 m, the average *NAIN* in the XD-District was 0.984, and when H_{a2} = 400 m, the average *NAIN* in the XD-District was 1.006. This indicated that the route decision nodes in the TP-District had better accessibility and were suitable for keeping. Increasing the height of landmarks, which increased the connection between decision nodes and landmarks, improved the *NAIN* values of nodes. The connection data indicated that the A2 landmark increased from 268 to 400 m with the addition of only three links, and the improvement in *NAIN* was not significant.

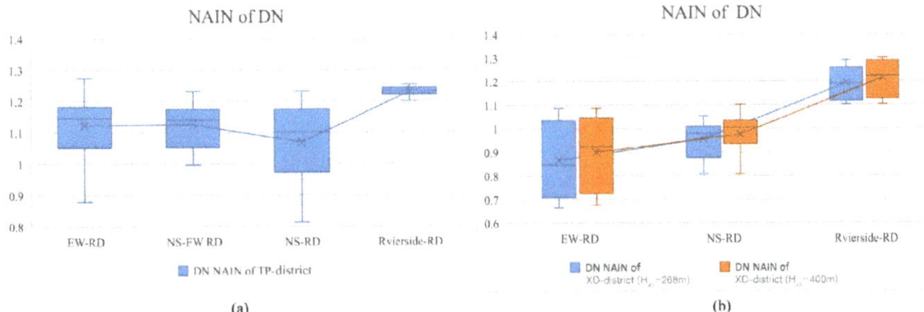

Figure 15. Boxplot analysis of *NAIN* values for various types of decision nodes. (**a**) TP-District; (**b**) XD-District.

The integration of decision nodes in the TP-District was relatively concentrated, and the grid was uniform. The box plot (Figure 15a) of the integration of the decision nodes and the standard deviation (Table 3) of various types of data demonstrated that the decision nodes along the Riverside-RD had a wide field of view on its west side, and their integration maintained the maximum value in the system, as well as the most concentrated numerical range. The values for the NS-EW-RD were relatively consistent and at a moderate level. The values for the NS-RD were relatively scattered, with low mean and median values.

Table 3. Standard deviation of *NAIN* values for each type of node.

Type	TP-District	XD-District (H_{a2} = 268 m)	XD-District (H_{a2} = 400 m)
Riverside RD	0.018	0.067	0.073
NS-EW RD	0.076	-	-
EW RD	0.098	0.160	0.158
NS RD	0.142	0.77	0.85
ALL	0.108	0.144	0.145

The *NAIN* values of decision nodes in the XD-District were relatively scattered. The box plot (Figure 15b) and the standard deviation (Table 3) of the data of each type of decision node indicated that the *NAIN* values in both cases were relatively scattered, and the data were located in a large range. The *NAIN* values of the path decision nodes along the Riverside-RD remained at the maximum value in the system but they were more scattered than in the TP-District. Although the decision nodes of the NS-RD had the largest number of nodes, their *NAIN* values were relatively concentrated, with mean and median values close to one. The *NAIN* values of the EW-RD were scattered, and most of the values were less than one; that is, the node space was relatively isolated.

After increasing the height of the A2 landmark in the new 3D NL GAM, the *NAIN* values of a few decision nodes remained unchanged, while the *NAIN* values of most decision nodes were improved. In particular, the *NAIN* values of the three nodes that gained visibility of the A2 landmark because of its elevation were significantly enhanced.

Comparing the landmark *NACH* values of the three 3D NL GAMs (Figure 16, Tables S2–S4), the TP-District still maintained a clear advantage. The average value of landmarks in the TP-District was 0.789; when H_{a2} = 268 m, it was 0.595 for the XD-District and when H_{a2} = 400 m, it was 0.573 for the XD-District. That is, the TP-District landmarks had higher *NACH* values, indicating that landmarks in this area were more guided. Increasing the height of the landmarks changed the structural relations of the grid, but the results indicated that the mean *NACH* value of the landmarks decreased instead.

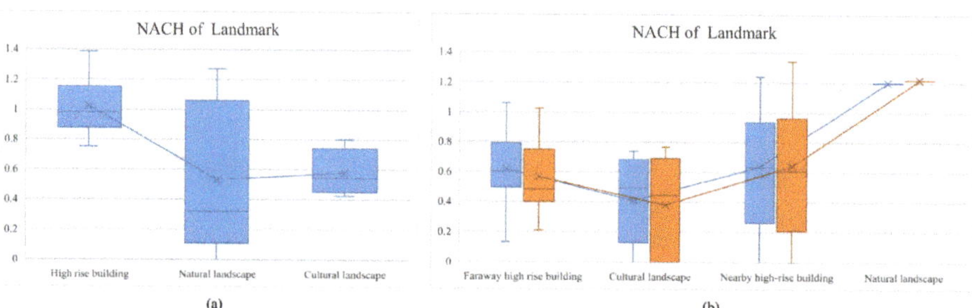

Figure 16. Boxplot analysis of *NACH* values for various types of landmarks. (**a**) TP-District; (**b**) XD-District.

The results of the box plot (Figure 16) and standard deviation (Table 4) indicated that the *NACH* data of the TP-District landmarks were more dispersed, and the guiding of landmarks had obvious variability. High-rise buildings maintained larger *NACH* values in the system and remained within a more concentrated range of values, with the IFS-West Tower having the highest *NACH* value of those in the grid. Although there were only five natural landscape landmarks, the *NACH* value interval of landmarks was large and the data dispersion was high, among which the *NACH* value of Yuelu Mountain was second only to that of the IFS-West Tower, but the *NACH* value of Orange Head was zero, the smallest value in the grid. The number of humanistic landmarks was smaller, and the connections with decision nodes were lower. Their landmark *NACH* values were all lower, and the data were more concentrated.

Table 4. Standard deviations of *NACH* values for each type of landmark.

Type	TP-District	XD-District (H_{a2} = 268 m)	XD-District (H_{a2} = 400 m)
Nearby high-rise building	0.193	0.424	0.419
Distant high-rise building	-	0.260	0.243
Cultural landscape	0.160	0.313	0.333
Natural landscape	0.519	-	-
ALL	0.379	0.363	0.366

The *NACH* values of the landmarks in the XD-District were more scattered in both cases. The Xiangjiang River was the only natural landscape landmark in the area, and its *NACH* value was second only to the Beichen-A2 landmark. For the neighboring high-rise landmarks, the data lay in a larger dispersion interval, and their mean and median values were only lower than those of the Xiangjiang River. For distant high-rise landmarks, the *NACH* value interval remained low, even though they were all ultra-high-rise buildings. Cultural landscape landmarks had lower *NACH* values and more scattered data.

After raising the height of the Baichen-A2 landmark, the *NACH* was first significantly increased, and the *NACH* values of most of the nearby high-rise landmarks were raised. However, this also lowered the *NACH* values of most of the distant high-rise buildings, as well as those of the area's cultural landmarks.

5.2. Structural Salience of Landmarks

The landmark visibility values at each node were calculated according to Equation (5), and the structural salience of the landmarks was then calculated by substituting Equation (6). The results of the TP-District (Figure 17, Table S2) and XD-District (Figures 18 and 19; Tables S3 and S4) expressed the structural salience of landmarks in the area with a combination of high- and low-rise building groups, and in the area dominated by high-rise building groups, respectively. The calculated results indicated that the mean value of landmark structure salience was 6.402 for the TP-District and 1.211 (H_{a2} = 268 m) and 1.291 (H_{a2} = 400 m) for the XD-District. This meant that the landmark salience of the TP-District was much larger than that of the XD-District, and raising the height of the landmark improved its own salience and that of the area.

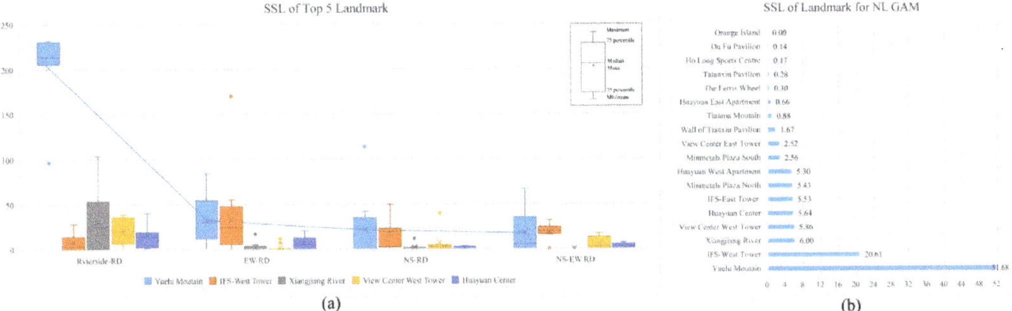

Figure 17. Structural salience of TP-district landmarks. (**a**) Boxplot of the structural salience of TOP5 landmarks corresponding to each decision point, (**b**) bar chart of the structural salience of landmarks in grid.

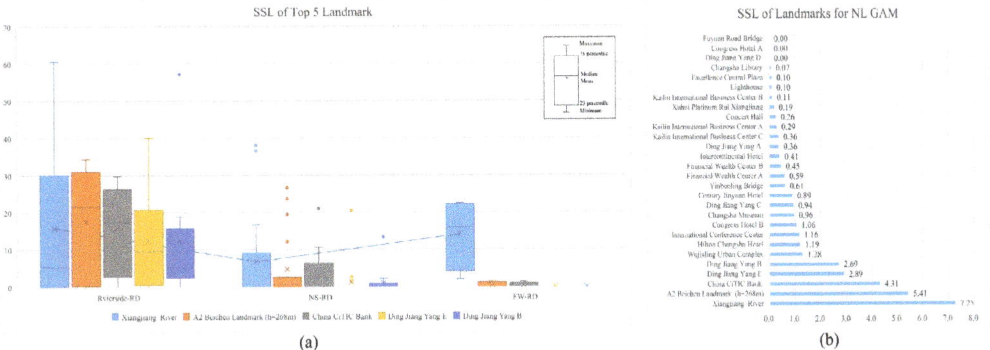

Figure 18. Structural salience of XD-district landmarks (H_{a2} = 268 m). (**a**) Boxplot of the structural salience of TOP5 landmarks corresponding to each decision point, (**b**) bar chart of the structural salience of landmarks in grid.

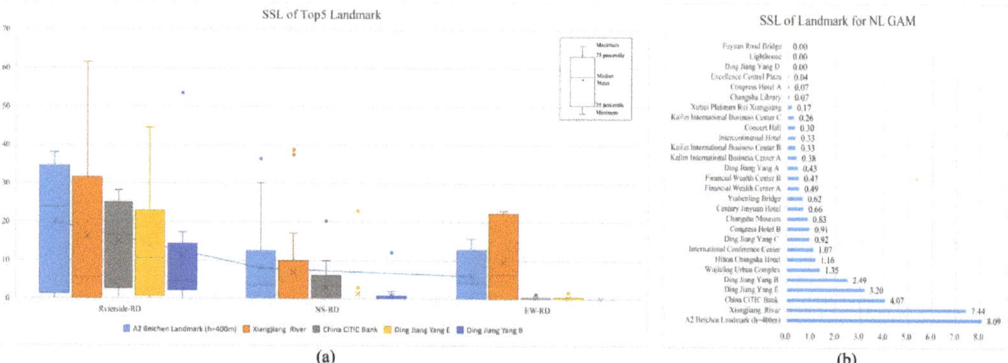

Figure 19. Structural salience of XD-district landmarks (H_{a2} = 400 m). (**a**) Boxplot of the structural salience of the Top 5 landmarks corresponding to each decision point, (**b**) bar chart of the structural salience of landmarks in grid.

In the TP-District, the structural salience of all landmarks had a large range of values, with a median of 2.54. The structural salience of Yuelu Mountain and IFS-West Tower was definitely dominant, with Yuelu Mountain being the largest. Yuelu Mountain could be seen from 29 of the 44 path nodes, mainly those along the EW-RD, including Xiangjiang Middle Road, Jiefang Road, Renmin Road, Chengnan Road, and Baisha Road. Although the IFS-West Tower was visible from 30 nodes, the building's visibility was not comparable to that of a large-scale natural mountain. Thus, the structural salience of the IFS-West Tower was the second-highest in GAM. The structural salience of the Xiangjiang River was much lower than that of the Yuelu Mountain, with only 15 nodes visible to the Xiangjiang River, ranking third. High-rise buildings had a high structural salience, with most of them ranking above the median. Cultural landscape landmarks had lower structural salience and ranked below the median.

In the XD-District, the structural landmark salience of the Xiangjiang River ranked first when the Baichen-A2 landmark H_{a2} = 268 m (Figure 18). There were 11 high-rise buildings in the neighboring landmarks, and although the top five architectural landmarks were all neighboring landmarks, four of them were also ranked below the median, indicating that the structural salience of high-rise buildings in the neighboring landmarks was obviously less advantageous than that in the TP-District. For the distant landmarks, there were four super-high-rises that ranked above the median. The Changsha Hilton Hotel was ranked seventh, with a height of only 241 m, but it had more structural salience than the Financial Center Block A (328 m, ranked fifth in Changsha) in the RiverSide New Town. The cultural landscape with a lower volume and height was ranked lower, but Changsha Museum was ranked 10th because of its location advantage in the grid.

After raising the building height of the Baichen-A2 landmark H_{a2} = 400 m (Figure 19), firstly, the structural significance of the A2 landmark overtook Xiangjiang, taking first place. Second, the structural salience values of 15 landmarks were raised while the others were lowered, resulting in a further broadening of the numerical range of structural salience values and a certain change in the ranking order. This suggested that changing the shape of a landmark building can change the entire grid system of structures in the area.

6. Discussion

6.1. Correlation between Structural Landmark Salience and Urban Intelligibility

The structural salience of landmarks can be considered an intelligibility indicator for urban districts. Claramunt and Winter synthesized the structural salience of landmarks from four levels of salience, district, path, edge, and node [12], which is a pure street grid structure idea that cannot express the visual relationship between decision points and landmarks. In this paper, using this reference, we obtained visibility measurements between decision nodes and landmarks using 3D isovist methods, such that the structural saliency of landmarks constituted a comprehensive quantification of each element of urban intent. From the relationship between the structural salience of landmarks and the other four elements of urban imagery (Table 5), the structural salience of landmarks was a function of the spatial attributes of landmark visibility and grid nodes.

Table 5. The relationship between the structural salience of landmarks and the other four elements of the city.

Elements of Urban Imagery	Structural Landmark Salience
district	The visibility measure expressed in terms of decision nodes and landmarks is the morphological feature of the region. Differences in the morphology of the region lead to differences in the shading relation between the landscape resources of the regional landmarks and the surrounding environment. It is difficult to obtain good structural salience for landmarks with excessive occlusion of the environment.
edge	The structural salience of landmarks has broadened the scope of the analysis of urban areas. The edges of the visual field are formed by distant landmarks, and the edges of human activity are peripheral streets or large-scale natural landscape dividers within the area.
path	The wayfinding decision nodes are grouped according to the street alignment, and the decision nodes themselves express the sequential relationship of paths. 3D NL GAM incorporates landmark directional guidance interventions into the path selection analysis.
node	The *NAIN* values of the nodes at the decision point locations express the influence of landmark structural salience on the aggregation and accessibility of people, and the *NACH* values of the landmark node locations express the influence of landmark structural salience on the guidance of the wayfinding directions.

The computation of landmark salience results reflects the synthesis of all elements of urban imagery within the field of view of the urban network. The realization of this synthesis benefits from the transformation of structural thinking regarding the space syntax model [56]. The 3D NL GAM extends the horizontal and vertical spatial extents constrained by the street network. In terms of horizontal spatial extent, the 3D NL GAM contains both accessible and visibility dual regions and dual edges. In terms of vertical spatial extent, it expresses the influence of the combination of heights of building clusters on spatial morphology [32]. Thus, the 3D NL GAM incorporates the effects of more agglomerative phenomena in the urban spatial environment, thus enabling an effective analysis of the intelligibility of urban areas.

It can be assumed that the values of landmark structural saliency have a positive correlation with the comprehensibility of the city. Firstly, landmark saliency includes the values of spatial attributes of grid nodes, i.e., *NAIN* for path decision points in the street grid and *NACH* for landmark locations in the field of view. The corresponding spatial attribute values in the two sample district cases show that TP-district has higher ones than those of the XH-district. Secondly, the fact that landmark saliency values include indicators of landmark visibility further enhances the description of the comprehensibility of the urban spatial environment. Since the combined pattern of high- and low-rise building clusters tends to result in better visibility of the ground's surface, the numerical results of landmark structural saliency for the two sample districts indicate that TP-district receives significantly higher values than XH-district. It is clear from the calculated results that TP-District is more comprehensible than XH-District.

Based on the comprehensive analytical capabilities of 3D NL GAM, analytical results have been validated in this project. First, the results of the 3D NL GAM and the traditional axial map analysis are compared. For the TP-District, the 3D NL GAM indicated that the *NAIN* values of decision points on Riverside-RD were higher than those of other streets, indicating that Riverside-RD had a higher integration of pedestrian flow, while the analysis of traditional axial maps reflected a higher integration of roads located in the center of the district (Figure 20). The comparison with the Baidu heat map [84] showed that the actual number of pedestrian gatherings on Riverside-RD was indeed the most aggregated place in the district, and the analysis of the actual pedestrian flow showed that the 3D NL GAM analysis map better correlated with the actual street pedestrian flow than the axial analysis map. Therefore, the use of the 3D NL GAM enabled more realistic wayfinding simulations and more accurate quantitative indicators of spatial node movement properties.

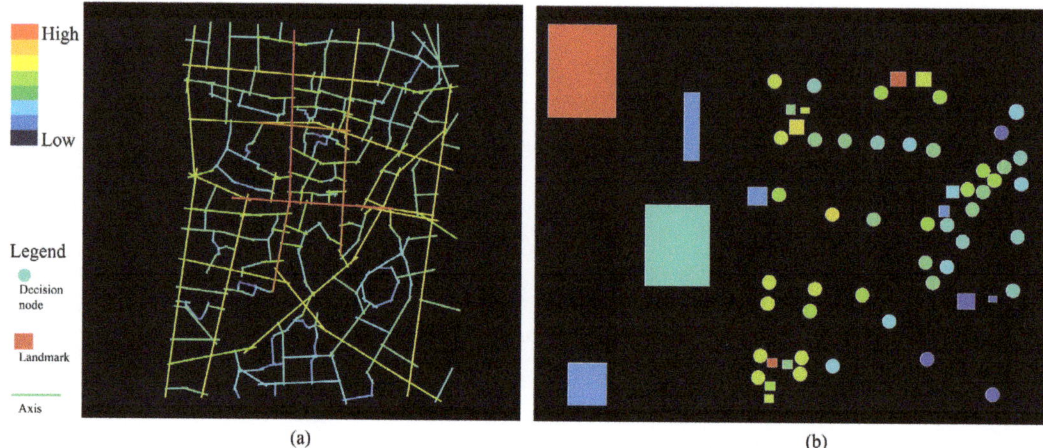

Figure 20. Comparing the calculation results of the node integration degree of the 3D NL GAM graph with the axial analysis graph in TP-District. (**a**) Axial analysis graph; (**b**) 3D NL GAM graph.

6.2. The Significance of Computing the Structural Salience of Landmarks for Urban Design

In addition to evaluating the legibility of built-up urban environments, structural landmark salience analysis has practical guidance for the spatial design of compact cities. It mainly manifests itself in the formulation of planning strategies for large-scale natural landscape resources, control of the combined morphology of building groups, site selection, and body shape control of important single landmark buildings.

The 3D NL GAM of the three sample cases demonstrated that large-scale natural landscapes typically have the highest landmark structural salience in urban areas, and mountainous landscapes also reflect absolute dominance. Therefore, the natural landscape of the mountain should be expanded as much as possible for the landscape corridor perpendicular to the natural landscape interface road. For example, the decision nodes on the EW-RD in the TP-District can see Mount Yuelu, so pedestrians can quickly identify their locations. In urban areas, the water surface of rivers is easily obstructed, but the roads immediately adjacent to the water boundary have individual open view spaces, and the analysis results of the sample cases confirmed that the riverside avenues tended to form the highest concentration of people, so that the river interface played a decisive role in the arrangement of the street grid.

In compact urban areas, the 3D NL GAM can guide urban design in determining building height control requirements for each area. Clearly, a combination of high-rise and low-rise groups in the TP-District is more likely to create significant visible permeable space than a single combination of high-rise groups in the XD-District. Therefore, the placement of large groups of high-rise buildings must fully consider the visible permeable space of the city. With the 3D NL GAM, it is possible to screen important decision point locations and landmark locations, based on the spatial attribute index of the decision points and landmarks, and to achieve as much visual connectivity of important decision points and important landmarks as possible.

The 3D NL GAM facilitates the siting, architectural form, and facade design of individual architectural landmarks. In the sample case, the height variation of the Beichen-A2 landmark in the XD-District demonstrated the interdependence of landmarks and decision nodes. First, landmark buildings should be located on sites where they can form line-of-sight connections to many important decision nodes. Second, optimizing the shape and facade of landmark buildings in the direction of important decision nodes is an important means to highlight the cognitive salience and visual salience of landmarks [13]. In fact, landmark salience is a combination of several salience indicators [24], and the image design of landmarks that is based on the most important line decision nodes is an important means of maintaining the structural prominence of landmarks.

6.3. Further Research Objectives

The current study still has certain limitations, including regarding the exploration of types of urban sample areas and the impact of changes in old and new urban environments. In addition, the construction of the 3D NL GAM is not intelligent enough. The visibility results of landmarks are not automatically passed to the spatial analysis software, and the link of wayfinding decision nodes and landmarks needs to be done manually. Thus, the computational procedure needs to be improved and optimized in the future.

The development of a legible city into an imageable city is the unremitting goal of urban design [10,50,85]. The structural landmarks mentioned in this paper expressed only the visibility and location of landmarks. However, only integrated landmarks can express the image of the city. When the 3D NL GAM confirms landmarks, it is essentially a subjective judgment, which often leads to an exaggerated landmark salience or neglect of landscape resources with more landmark effects in the city. So, is it possible to use 3D NL GAM for inverse derivation of landmark salience for all landscape objects and buildings spanning an urban area?

In addition to the visual, cognitive, and structural salience of landmarks proposed by Sorrows and Hirtle [13], color is also an important factor in landscape architectural identity itself [86]. In a study on the semantic significance of landmarks, Bartie used semantic clustering technology to study the semantic aspect of landmark significance [87,88]. Landmark salience also includes the impact of landmark facade images on cognitive affect [23,87,89]. The literature presents more landmark salience-related factors for 3D NL GAM, which can further enhance the intelligibility evaluation of cities. Therefore, how to deeply interpret the perception of natural landscapes and landmarks is the main research objective for future work on this topic. Furthermore, the 3D NL GAM can be further developed as a tool for assessing and predicting landmark salience, with the integration of multiple factors.

On the one hand, the current implementation of smart cities hopes to identify landmarks in the urban network by machines [24,64]. As applied to a wider range of domains, the analysis of more types of landmark association factors could provide more detailed landmark data to facilitate the establishment of machine learning, and thus improve intelligent landmark navigation technique. On the other hand, compact cities are changing the urban form of the past. Most of the relevant literature is focused on the context of landmark distinctiveness identification in built environments, and it is difficult to form guidelines for landmark architecture design in sheltered urban environments. With increasing computational power, 3D simulation analysis has been able to achieve deeper and more accurate visibility measurements. This facilitates quantitative analysis of the combined indicators of the visual, cognitive, and emotional significance of landmarks, which can be used as design guidance for architects and planners.

7. Conclusions

In this paper, we developed a 3D NL GAM to compute the structural salience of landmarks in compact urban tracts, from the perspective of improving the intelligibility of compact cities. The 3D NL GAM is able to quantitatively describe the spatial properties of urban nodes, their visibility and the structural salience of landmarks with different urban spatial morphological types. Thus, the intelligibility of cities can be deciphered and used to guide urban design for the sustainable development of compact cities.

Currently, cities around the world are in a period of sustainable transition, and there is a need to further strengthen the research regarding the development, planning, and design of urban spaces, in order to cope with the pressures of urban population growth [2]. 3D NL GAM is an effective means to help designers enhance their knowledge of cities from the level of spatial composition and comprehensively interpret the connectivity of urban spatial nodes. It also provides a comprehensive explanation of the connectivity of spatial nodes in the city and extends to the directional guidance of the spatial locations of landmarks, from which 3D NL GAM accurately quantifies the visibility and structural saliency measures of landmarks. Therefore, this study has practical implications for urban studies.

Reinforcing the interactive interpretation of humans and the environment is a fundamental objective of space syntax theory, and the 3D NL GAM reflects that space syntax theory is also highly scalable for the study of urban environments. The computational results of the study show that large-scale natural landscape resources have outstanding structural salience in urban environments. Tall buildings can only reflect a higher structural salience in urban forms with a combination of high- and low-rise building groups, but in forms dominated by purely tall buildings, the role of supertall buildings in enhancing urban intelligibility is limited. These results can help designers effectively deal with urban planning, architectural design, and the conservation of natural landscape resources.

The 3D NL GAM extends the analysis range of urban tracts and upgrades the traditional analysis of the grid from 2D to 3D, thus expanding the application of space syntax theory. In fact, the expansion of the 3D analysis of space syntax has achieved results, with studies investigating the distribution of spatial morphological openness [7,32], the simulation of intelligent bodies [53], etc., which are the basis for forming the concept of this study. The most important contribution of this work is the response of the 3D NL GAM, constructed by expanding the visual perception of the city in horizontal and vertical ranges, and the interpretation of the urban structure in terms of the visibility and saliency of landmarks. This study is limited in terms of the number of samples analyzed, and more sample studies are needed to explore the interaction between humans and urban environments. The construction procedure of 3DNLGAM should be continuously improved in the future, including an intelligent construction of the computational software and the loading of factors for urban environment analysis, so as to expand the applicability of 3D NL GAM in urban environment analysis and to gradually realize the extent of the findings garnered with this method.

Supplementary Materials: The following supporting information can be downloaded at: https://www.mdpi.com/article/10.3390/buildings13041024/s1, Table S1. Spatial attribute of nodes in standard city block when linked and unlinked with landmarks; Table S2. Computation of Structural salience of landmarks in TP-District; Table S3. Computation of Structural salience of landmarks in XD-District (Hight of Beichen-A2 Landmark is 268 m); Table S4. Computation of Structural salience of landmarks in XD-District (Hight of Beichen-A2 Landmark is 400 m).

Author Contributions: Conceptualization, Y.G. and X.H.; methodology, Y.G.; software, Y.G. and J.T.; validation, Y.G. and J.T.; formal analysis, Y.G.; investigation, Y.G.; resources, Y.G.; data curation, Y.G.; writing-original draft preparation, Y.G.; writing-review and editing, Y.G.; visualization, Y.G.; supervision, X.H.; project administration, Y.G.; funding acquisition, Y.G. and X.H. All authors have read and agreed to the published version of the manuscript.

Funding: This work was supported by the Key Disciplines of State Forestry Administration of China (No. 21 of Forest Ren Fa, 2016); Hunan Province "Double First-class" Cultivation discipline of China (No. 469 of Xiang Jiao Tong, 2018).

Data Availability Statement: The data that support the findings of this study are available from the authors upon reasonable request.

Acknowledgments: The authors would like to thank Ke Huang, a planner at the Changsha Planning Information Center, for his suggestions in selecting the landmarks, and the 32 architecture students from the Central South University of Forestry and Technology in 2016 for their on-site crowd data collection. We would like to thank Jikuang Yang from Chalmers University for suggesting this paper.

Conflicts of Interest: The authors declare no conflict of interest.

References

1. Saaty, T.; De Paola, P. Rethinking Design and Urban Planning for the Cities of the Future. *Buildings* **2017**, *7*, 76. [CrossRef]
2. Secretariat, H.I. *New Urban Agenda*; United Nations: New York, NY, USA, 2017.
3. Showkatbakhsh, M.; Makki, M. Multi-Objective Optimisation of Urban Form: A Framework for Selecting the Optimal Solution. *Buildings* **2022**, *12*, 1473. [CrossRef]
4. Programme, U.P.F.G.; Office, U.H.F.U. *SDG Project Assessment Tool Vol1: General Framework*; United Nations Human Settlements Programme: Nairobi, Kenya, 2020; p. 13. Available online: https://unhabitat.org/sites/default/files/2020/07/sdg_tool_general_framework_jan_2020.pdf (accessed on 2 November 2022).
5. Carmona, M.; Tiesdell, S.; Heath, T.; Oc, T. *Public Places—Urban Spaces: The Dimensions of Urban Design*, 2nd ed.; Architectural Press of Elsevier: Burlington, MA, USA, 2010.
6. Rapoport, A. Toward a Redefinition of Density. *Environ. Behav.* **1975**, *7*, 133–158. [CrossRef]
7. Fisher-Gewirtzman, D.; Wagner, I.A. Spatial Openness as a Practical Metric for Evaluating Built-up Environments. *Environ. Plan. B Plan. Des.* **2003**, *30*, 37–49. [CrossRef]
8. Fisher-Gewirtzman, D. The association between perceived density in minimum apartments and spatial openness index three-dimensional visual analysis. *Environ. Plan. B Urban Anal. City Sci.* **2017**, *44*, 764–795. [CrossRef]
9. Dalton, R.C.; Bafna, S. The syntactical image of the city: A reciprocal definition of spatial elements and spatial syntaxes. In Proceedings of the 4th International Space Syntax Symposium, London, UK, 17–19 June 2003.
10. Stamps, A.E. Mystery, complexity, legibility and coherence: A meta-analysis. *J. Environ. Psychol.* **2004**, *24*, 1–16. [CrossRef]
11. Lynch, K. *The Image of the City*; The MIT Press: Cambridge, MA, USA, 1960.
12. Claramunt, C.; Winter, S. Structural Salience of Elements of the City. *Environ. Plan. B Plan. Des.* **2007**, *34*, 1030–1050. [CrossRef]
13. Sorrows, M.E.; Hirtle, S.C. *The Nature of Landmarks for Real and Electronic Spaces*; Springer: Berlin/Heidelberg, Germany, 1999; pp. 37–50.
14. Caduff, D.; Timpf, S. On the assessment of landmark salience for human navigation. *Cogn. Process.* **2008**, *9*, 249–267. [CrossRef] [PubMed]
15. Steck, S.D.; Mallot, H.A. The Role of Global and Local Landmarks in Virtual Environment Navigation. *Presence Teleoperators Virtual Environ.* **2000**, *9*, 69–83. [CrossRef]
16. Röser, F.; Hamburger, K.; Krumnack, A.; Knauff, M. The structural salience of landmarks: Results from an on-line study and a virtual environment experiment. *J. Spat. Sci.* **2012**, *57*, 37–50. [CrossRef]
17. Caduff, D.; Timpf, S. *The Landmark Spider: Representing Landmark Knowledge for Wayfinding Tasks, 2005 AAAI Spring Symposium*; Stanford: San Jose, CA, USA, 2005.
18. Nuhn, E.; König, F.; Timpf, S. "Landmark Route": A Comparison to the Shortest Route. *AGILE GIScience Ser.* **2022**, *3*, 12. [CrossRef]
19. Filomena, G.; Verstegen, J.A. Modelling the effect of landmarks on pedestrian dynamics in urban environments. *Comput. Environ. Urban Syst.* **2021**, *86*, 101573. [CrossRef]
20. Yesiltepe, D.; Dalton, R.; Ozbil, A.; Dalton, N.; Noble, S.; Hornberger, M.; Coutrot, A.; Spiers, H. Usage of landmarks in virtual environments for wayfinding: Research on the influence of global landmarks. In Proceedings of the 12th Space Syntax Symposium, Beijing, China, 8–13 July 2019.
21. Ruddle, R.A.; Volkova, E.; Mohler, B.; Bülthoff, H.H. The effect of landmark and body-based sensory information on route knowledge. *Mem. Cogn.* **2011**, *39*, 686–699. [CrossRef]
22. Kuipers, B. Modeling Spatial Knowledge. *Cogn. Sci.* **1978**, *2*, 129–153. [CrossRef]
23. Yesiltepe, D.; Conroy Dalton, R.; Ozbil Torun, A. Landmarks in wayfinding: A review of the existing literature. *Cogn. Process.* **2021**, *22*, 369–410. [CrossRef]
24. Richter, K.; Winter, S. *Landmarks: GIScience for Intelligent Services*; Springer International Publishing: Cham, Switzerland; Heidelberg, Germany; New York, NY, USA; Dordrecht, The Netherlands; London, UK, 2014.
25. Fischer, L.F.; Mojica Soto-Albors, R.; Buck, F.; Harnett, M.T. Representation of visual landmarks in retrosplenial cortex. *Elife* **2020**, *9*, e51458. [CrossRef] [PubMed]

26. Klippel, A.; Winter, S. *Structural Salience of Landmarks for Route Directions*; Springer: Berlin/Heidelberg, Germany, 2005; pp. 347–362.
27. Winter, S. Route adaptive selection of salient features. In *Spatial Information Theory, Lecture Notes in Computer Science*; Kuhn, W., Worboys, M., Timpf, S., Eds.; Springer: New York, NY, USA, 2003; pp. 349–361.
28. Appleyard, D. Why Buildings Are Known: A Predictive Tool for Architects and Planners. *Environ. Behav.* **1969**, *1*, 131–156. [CrossRef]
29. Hillier, B. *Space is the Machine*, Electronic ed.; Space Syntax: London, UK, 2007.
30. Ostwald, M.J. The Mathematics of Spatial Configuration: Revisiting, Revising and Critiquing Justified Plan Graph Theory. *Nexus Netw. J.* **2011**, *13*, 445–470. [CrossRef]
31. Morello, E.; Ratti, C. A digital image of the city: 3D isovists in Lynch's urban analysis. *Environ. Plan. B Plan. Des.* **2009**, *36*, 837–853. [CrossRef]
32. Kim, G.; Kim, A.; Kim, Y. A new 3D space syntax metric based on 3D isovist capture in urban space using remote sensing technology. *Comput. Environ. Urban Syst.* **2019**, *74*, 74–87. [CrossRef]
33. Morais, F.; Vaz, J.; Viana, D.L.; Carvalho, I.C. 3D Space Syntax Analysis—Case Study 'Casa da Música'. In Proceedings of the 11th Space Syntax Symposium, Lisbon, Portugal, 3–7 July 2017.
34. Ascensão, A.; Costa, L.; Fernandes, C.; Morais, F. 3D Space Syntax Analysis: Attributes to be Applied in Landscape Architecture Projects. *Urban Sci.* **2019**, *3*, 20. [CrossRef]
35. Dalton, R.C.; Dalton, N.S. The problem of representation of 3D isovists. In Proceedings of the 10th International Space Syntax Symposium, London, UK, 13–17 July 2015.
36. Tara, A.; Lawson, G.; Renata, A. Measuring magnitude of change by high-rise buildings in visual amenity conflicts in Brisbane. *Landsc. Urban Plan.* **2021**, *205*, 103930. [CrossRef]
37. Karimi, K. A configurational approach to analytical urban design: 'Space syntax' methodology. *Urban Des. Int.* **2012**, *17*, 297–318. [CrossRef]
38. Yamu, C.; van Nes, A.; Garau, C. Bill Hillier's Legacy: Space Syntax—A Synopsis of Basic Concepts, Measures, and Empirical Application. *Sustainability* **2021**, *13*, 3394. [CrossRef]
39. Hiller, B.; Leaman, A. The man-environment paradigm and its paradoxes. *Arch. Des.* **1973**, *78*, 507–511.
40. Karimi, K. Space syntax: Consolidation and transformation of an urban research field. *J. Urban Des.* **2018**, *23*, 1–4. [CrossRef]
41. Batty, M. *The New Science of Cities*; The MIT Press: Cambridge, MA, USA; London, UK, 2013.
42. Natapov, A.; Fisher-Gewirtzman, D. Visibility of urban activities and pedestrian routes: An experiment in a virtual environment. *Comput. Environ. Urban Syst.* **2016**, *58*, 60–70. [CrossRef]
43. Alkamali, N.; Alhadhrami, N.; Alalouch, C. Muscat City Expansion and Accessibility to the Historical Core: Space Syntax Analysis. *Energy Procedia* **2017**, *115*, 480–486. [CrossRef]
44. Koohsari, M.J.; Oka, K.; Owen, N.; Sugiyama, T. Natural movement: A space syntax theory linking urban form and function with walking for transport. *Health Place* **2019**, *58*, 102072. [CrossRef]
45. Shen, Y.; Karimi, K. Urban evolution as a spatio-functional interaction process: The case of central Shanghai. *J. Urban Des.* **2018**, *23*, 42–70. [CrossRef]
46. Liao, P.; Yu, R.; Gu, N.; Soltani, S. A Syntactical Spatio-Functional Analysis of Four Typical Historic Chinese Towns from a Heritage Tourism Perspective. *Land* **2022**, *11*, 2181. [CrossRef]
47. Bill Hiller, J.H. *The Social Logic of Space*; Cambridge University Press: New York, NY, USA, 1984.
48. Turner, A.; Penn, A.; Hillier, B. An Algorithmic Definition of the Axial Map. *Environ. Plan. B Plan. Des.* **2005**, *32*, 425–444. [CrossRef]
49. Varoudis, T.; Penn, A. Visibility, accessibility and beyond: Next generation visibility graph analysis. In Proceedings of the 10th International Space Syntax Symposium, London, UK, 13–17 July 2015.
50. Lee, J.H.; Ostwald, M.J.; Lee, H. Measuring the spatial and social characteristics of the architectural plans of aged care facilities. *Front. Archit. Res.* **2017**, *6*, 431–441. [CrossRef]
51. Hillier, B.; Iida, S. *Network and Psychological Effects in Urban Movement*; Springer: Berlin/Heidelberg, Germany, 2005; Volume 3693, pp. 475–490.
52. Hillier, B. The architectures of seeing and going: Or, are cities shaped by bodies or minds? And is there a syntax of spatial cognition? In Proceedings of the 4th International Space Syntax Symposium, London, UK, 17–19 June 2003.
53. van Nes, A.; Yamu, C. *Introduction to Space Syntax in Urban Studies*; Springer: Cham, Switzerland, 2021.
54. Benedikt, M.L. To take hold of space: Isovists and isovist fields. *Environ. Plan. B Plan. Des.* **1979**, *6*, 47–65. [CrossRef]
55. Batty, M. Exploring Isovist Fields: Space and Shape in Architectural and Urban Morphology. *Environ. Plan. B Plan. Des.* **2001**, *28*, 123–150. [CrossRef]
56. Batty, M.; Rana, S. The Automatic Definition and Generation of Axial Lines and Axial Maps. *Environ. Plan. B Plan. Des.* **2003**, *31*, 615–640. [CrossRef]
57. Zolfagharkhani, M.; Ostwald, M.J. The Spatial Structure of Yazd Courtyard Houses: A Space Syntax Analysis of the Topological Characteristics of the Courtyard. *Buildings* **2021**, *11*, 262. [CrossRef]
58. Hillier, B.; Yang, T.; Turner, A. Normalising least angle choice in Depthmap and how it opens up new perspectives on the global and local. *J. Space Syntax* **2012**, *3*, 155–193.

59. Yang, T.; Li, M.; Shen, Z. Between morphology and function: How syntactic centers of the Beijing city are defined. *J. Urban Manag.* **2015**, *4*, 125–134. [CrossRef]
60. Turner, A. Depthmap-a program to perform visibility graph analysis. In Proceedings of the 3rd International Space Syntax Symposium, Atlanta, GA, USA, 7–11 May 2001.
61. Pinelo, J.; Turner, A. *Introduction to UCL Depthmap 10 Version 10.08.00r*; UCL Press: London, UK, 2010.
62. Abdulmawla, A.; Bielik, M.; Buš, P.; Chang, M.-C.; Dennemark, M.; Fuchkina, E.; Miao, Y.; Knecht, K.; Koenig, R.; Schneider, S. DeCodingSpaces Toolbox for Grasshopper: Computational Analysis and Generation of STREET NETWORK, PLOTS and BUILDINGS. Available online: https://kar.kent.ac.uk/id/eprint/68953 (accessed on 10 April 2023).
63. Natapov, A.; Czamanski, D.; Fisher-Gewirtzman, D. Can visibility predict location? Visibility graph of food and drink facilities in the city. *Surv. Rev.* **2013**, *45*, 462–471. [CrossRef]
64. Natapov, A.; Fisher-Gewirtzman, D. Cities as Visuospatial Networks. In *Smart City Networks*; Rassia, S.T., Pardalos, M., Eds.; Springer: Cham, Switzerland, 2017; pp. 191–205.
65. Bielik, M.; König, R.; Fuchkina, E.; Schneider, S.; Abdulmawla, A. Evolving Configurational Properties: Simulating multiplier effects between land use and movement patterns. In Proceedings of the 12th Space Syntax Symposium, Beijing, China, 8–13 July 2019.
66. Hillier, B. Cities as movement economies. *Urban Des. Int.* **1996**, *1*, 41–60. [CrossRef]
67. Hillier, B.; Penn, A.; Hanson, J.; Ki, T.G.; Xu, J. Natural movement: Or, configuration and attraction in urban pedestrian movement. *Environ. Plan. B Plan. Des.* **1993**, *20*, 29–66. [CrossRef]
68. Nuhn, E.; Timpf, S. Do people prefer a landmark route over a shortest route? *Cartogr. Geogr. Inf. Sci.* **2022**, *49*, 407–425. [CrossRef]
69. Nuhn, E.; Timpf, S. Landmark weights—An alternative to spatial distances in shortest route algorithms. *Spat. Cogn. Comput.* **2022**, 1–27, *ahead-of-print*. [CrossRef]
70. De Floriani, L.; Marzano, P.; Puppo, E. Line-of-Sight Communication on Terrain Models. *Geogr. Inf. Syst.* **1994**, *8*, 329–342. [CrossRef]
71. Suleiman, W.; Joliveau, T.; Favier, E. A New Algorithm for 3D Isovists. In *Advances in Spatial Data Handling*; Richardson, D., Oosterom, P., Eds.; Springer: Berlin/Heidelberg, Germany, 2012; pp. 157–173.
72. Fisher-Gewirtzman, D. Can 3D Visibility Calculations along a Path Predict the Perceived Density of Participants Immersed in a Virtual Reality Environment. In Proceedings of the 11th Space Syntax Symposium, Lisbon, Portugal, 3–7 July 2017; Heitor, T., Serra, M., Silva, J.P., Bacharel, M., Silva, L.C.D., Eds.; Departamento de Engenharia Civil, Arquitetura e Georrecursos, Instituto Superior Técnico: Lisboa, Portugal, 2017.
73. Bhatia, S.; Chalup, S.K.; Ostwald, M.J. Wayfinding: A method for the empirical evaluation of structural saliency using 3D Isovists. *Archit. Sci. Rev.* **2013**, *56*, 220–231. [CrossRef]
74. Fisher-Gewirtzman, D.; Bruchim, E. Considering Variant Movement Velocities on the 3D Dynamic Visibility Analysis (DVA)—Simulating the perception of urban users: Pedestrians, cyclists and car drivers, Computing for a better tomorrow. In Proceedings of the 36th eCAADe Conference, 17–21 September 2018; Kepczynska-Walczak, A.B.S., Ed.; Lodz University of Technology: Lodz, Poland, 2018; pp. 569–576.
75. Bielik, M.; Fuchkina, E.; Schneider, S.; Koenig, R. 2D and 3D Isovists for Visibility Analysis, DeCodingSpaces Toolbox for Grasshopper. 2019. Available online: https://toolbox.decodingspaces.net/tutorial-2d-and-3d-isovists-for-visibility-analysis/ (accessed on 10 April 2023).
76. Oxman, R.; Gu, N. Theories and Models of Parametric Design Thinking. In Proceedings of the 33rd International Conference of ECAADE Conference, Vienna, Austria, 16–18 September 2015; pp. 2–6.
77. Payne, A.; McNeel, B.; Davidson, S. *The Grasshopper Primer*, 3rd ed.; GitBook: Lyon, France, 2016.
78. Issa, R. *Essential Algorithms and Data Structures for Computational Design in Grasshopper*, 1st ed.; Robert McNeel & Associates: Seattle, WA, USA, 2020.
79. von Richthofen, A.; Knecht, K.; Miao, Y.; König, R. The 'Urban Elements' method for teaching parametric urban design to professionals. *Front. Archit. Res.* **2018**, *7*, 573–587. [CrossRef]
80. Bielik, M.; Schneider, S.; König, R. Parametric Urban Patterns—Exploring and integrating graph-based spatial properties in parametric urban modelling. In Proceedings of the eCAADe 2012, Prague, Czech Republic, 12–14 September 2012.
81. Gehl, J. *Cities for People*; ISLAND PRESS: Washington, DC, USA, 2010.
82. Garnero, G.; Fabrizio, E. Visibility analysis in urban spaces: A raster-based approach and case studies. *Environ. Plan. B Plan. Des.* **2015**, *42*, 688–707. [CrossRef]
83. Sayed, K.A.; Turner, A.; Hillier, B.; Iida, S.; Penn, A. *Space Syntax Methodology*; Bartlett School of Architecture, UCL: London, UK, 2014.
84. Bao, W.; Gong, A.; Zhang, T.; Zhao, Y.; Li, B.; Chen, S. Mapping Population Distribution with High Spatiotemporal Resolution in Beijing Using Baidu Heat Map Data. *Remote Sens.* **2023**, *15*, 458. [CrossRef]
85. Jacobs, A.; Appleyard, D. Toward an Urban Design Manifesto. *J. Am. Plan. Assoc.* **1987**, *53*, 112–120. [CrossRef]
86. Zhang, L.; Cheng, Y.; Ma, L. A Quantitative Study on the Colour of City Landmark Landscape Architectures. *J. Phys. Conf. Ser.* **2019**, *1288*, 12011. [CrossRef]
87. Balaban, C.Z.; Karimpur, H.; Röser, F.; Hamburger, K. Turn left where you felt unhappy: How affect influences landmark-based wayfinding. *Cogn. Process.* **2017**, *18*, 135–144. [CrossRef] [PubMed]

88. Bartie, P.; Mackaness, W.; Petrenz, P.; Dickinson, A. Identifying related landmark tags in urban scenes using spatial and semantic clustering. *Comput. Environ. Urban Syst.* **2015**, *52*, 48–57. [CrossRef]
89. Palmiero, M.; Piccardi, L. The Role of Emotional Landmarks on Topographical Memory. *Front. Psychol.* **2017**, *8*, 763. [CrossRef] [PubMed]

buildings

Review

Socio-Spatial Experience in Space Syntax Research: A PRISMA-Compliant Review

Ju Hyun Lee *, Michael J. Ostwald and Ling Zhou

School of Built Environment, Faculty of Arts, Design & Architecture, The University of New South Wales, Sydney, NSW 2052, Australia
* Correspondence: juhyun.lee@unsw.edu.au

Abstract: Characterising and predicting socio-spatial experience has long been a key research question in space syntax research. Due to the lack of synthesised knowledge about it, this review conducts the first systematic scoping review of space syntax research on the relationships between spatial properties and experiential values. Adopting the "Preferred Reporting Items for Systematic reviews and Meta-Analyses" (PRISMA) framework, this review of space syntax research identifies 38 studies that examine socio-spatial experiences in architectural, medical, and urban spaces. The data arising from this systematic review are used to identify trends in this sub-field of research, including the growth of socio-spatial methods and applications in urban analytics since 2016 and key methodological approaches, characteristics, and factors in space syntax research about socio-spatial experience. The research identified using the systematic framework employs a mixture of descriptive, correlation, and regression methods to examine the dynamic effects of spatial configurations on human experiences. Arising from the results of the review, the article further identifies a collective, predictive model consisting of five syntactic predictors and three categories of experiential values. This article, finally, examines research gaps and limitations in the body of knowledge and suggests future research directions.

Keywords: systematic review; space syntax; PRISMA; PECO framework; socio-spatial experience

Citation: Lee, J.H.; Ostwald, M.J.; Zhou, L. Socio-Spatial Experience in Space Syntax Research: A PRISMA-Compliant Review. *Buildings* **2023**, *13*, 644. https://doi.org/10.3390/buildings13030644

Academic Editor: Lucia Della Spina

Received: 27 January 2023
Revised: 17 February 2023
Accepted: 24 February 2023
Published: 28 February 2023

1. Introduction

Space syntax theorises the relationship between spatial patterns and human behaviours. For example, the space syntax method axial line analysis (ALA) identifies and examines a network of "longest and fewest" lines of sight [1] and access in a plan of the built environment, which also represent human behavioural characteristics such as movement paths and navigational choices. A second space syntax technique, convex space analysis (CSA), develops and then uses a map that describes visibly defined or enclosed spaces and their adjacent accesses in terms of social interaction and inhabitation. Such a map can be used to measure the ways architectural and urban spaces both limit and enable human movement and perception [2]. A third technique, isovist analysis (ISA), is based on isovist fields or viewsheds that geometrically represent the spatio-visual properties of locations in an environment. This method can be used for predicting movement as well as controlling observation [3]. Additional space syntax techniques such as justified plan graph (JPG) analysis, visibility graph analysis (VGA), and agent-based simulation (ABS) [3] have also been used to study the social, cognitive, and behavioural properties of spaces. From these studies, multiple space syntax measures and indexes have been proposed and applied for measuring and understanding socio-spatial properties.

Using these various syntactic measures, space syntax has been widely applied to explore the relations between spatial organisations and social effects in the built environment. Furthermore, multiple spatial properties are measured to identify spatial topologies and social relations, regardless of the forms that contain them [4]. The complicated statistics of spatial configurations have enabled the characterisation and prediction of the socio-spatial experience in the built environment [5]. However, while hundreds of space syntax studies have been published in a range of fields over the last 40 years, systematic literature reviews of space syntax methods and results have only rarely been conducted. There are a few notable exceptions [6–8], including Nes's and Yamu's *"Introduction to Space Syntax in Urban Studies"* [9]. Such past reviews typically employ descriptive or narrative approaches, and only rarely consider specific research applications, the validity of results, or methodological limitations. An exception is the work of Sharmin and Kamruzzaman [7], who conducted a meta-analysis of the relationship between syntactic measures and pedestrian movements, providing a statistical analysis of the findings and a synthesis of the evidence. Their approach, which both is focussed on a particular application and includes a critical review of the evidence, is one of the few examples of its type. It highlights the lack of other similar systematic reviews in this field and the need for research which critically syntheses complex space syntax methods, indices, and results about the relationships between syntactic properties and human experiences. As such, a major research question in this review is: "how has space syntax research characterised and predicted socio-spatial experience in the built environment?" To address this research problem, this article conducts the first systematic scoping review of space syntax research on socio-spatial relations—which is a sub-set of the much larger body of space syntax applications. This article aims to develop a collective understanding of socio-spatial approaches and factors in space syntax research by identifying and mapping the available evidence in the literature and discussing the relationships between syntactic values and human experiences.

The primary approach used in this article, a systematic literature review, is a rigorous, reproducible methodology that identifies, selects, appraises, and synthesises the relevant studies and evidence [10]. In accordance with the best practice in the field, this research adopts both the Preferred Reporting Items for Systematic reviews and Meta-Analyses (PRISMA) standard [10] and the PECO framework (population or problem, exposure or environment, comparator, and outcome) [11] to frame the research questions and goals of the review. Although these systematic review frameworks were largely developed and applied in medical research, a PRISMA-compliant review has been widely used in a variety of domains as it is an accepted approach for synthesising knowledge about large research fields and specific applications. It has been used, for example, for conceptualising and categorising smart-city projects [12], identifying trends and perspectives in smart-cities research [13], and introducing the concepts and techniques of machine learning in urban studies [14]. Most of these reviews are, however, limited to identifying the types of available evidence and key characteristics or factors. Thus, they could be considered 'systematic scoping reviews'. In contrast, a full systematic review should identify and synthesis the evidence [15], but due to the complexity in the research scales and measures, only a few studies in architectural and urban sciences have fulfilled the required conditions of a systematic review. For this reason, the present article conducts a systematic scoping review, a precursor to a full systematic review, using the PRISMA framework to provide a holistic overview of experiential variables in space syntax research. In accordance with the principles of meta-analysis, the experiential data, comparative values, and syntactic parameters considered in this article are synthesised and explored to capture the key characteristics and socio-spatial factors in space syntax research. In addition, as the problem addressed in this article is socio-spatial experience in space syntax research, the relevant outcomes identified through this scoping review are focused on the relationship between syntactic and experiential data.

The following section highlights three methodological stages in a PRISMA-compliant review: (1) finding relevant data, (2) data selection, and (3) synthesis. This article then reports the synthesised characteristics and outcomes of the identified space syntax studies and considers their implications. This article concludes with a discussion about the research limitations and contributions.

2. Materials and Methods

2.1. Finding Relevant Data

In accordance with the PECO framework for systematic reviews, the scope of this review has four elements. First, it is concerned with socio-spatial experience in space syntax research (P: problem). Second, it is focused on syntactic properties of the built environment that were measured using space syntax techniques (E: Environment). Experiential or experimental properties are identified as comparative values to the syntactic parameters (C: Comparison). The last element highlights the relationship between syntactic and experiential data (O: Outcome). From these criteria, search terms are developed for finding relevant articles. These terms are (i) space syntax, (ii) experience, and (iii) experiment or test. This review used a specific search string:

"Space syntax" AND experience AND (experiment OR test)

This review developed the above search string to produce a set of articles with an appropriate degree of relevance. Nine academic databases—Science Direct, Wiley Online Library, Taylor & Francis Online, SAGE Journals, JSTOR, Web of Science, UCL Discovery, Emerald Insight, and MDPI—were used to identify relevant publications (refereed journal articles only) in late 2021 (November 2021 last search). In addition to the online database search, both forward and backward citation searches were run using Google Scholar.

2.2. Data Selection

This review used two rounds of data-selection workflows, online database search, and citation search in Figure 1, extended from the conventional PRISMA templates [10,16]. The first, *online database search*, starts by identifying articles from the nine databases with the chosen search string (n = 1317). After excluding duplicate records and inappropriate formats of publications (conference papers, reports, and chapters), 969 articles were screened by title and abstract using the PECO selection criteria framework. Two authors assessed the quality of the resulting 42 articles for eligibility using an independent full-text review structured around the four PECO questions. A simple rating system (Yes = 1; No = 0; partially = 0.5) was used for the full-text screening and 26 articles from the database search that scored greater than or equal to 3.5 points by each assessor were included in the online database search.

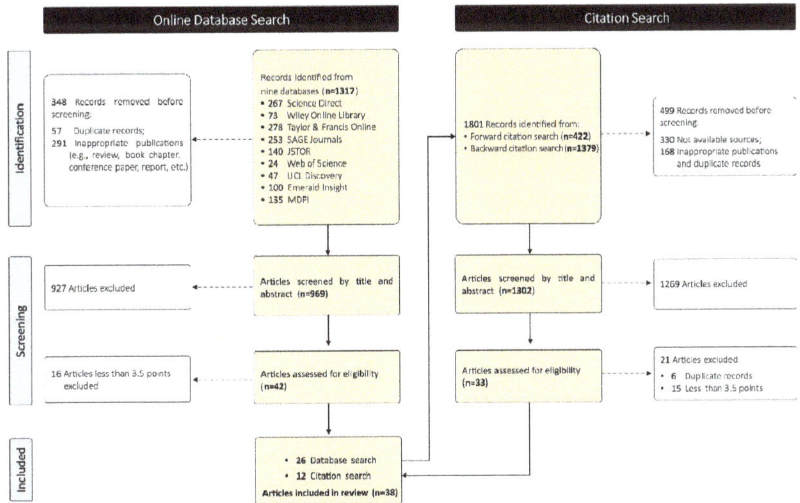

Figure 1. Two rounds of data-selection flows based on the PRISMA templates [10,16].

The second round of data selection, *citation search*, involved forward and backward citation searches. The former identified research articles that cited the 26 articles selected in the online database search (n = 422), while the latter selected all the references in the 26 articles (n = 1379). In the screening phase, 1302 articles were screened by title and abstract and 33 articles were then assessed for eligibility by the two authors. Finally, 12 articles were included in the citation search, leading to a total of 38 articles being selected for this scoping review (see also Table 1).

2.3. Synthesis

Space syntax research on socio-spatial experience typically consists of three research components, (i) syntactic (or topological map) analysis and (ii) experiential (or experimental) data analysis, closely relating to computational and social research, respectively. The two analytic results are then mapped and compared in (iii) relational analysis such as correlation, where conventional statistical analysis is often applied. Collectively, this review takes a step-by-step synthesis to the three components. Furthermore, based on the PRISMA checklist, the synthesis of the data highlights three aspects of the findings of the 38 articles, (i) syntactic properties of the built environment, (ii) experiential values, and (iii) the relationships between spatial properties and experiential values. The first synthesis addresses specific space syntax techniques (ALA, CSA, ISA, JPG, VGA, ABS, SA) and syntactic parameters or measures, e.g., *integration, connectivity, intelligibility, control value, mean depth, choice, isovist properties, visibility,* and *synergy*. Second, experiential values are synthesised with a focus on experimental methods and their results recorded for factors such as behavioural pattern, movement, spatial choice, and human perception. Lastly, this review categorises and discusses the findings of the articles in terms of the relationships between syntactic and experiential data. In addition, this review summarises the spatial properties impacting on socio-spatial experience in the research.

Table 1. Thirty-eight articles included in two rounds of data selection.

Author	Year	Title	Research Area	Source
26 articles included in the online database search				
Aknar and Atun	2017	Predicting movement in architectural space	Architecture	[17]
Alalouch and Aspinall	2007	Spatial attributes of hospital multi-bed wards and preferences for privacy	Medical space	[18]
Askarizad and Safari	2020	The influence of social interactions on the behavioral patterns of the people in urban spaces (case study: The pedestrian zone of Rasht Municipality Square, Iran)	Urban space	[19]
Can and Heath	2016	In-between spaces and social interaction: a morphological analysis of Izmir using space syntax	Urban space	[20]
Chiang and Li	2019	Metric or topological proximity? The associations among proximity to parks, the frequency of residents' visits to parks, and perceived stress	Urban space	[21]
El-Hadedy and El-Husseiny	2021	Evidence-Based Design for Workplace Violence Prevention in Emergency Departments Utilizing CPTED and Space Syntax Analyses	Medical space	[22]
Elshater et al.	2019	What makes livable cities of today alike? Revisiting the criterion of singularity through two case studies	Urban space	[23]
Esposito and Camarda	2020	Agent-Based Analysis of Urban Spaces Using Space Syntax and Spatial Cognition Approaches: A Case Study in Bari, Italy	Urban space	[24]
Ferdous and Moore	2015	Field Observations into the Environmental Soul	Medical space	[25]
Geng et al.	2021	Comparative analysis of hospital environments in Australia and China using the space syntax approach	Medical space	[26]

Table 1. *Cont.*

Author	Year	Title	Research Area	Source
Güngör and Harman	2020	Defining urban design strategies: an analysis of Iskenderun city center's imageability	Urban space	[27]
Hidayati et al.	2020	Realised pedestrian accessibility of an informal settlement in Jakarta, Indonesia	Urban space	[28]
Hölscher et al.	2012	Challenges in Multilevel Wayfinding: A Case Study with the Space Syntax Technique	Architecture	[29]
Knöll et al.	2018	A tool to predict perceived urban stress in open public spaces	Urban space	[30]
Li and Klippel	2012	Wayfinding in Libraries: Can Problems Be Predicted?	Architecture	[31]
Liu et al.	2018	Spatial Configuration and Online Attention: A Space Syntax Perspective	Urban space	[32]
Mansouri and Ujang	2017	Space syntax analysis of tourists' movement patterns in the historical district of Kuala Lumpur, Malaysia	Urban space	[33]
Marquardt et al.	2011	Association of the Spatial Layout of the Home and ADL Abilities Among Older Adults with Dementia	Medical space	[34]
Mohamed and Stanek	2020	The influence of street network configuration on sexual harassment patterns in Cairo	Urban space	[35]
Nubani et al.	2018	Measuring the Effect of Visual Exposure and Saliency of Museum Exhibits on Visitors' Level of Contact and Engagement	Architecture	[36]
O'Hara et al.	2018	Macrocognition in the Healthcare Built Environment (mHCBE): A Focused Ethnographic Study of "Neighborhoods" in a Pediatric Intensive Care Unit	Medical space	[37]
Omer and Goldblatt	2017	Using space syntax and Q-analysis for investigating movement patterns in buildings: The case of shopping malls	Architecture	[38]
Ozbil et al.	2021	Children's Active School Travel: Examining the Combined Perceived and Objective Built-Environment Factors from Space Syntax	Urban space	[39]
Rashid et al.	2014	Network of Spaces and Interaction-Related Behaviors in Adult Intensive Care Units	Medical space	[40]
Tzeng and Huang	2009	Spatial Forms and Signage in Wayfinding Decision Points for Hospital Outpatient Services	Medical space	[41]
Zeng et al.	2020	Inheritance or variation? Spatial regeneration and acculturation via implantation of cultural and creative industries in Beijing's traditional compounds	Architecture	[42]
Zhai and Baran	2016	Do configurational attributes matter in context of urban parks? Park pathway configurational attributes and senior walking	Urban space	[43]
12 articles included in the citation search				
Domènech et al.	2020	Built environment and urban cruise tourists' mobility	Urban space	[44]
Keszei et al.	2019	Space Syntax's Relation to Seating Choices from an Evolutionary Approach	Architecture	[45]
Koohsari et al.	2016	Walkability and walking for transport: characterizing the built environment using space syntax	Urban space	[46]
Neo and Sagha-Zadeh	2017	The influence of spatial configuration on the frequency of use of hand sanitizing stations in health care environments	Medical space	[47]
Pagkratidou et al.	2020	Do environmental characteristics predict spatial memory about unfamiliar environments?	Urban space	[48]

Table 1. *Cont.*

Author	Year	Title	Research Area	Source
Rashid et al.	2016	Physical and Visual Accessibilities in Intensive Care Units: A Comparative Study of Open-Plan and Racetrack Units	Medical space	[49]
Shatu et al.	2019	Shortest path distance vs. least directional change: Empirical testing of space syntax and geographic theories concerning pedestrian route choice behaviour	Urban space	[50]
Sheng et al.	2021	Effect of Space Configurational Attributes on Social Interactions in Urban Parks	Urban space	[51]
Zerouati and Bellal	2020	Evaluating the impact of mass housings' in-between spaces' spatial configuration on users' social interaction	Urban space	[52]
Zhai et al.	2018	Can trail spatial attributes predict trail use level in urban forest park? An examination integrating GPS data and space syntax theory	Urban space	[53]
Zhang et al.	2020	Combining GPS and space syntax analysis to improve understanding of visitor temporal–spatial behaviour: a case study of the Lion Grove in China	Urban space	[54]

3. Results

3.1. General Characteristics

As shown in Figure 2 (see also Table 1), 20 of the collected 38 articles have a focus on 'urban space' (52.63%), ten address 'medical space' (26.32%), and seven 'architecture' (18.42%). One article investigating mountainous sites was not included in these three categories. The two dominant research areas were neighbourhoods [20,24,28,39,52] and parks [21,43,51,53]. The 38 research articles were published between 2007 and 2021. The number of articles records a spike in 2020, largely published about 'urban space'.

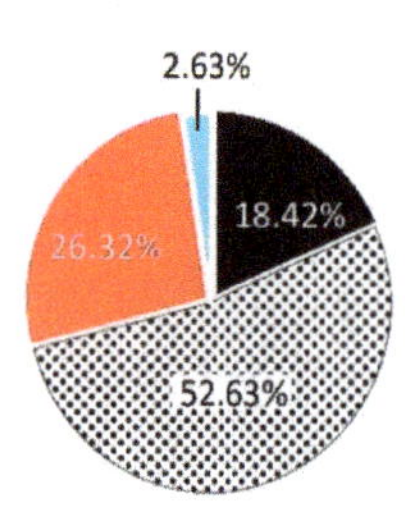

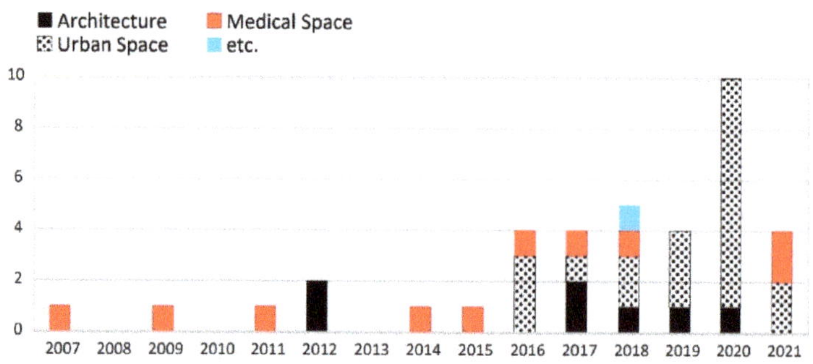

Figure 2. The percentage of articles by research area and the number of articles by publication year.

The review suggests that there is a growth in the number of relevant publications, although this may also be a by-product of database indexing time. The increasing availability of geographical information systems (GIS), global positioning systems (GPS), and state-of-the-art tracking technologies may also be shaping the increase in publications in urban space since 2016 in Figure 2. Each of three journals, *Cities*, *Environment and Planning B: Urban Analytics and City Science*, and *Sustainability*, published three articles. These journals have also regularly published space syntax research on architectural and urban spaces. The other journals, e.g., *Journal of Urbanism*, *Habitat International*, *Frontiers of Architectural Research*, and *Open House International*, published one or two articles on this topic. Interestingly, space syntax research on "medical space" (e.g., hospital planning, intensive care units (ICU), etc.) has been published at a steady rate of one or two articles per year

(Figure 2). As such, health-related journals, e.g., *HERD: Health Environments Research & Design Journal, Behavioral Sciences, American Journal of Alzheimer's Disease & Other Dementias,* and *International Journal of Environmental Research and Public Health,* have an interest in space syntax, with ten articles identified in this review. The relationship between syntactic measures of medical spaces and human experiences or responses has become a valuable research topic in recent years [3].

3.2. Space Syntax Analysis

3.2.1. Methodological Approaches

Of the 38 articles, space syntax research about urban space generally used ALA. As such, the collected articles typically generated the syntactic properties of cities using axial maps, focusing on movement paths (typically roads). A wide range of sites was explored in this research, including city areas [23,46], public open spaces [19,30], city centres or central business districts [27,50], tourist destinations [44], and historical sites [33]. Several different ranges of radius measures were used for the analysis of movement: an 800 m radius measure for a ten-minute walk [27,39], a 400 m for a five-minute walk [35,44], and a combined measure of 1000 m and 400 m for topological proximity [21]. One lesson from these studies is that a buffer-zone radius measure should be carefully defined before the development of a topological map.

UCL DepthmapX (version 0.8.0) software was typically used for the ALA on urban spaces, often in parallel with *ArcGIS* [21,27,53,54]. The *QGIS* space syntax toolkit was also developed for this purpose [44]. ALA provided the dominant technique to measure *global integration* and *local integration* from traditional street segments, while it was applied to develop angular syntactic properties in urban networks [27,28,50,51]. The 20 urban studies in the collected data tended to use such *single* syntactic (ALA or SA) methods focusing on natural movements. In contrast, three studies combined it with visuo-spatial analysis (i.e., VGA and ISA) to address social interaction [19], cognitive aspects of a city's imageability [27], and even perceived urban stress [30]. Interestingly, instead of ALA, the configurational characteristics of park pathways were captured using a modified CSA focusing on the depth of a space [43]. Density, an influential factor on human experience in urban space, was also measured to improve the predictive power of spatial analytic data [30,32,46]. Beyond these syntactic approaches, geographic information was developed using Google Earth [23], OpenStreetMap [30,35], and GIS [44,53,54].

The remainder (18 articles) conducted syntactic analyses of plan layouts in architecture (medical and architectural spaces). ALA tended to be used to investigate indoor movement (eight articles), while other architectural studies developed visuo-spatial properties from visibility graph measures such as VGA (11 articles) and ISA (four articles). A few articles used VGA or ISA as a *sole* technique for spatial analysis to explore psychological seating choices in a three-dimensional (3D) virtual model [45], the visibility and accessibility of medical spaces [18,25,47], and macro-cognitive interactions [37]. Interestingly, Alalouch and Aspinall [18] used a small grid size (200 mm × 200 mm) for VGA in a possible attempt to increase the statistical significance in their relational analysis. However, there is no consensus about the spacing across the set of works. For example, one study used 1 foot [36] and another used 1 m [55], while a "human-scale" grid was typically recommended [56]. Furthermore, in terms of the combination of VGA and ABS, a Fibonacci retracement calculator was applied to calculate "ideal" *integration* values and predict human movements in a restaurant [17].

In the set of articles, the combination of ALA and VGA (or ISA) was often used to investigate wayfinding [29,31,41], visual accessibility [22,49], and spatial memory [48]. Several architectural studies developed syntactic properties from convex maps (CSA) or plan graphs (JPG). Marquardt et al. [34] addressed the convexity of homes along with clinical variables, and Zeng et al. [42] measured "symmetry–asymmetry" using JPG. To examine both axial and node integrations in hospitals, JPG was associated with ALA [22,40]. Furthermore, as a complementary approach to space syntax techniques, Omer and Goldblatt [38] presented "Q-analysis" using a connectivity graph where axial maps for each building floor were transformed into segment maps. Their method could be useful for investigating the multidimensional structure of movement patterns in architecture.

In summary, the collected data revealed that the conventional alpha analysis, along with geographic information, was widely used in urban studies. In contrast, architectural studies tended to adopt more diverse space syntax techniques, highlighting visuo-spatial properties in the built environment. *UCL Depthmap* was dominantly used in 27 articles (71.05%) in the data, and it is clear that space syntax techniques and approaches have continued to evolve across the body of research captured in this analysis.

3.2.2. Syntactic Properties

The syntactic properties of the built environment can be understood as pertaining to either "environment" or "exposure" (E) in the PECO framework. Figure 3 describes the percentage of articles using specific space syntax techniques and syntactic parameters and their relationships. ALA was a dominant technique, accounting for 23 out of 38 articles or 60.53%. This result aligns with the dominant research area, "urban space", in Figure 2, because it was usually explored with ALA. VGA was the second-most-frequently (36.84%) used to examine indoor spaces. ISA is the third (18.42%) in Figure 3a. Each of CSA, JPG, and SA accounted for 13.16%, while only three articles used ABS. It should be acknowledged, however, that many articles use more than one space syntax technique. VGA is, for example, combined with ALA [19,27,29,31,48], ISA [19,26,27], ABS [17,26], and JPG [22]. There are also combinations of three techniques—ALA, ISA, and VGA [19]; ALA, JPG, and VGA [22]; ISA, VGA, and ABS [26]. Lastly, Güngör and Harman Aslan [27] present a combined method of four techniques, ALA, ISA, VGA, and SA, as a holistic way to define urban design strategies from cognitive and syntactic properties.

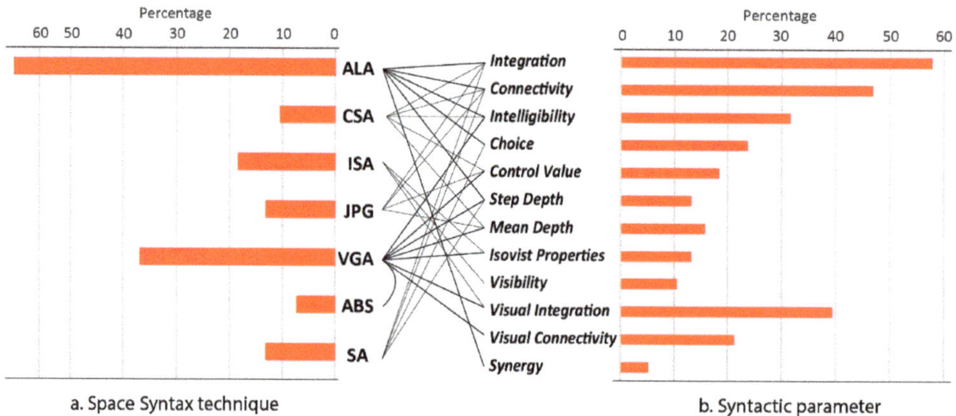

Figure 3. The percentage of articles using specific space syntax techniques and syntactic parameters. (KEY—ALA: Axial Line Analysis, CSA: Convex Space Analysis, ISA: Isovist Analysis, JPG: Justified Plan Graph Analysis, VGA: Visibility Graph Analysis, ABS: Agent-based Simulation, and SA: Segment Analysis).

As shown in Figure 3b, each space syntax technique produced several syntactic parameters. Among them, *integration*, including *global integration* and *local integration*, was measured in 22 out of 38 articles or 57.89%. Since *visual integration* (39.47%) of VGA is a variation of the same parameter, the number of articles addressing the dominant syntactic parameter could be more than 22. However, this review distinguishes the two, as they use different maps, axial map and visibility map, respectively. The second-most-common parameter was *connectivity* (47.37%), the third *visual integration*, and the fourth *intelligibility* (31.58%). *Integration, connectivity,* and *intelligibility* should be fundamental values because research suggests that they are closely related to behaviour or movement patterns in the built environment. In addition to these dominant parameters, *choice* (23.68%) was frequently used in ALA or SA. *Synergy*, as a second-order measure like *intelligibility*, used in ALA, accounted for 5.26%. SA, as an alternative to ALA, also produced these parameters.

In contrast, CSA and JPG developed relatively more *control value* and *mean depth* measures but both techniques also weighted *integration* and *connectivity*. Again, VGA produced *visual integration* (39.47%) and *visual connectivity* (21.05%), while ISA developed several isovist properties (13.16%) such as *isovist area* and *perimeter*. Two articles used ABS along with VGA. Interestingly, *visibility* accounted for 10.53%, but the definitions of *visibility* were slightly different across the articles. For example, *visibility* could be regarded as either *visual integration* or *visual access* [31], the relative size of isovist areas [30], line of sight [37], and even *visual connectivity* [47]. A few additional parameters—*convexity* [34], *layout complexity* [31], and *openness* [37]—were derived only once in the set of work and are not counted in Figure 3b. Acknowledging these specific parameters, the relationship between space syntax techniques and syntactic parameters in the figure and the synthesised, holistic view should be very useful for space syntax research with a focus on socio-spatial experience.

3.2.3. Discussion of Syntactic Analysis

Although this review has focused on the synthesis of syntactic properties, there is a level of complexity in these which is important to note. All space syntax variables have their own definitions and purposes in terms of socio-spatial epistemology [5]. For example, past research has clearly distinguished *local integration (R3)* from *global integration (Rn)* [21,23,32,36,57,58]. The former is used as the best predictor of "smaller-scale movement" or "pedestrian movement", while the latter is better for "larger-scale movement" or "vehicular movement" [57]. Furthermore, *normalized angular integration* [51,59,60] developed by SA, could be more useful for an accurate prediction of movement than topological *integration* [61–63]. *Visual integration* (local and global) of VGA was also different from *integration* measured with ALA. That is, different space syntax maps or graphs could produce measures with the same name but meaning different things in different contexts. In addition, *integration* generally indicates the accessibility of the system, but *choice* is also sometimes used for this purpose because it can be regarded as an improved measure for human movement prediction [64]. Thus, like *visibility* in this review, each syntactic variable should be clearly defined, even of it is widely used in this domain.

Due to the heterogeneity and complexity of socio-spatial properties, it is reasonable that space syntax research uses more than one syntactic technique and variable. Specifically, many articles combined VGA with another syntactic technique. This is possible because, regardless of whether the topological graph (of urban space or of a large building) is partly or fully manual, the *UCL Depthmap* software automatically extracts both ALA and VGA from it. Another reason why VGA was preferred by space syntax researchers is the relationship between "visibility and permeability" [65,66]. That is, ALA and CSA describe the permeability of the system, while VGA identifies the visibility properties. In this way, their combination could capture the hybrid impact of spatial perceptions and structures on movement and wayfinding in architectural and urban design.

3.3. Experiential Data Analysis

3.3.1. Experiential Values

The experiential or experimental data in the 38 articles are now classified into four types: (i) movement, (ii) behaviour, (iii) perception, and (iv) spatial preference. Fifteen articles (39.47%) examined movement patterns that were identified by route, movement volume, time, stopping, distance, and speed. These parameters were developed from GPS or electronic tracking [44,47,53,54], questionnaire and observation [24,32,39,43,46,50], video recording [29,41], gate counting [33,38], and observation tracking [36]. Wayfinding behaviours [29,31,36,41] were also regarded as part of movement patterns.

Secondly, ten articles (26.32%) produced behavioural data using questionnaires, interviews, and observations. For example, Sheng et al. [51] adopted field observations to count personal and social activities in urban parks. Askarizad and Safari [19] used gate counting and snapshot observation to identify both movement and behavioural data (e.g., walking, talking, and standing). Snapshot observation was also used in two more articles [20,52]. Marquardt et al. [34], furthermore, used an in-home assessment to capture activities of daily living (ADL). In contrast, seven articles (18.42%) dealt with participants' perceptions of built environments. These types of experiential values were also mostly captured with questionnaires and interviews. Interestingly, O'Hara et al. [37] addressed situated macro-cognitive activities in a paediatric ICU and El-Hadedy and El-Husseiny [22] observed workplace violence (WPV) locations to apply Crime Prevention Through Environmental Design (CPTED). Lastly, three articles addressed spatial preferences or choices to examine social activity [17], privacy [18], and prospect-refuge theory [45]. Two articles that were not categorised in this review produced specific data about spatial memory [48] and cultural identity [42].

3.3.2. Sampling

The development of human experiential data—of physical sites, virtual, or laboratory settings—typically adopts social science techniques. Whilst human data could be automatically collected using tracking technologies such as GPS, most research in this review employed convenient or purposive sampling because of the difficulty of developing population-level data in space syntax research. For example, human data were collected by tracking park visitors [53], counting daily visitors entering a garden [54], interviewing people on a street [28], and conducting visitor surveys [21,32]. Students were also conveniently recruited from universities [24,30,48]. In addition, questionnaire and observation tended to use self-selection or volunteer sampling [20,23,46,49] and systematic sampling [50]. In contrast, only two studies used probability sampling, for example, simple random sampling [44] and stratified random sampling [52], because they knew the population of their research, cruise passengers and housing neighbourhoods, respectively.

To effectively develop human data, the sample size is a key element of the research design. However, the number of participants in the selected articles varied widely from eight to 7319, while the median was 105. These numbers might depend not only on sampling targets, but also on research methods. For example, research using interviews ranged from eight for staff interviews [17] to 75 for street interviews [28]. In contrast, field observations recorded the higher numbers of participants, such as 7319 [43] and 4983 [51]. GPS tracking also tended to recruit more than 300 participants [53,54]. A face-to-face survey or questionnaire that produced movement or behavioural patterns also collected hundreds of samples. For example, Koohsari et al. [46] analysed data from 2591 participants for the Physical Activity in Localities and Community Environments (PLACE) study. Chiang and Li [21] used 964 residents' responses and Ozbil et al. [39] 843 parental surveys. Liu et al. [32] also collected 510 route samples through questionnaires. However, the studies of wayfinding tended to use smaller samples because they recorded routes as well as navigation behaviours. For example, Hölscher et al. [29] conduced wayfinding experiments with 12 participants and Li and Klippel [31] assessed eight participants' wayfinding behaviours in terms of time, distance, and pointing errors. In addition, an in-depth survey used medium-sized samples. For example, Marquardt et al. [34] collected 82 in-home assessments and Rashid et al. [49] 81 staff's perceptions of interaction and communication. In summary, research on movement patterns tended to recruit hundreds of participants, but experimental data such as perceptions and spatial preferences were often developed by small cases. In general, few articles described a clear rationale for determining an effective sample size.

3.3.3. Discussion of Human Data Analysis

A detailed hierarchy of socio-spatial experience, based on the four categories in this review, can be used to support a systematic identification of the appropriate types of syntactic techniques for future research. One of the first considerations when doing this relates to 2D and 3D space. For example, Lu et al.'s work [67] presents 3D visibility graphs associated with social network theory. Despite this example, most space syntax methods are applied to 2D human patterns, although some research captures complex activities [19,34]. While it could be argued that 2D "movement" (that is, horizontal rather than vertical) was the dominant category of experiential data, a range of approaches and digital technologies were also used, led by observation studies and questionnaires. In other words, human experience was developed from multi-dimensional factors or "complex dimensions and elements" [68]. This is also why multiple social and environmental variables have been considered as predictors in the present article [30,34,39,43]. Regardless of the research area, human experience is collectively shaped by many social, economic, cultural, and environmental factors.

Space syntax studies have traditionally addressed the *innate* or *natural* socio-spatial properties of architectural and urban space. Thus, a large sample size is essential for this purpose [21,43,46,51,53,54]. In contrast, relatively small sample sizes were employed for the in-depth analysis of behavioural or perceptual patterns [29,31,34]. That is, the sample size for this comparator data was dependent on the applied syntactic technique(s) as well as the varied quality of data. Furthermore, this review reveals that space syntax research on human preferences, which might need an intensive body of data to study, has been relatively poorly addressed so far. The quality of spatial experiences, probably led by a higher level of cognitive operations [69], interactive aspects of space [68], and socio-economic forces [70], would be further investigated in space syntax communities.

3.4. The Relationships between Syntactic and Experiential Data

The findings of the 38 articles can be categorised into three analytic relationships, (i) descriptive, (ii) correlated, and (iii) predictive relationships (regression). The first addresses the descriptive statistics of syntactic and experiential data, separately, and discusses the connections between their results. The correlation and regression methods use statistical relationships, with the latter including the estimation of the relationship through a model in statistics. Generally, the correlations in these articles tend to include descriptive discussions whereas the regression applications are often presented with correlational analysis.

3.4.1. Descriptive Relationship

Articles dealing with descriptive relationships tended to have a small sample size [17,29] or frame an integrated discussion about several types of experiential data describing behavioural patterns [19,20,52] and perceptions [22,23,27,37]. For example, Aknar and Atun [17] argue that *integration* values are positively related to active tables in a restaurant and can be used for predicting social activity, based on eight manually compiled seating charts. Zerouati and Bellal [52] also find that syntactic values (*visual integration, visual connectivity,* and *intelligibility*), are important indicators of social activities, which are related to the frequencies of actions (sitting, standing, playing, buying/selling, and chatting) captured through snapshot observations. As such, *connectivity* is positively related to long-duration social activities in streets [20]. There is also a positive relationship between spatial properties (*integration* and *connectivity*) and perceived WPV locations in a healthcare setting, because WPV is largely shaped by patients and their relatives [22]. *Openness, connectivity,* and *visibility* also had an intricate link to macro-cognitive interactions in healthcare design [37]. Syntactic properties are also related to perceptions of Lynch's elements, specifically, paths and nodes, shaping the mental image of the city [27]. In contrast, the syntactic data did not always have a relationship with observational data in urban spaces, because pedestrian behaviours are affected by various social, commercial, and environmental factors [19]. Human cognition is also too complex to be predicted by space syntax approaches [27]. Other design strategies may also be superior to spatial configurations for specific purposes, such as CPTED [22].

Wayfinding research [29,31,41] often employed descriptive analytical methods for human movement and behavioural patterns. For example, novices' routes tended to show relatively higher values of *integration, connectivity,* and *step depth,* following a central point strategy, while experienced users employed a more effective plan to navigate environments [29]. *Integration, intelligibility,* and *choice* were also used to predict movement patterns captured by spatial cognition experiments [24]. Some of the findings suggest that low *visibility* could cause participants' wrong turns and hesitations while wayfinding, and *layout complexity* is related to wayfinding time and pointing errors [31]. It was, however, also suggested that wayfinding behaviours seemed to be impacted more by signage and forms such as "open L-type" spatial plans [41]. In summary, research adopting descriptive analysis relied on both syntactic properties and experimental data to examine spatial configurations.

3.4.2. Correlational Relationship

Correlation is the fundamental analytical approach to examining the relationship between variables. Eight articles used this approach, including the Pearson correlation coefficient [18,25,28,33,45], describing r values. For example, considering Appleton's prospect-refuge theory in architecture, Keszei et al. [45] discovered that there were positive correlations between seating choice and *visual integration* and *intervisibility* in simulated heritage buildings, e.g., $r = 0.351$, $p < 0.01$, and $r = 0.381$, $p < 0.01$, respectively, where participants tried to be seen. In a study of movement patterns in shopping malls, Omer and Goldblatt [38] calculate *connectivity, topological mean depth, topological choice,* and *intelligibility* at the axial line level and *connectivity, topological and metric mean depth,* and *topological and metric choice* at the segment level. Using this approach, they identify multiple correlations between syntactic properties and behaviours. For example, the aggregate movement flow in a shopping mall is found to be significantly ($p < 0.05$) correlated with *topological choice* ($r = 0.510$), *connectivity* ($r = 0.914$), *metric choice* ($r = 0.797$), and *mean depth* ($r = -0.881$). In terms of wayfinding behaviours [29], *step depth* shows a strong correlation with other performance measures (time, stop, getting lost, distance, way/shortest way, speed), ranging from $r = 0.78$, $p < 0.10$ (getting lost) to $r = -0.87$, $p < 0.05$ (speed).

The architectural studies dealing with "medical" spaces identified particular relationships between syntactic and experiential data. For example, Alalouch and Aspinall [18] reveal a strong negative relationship between preferences and *visual integration* in hospital wards, $r = -0.957$, $p < 0.05$, and a strong positive relationship between ward preference and *control value,* $r = 0.913$, $p < 0.05$, in terms of participants' chosen locations for privacy. Ferdous and Moore [25] also observe that in care facilities, high-level interactions and *proximity* (or *integration*) and *visibility* (or *isovist area*) are negatively correlated, $r = -0.565$, $p < 0.01$ and $r = -0.538$, $p < 0.01$, respectively. They argue that the syntactic properties might be positively correlated to low levels of interaction only. The high-level social interactions were better predicted using privacy parameters. Interestingly, the findings of Rashid et al. [40] include significant correlations between four interaction-related behaviours (walking, walking and interacting, standing and interacting, and sitting and interacting), three groups of users (nurse, physician, and others) and three syntactic properties (*axial integration, node integration,* and *control value*) in ICUs. For example, *axial integration* shows a negative correlation with "walking and interacting", $r(19) = -0.633$, $p = 0.004$, and a positive correlation with "sitting and interacting", $r(19) = 0.575$, $p = 0.016$, among nurses in nurse stations. It also has a significant positive correlation with "sitting and interacting", $r(148) = 0.387$, $p = 0.000$, among physicians at the unit level. In contrast, all observed behaviours were found to be negatively correlated with *node integration,* $r(53) = -0.335$, $p = 0.014$, and *control value,* $r(53) = -0.345$, $p = 0.011$, among nurses in patient rooms.

Urban studies, compared to architectural studies, tended to find only limited correlations. In Jakarta, Hidayati et al. [28] revealed that potential vehicular accessibility (normalised angular *choice*) and three locations (preschool, primary, and tertiary schools) were found to be correlated, $r = -0.109$, $p < 0.05$, $r = -0.143$, $p < 0.01$, and $r = 0.226$, $p < 0.01$, respectively. However, in addition to such weak correlations, their cases did not show any relationships between the realised pedestrian accessibility and school locations. As a particular case using the Spearman correlation, Zhang et al. [54] found that there were multiple significant relationships between temporal–spatial behaviours (first visiting proportion, revisiting proportion, average time, and average speed) and four syntactic indicators (*connectivity*, *mean depth*, *integration*, and *intelligibility*) in a Chinese garden, ranging from $\rho = -0.248$, $p < 0.05$ (average speed and *intelligibility* in the visible layer) to $\rho = 0.525$, $p < 0.01$ (first visiting proportion and *connectivity* in the passable layer). While there was no common consensus about the range of correlation coefficients, such weak correlations were discussed in some cases. There were also some formulas used to collectively quantify human experience such as social and spatial interactions. Nonetheless, *integration* and *connectivity* are fundamental syntactic parameters for correlation. *Choice* and simple topological *depth* also show significant correlations with many experiential variables. In summary, the correlational analysis addressed statistically significant relationships between these conventional syntactic properties and various types of experiential data (e.g., movement, behaviour, perception, and spatial preference).

3.4.3. Predictive Relationship (Regression)

The regression analysis approaches in this review include bivariate regression [53], stepwise regression [43,51], binomial regression [35], nominal logistic regression [39], ordinary least squares (OLS) and/or generalized least squares (GLS) regression [30,43,48,53], negative binomial regression [36,46], conditional (fixed-effects) logistic regression [50], mixed-effects model [48], and univariate and multivariate linear regression model [34]. Seventeen articles described their predictive models using regression analyses. Although reporting the regression results could start with the overall regression (F and p values) with R^2 values, many articles in this review addressed estimated coefficients of variables (β-value) with R^2.

In contrast, a few articles described regression equations. For example, Liu et al. [32] reveal that syntactic values (*integration* and *control value*) affect tourist trail flows, $y = -75.120 + 20.217S - 191.678L + 716.984G$ ($R^2 = 0.727$, $F = 22.321$, $p = 0.001$), and tourist visits, $y = 1253.037 + 1287.307C - 1567.360L + 388.846S$ ($R^2 = 0.331$, $F = 5.779$, $p = 0.004$), where S represents "online scenic attention kernel density value", L *local integration*, G *global integration*, and C *control value*. Zeng et al. [42] also found a linear equation, $y = 0.6619x - 2.7874$ ($R^2 = 0.3124$, $p < 0.05$), describing the influence of syntactic characters of the traditional houses on residents' acculturation results in Beijing. The syntactic characters are quantified from *integration* and *relative ringiness* of typical traditional spaces. In contrast, Rashid et al. [49] report only the overall regression (R^2 values with F and p values) to show the significant, positive impact of visual accessibilities (e.g., *visual connectivity* and *visual integration*) on interaction behaviours in ICUs. The *visibility* scores also predict the frequency of use of hand sanitising stations (HSS), adjusted $R^2 = 0.479$, $F(5, 65) = 13.877$, $p < 0.001$ [47].

Again, the dominant reporting format of regression was presenting predictive relationships with R^2 and β-values. The estimated coefficients of variables could be further explained with confidence intervals [46], odds ratios [50], and Wald x^2 tests [39], but were usually reported using t-statistics. Due to the complex nature of these values, however, this review highlights the significant links between predictor and response variables that could collectively generate social-spatial experience. Based on their relationships, this review identifies a collective model that consists of five syntactic predictors (*integration*, *connectivity*, *control value*, *choice*, and *step depth*) and three dependent categories (movement, behaviour, and perception). Whilst all predictors were presented in more than one article, *integration* and *connectivity* were dominantly examined in 13 and eight articles, respectively. The remainder were used only in two or three articles. Notably, *integration* was largely associated with both movement and behaviour, while *connectivity* tended to influence only movement. *Choice* and *step depth* were also regarded as a predictor of movement.

In the regression analysis set, there were no articles examining the fourth experiential value, "preference". Thus, all response variables (experiential values) were classified into three categories: movement, behaviour, and perception (see Figure 4). In eight articles, syntactic properties were considered influential factors on movement, e.g., trail and visit pattern [32,36,39,44,53], route choice [50], and frequency of use [43,47]. In addition, all five syntactic predictors were related to "movement". In contrast, five articles addressed behavioural responses such as social interaction activities [51] and behavioural patterns [21,35,46]. Lastly, only two articles revealed the predictive relationship of *visual integration* and *visual connectivity* to perception [30,49]. Importantly, like the negative correlations [18,25,40], syntactic variables had not only significant positive effects on experiential variables, but also negative coefficients as well [35,48,50]. Thus, the estimated coefficients of variables should be carefully interpreted, along with reference to social theories and phenomena. Despite this caveat, these predictive models using regression analysis can be applied to develop architectural, medical, and urban spaces, promoting positive socio-spatial experiences.

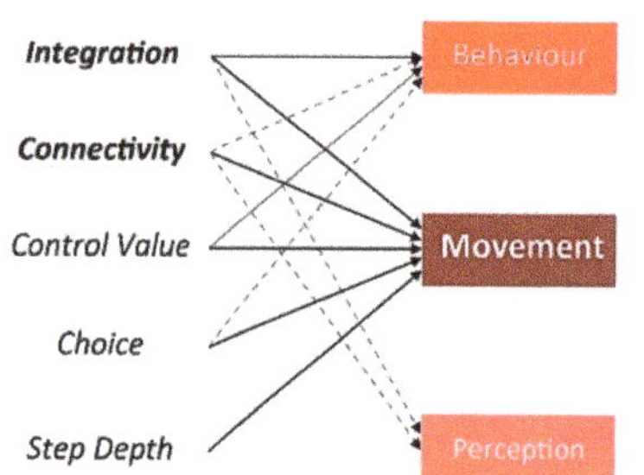

Figure 4. A collective model of the predictive relationships between syntactic and experiential variables [21,30,32,35,36,39,43,44,46,47,49–51,53].

3.4.4. Discussion of the Relationships between Syntactic and Experiential Data

It was not unexpected that the collective, predictive model in Figure 4 identified three experiential values (movement, simple behaviour, and perception) because they are closely related to the three socio-spatial phenomena addressed in many space syntax studies [3,71]. However, most of the 38 studies identified two or more syntactic predictors in their models, for example, *connectivity, choice, integration,* and *step depth* [44] and *integration, control value,* and *mean depths* [53]. Thus, there was no single syntactic parameter predicting social-spatial experience. A few syntactic properties were simultaneously measured in a study, and they were jointly functioning as predictor variables in the regression model(s). Thus, the dynamic effects of syntactic variables on social experience should be further examined in a future study synthesising β-values. That is, future space syntax studies need to better understand the multi-dimensional socio-spatial relationships identified in this review.

Statistics derived from spatial geometries or forms, from *segregation* to *integration*, have enabled the investigation of both the social logic and the consequences of spaces [5]. For example, in the research examined in this review, the *integration* or *connectivity* values of urban spaces had a strong correlation with human movements such as favourite routes and the frequencies of visits and use. However, the statistical analysis combining syntactic properties and experiential data often revealed the innate limitations of space syntax methods, due to reduced variables, possibly delivering unjustified conclusions about causal relationships. Of course, the original proposition was that syntactic accessibility could reflect the evolution of the system by way of a dynamic feedback cycle, and specifically, land uses and densities in an "organic" city [57,63,72–74]. However, the nature of socio-spatial phenomena, e.g., society–space relations, social interfaces and encounter, and the temporality of human experience [70], might not be encapsulated in quantitative values only. Thus, an additional in-depth or qualitative survey could be revealing for this purpose.

4. Discussion

4.1. A Systematic Review on Socio-Spatial Experiences in Space Syntax Research

Hillier and his co-authors argue that spatial configuration and morphology can supersede cognitive, economic, social, and cultural processes in this regard [57,63,74,75], although the reductionism of space syntax theory has been criticised [70,76,77]. *Syntactic* structures (topological) cannot capture the *semantic* meanings (cognitive and informational) of spaces, events, and acts [70]. Furthermore, without a deeper discussion about the complex, recursive phenomena occurring between human and environment, many space syntax researchers have relied on the hybrid theory (social and spatial) and its descriptive statistics, and have simply acknowledged syntactic, epistemological limitations. Likewise, the collected data in this review might be insufficient to demonstrate the quality or complexity of socio-spatial experiences in the built environment.

This review, furthermore, highlights the relationships between syntactic and experiential parameters, systematically examining socio-spatial factors in space syntax research. However, the formalised and codified evaluative framework proposed in this article may be limited to synthesising "nonlocal, or extrinsic properties of space" [72]. It is not equally useful for examining the intrinsic, cognitive, and interpretative operations characterising spatial cognition and human experience in a comparative evaluation, because of the complexity of various multi-relational, multi-dimensional factors. That is why a similar meta-analysis [7] was focused on the relationship of just four syntactic measures (*integration, choice, connectivity,* and *control*) and pedestrian movement only. Whilst acknowledging this, the limited scope of meta-analysis itself needs additional evaluation components. For example, a heuristic approach such as an analytic hierarchy process (AHP) or an in-depth analysis can be added to the existing framework. Particularly, the AHP can capture both the syntax and semantics of spatial experience comprising space, event, and activity, and thereby develop a hierarchical structure of related components, dimensions, and elements [68]. Thus, it can be further elaborated from cognitive and social theories, specifically, "visuospatial cognition" or "spatial cognition' [24,27,69].

4.2. Limitations of the Data

The findings of the studies reported in this review have multiple limitations associated with sample designs and data collection, e.g., convenience sampling [21,45], limited sampling [18,30,43], limited time of data collection [33,47,52,54], or cross-sectional data [39,43,53]. Some studies collected very small samples, e.g., eight participants [31] or 12 [29], although their data were in-depth and rich enough to address their research goals. The sampling of a few studies might also be too limited to generalise their findings [37,39,43,46,53]. Lastly, some researchers reported unexpected biases such as participants' environmental changes [34] and GPS data errors [54]. These limitations might be common challenges and unavoidable, but the sample design and data collection method should be carefully devised to conduct research particularly on socio-spatial experience. A potential approach to overcoming a limitation of the sample design and data collection would be using a combined and complementary method, where each part enhances the qualities of the other, e.g., observation and questionnaire or interview. An in-depth analysis on a rich data set can also be a promising strategy to examine complex socio-spatial behaviours and perceptions.

4.3. Future Research in Socio-Spatial Experience

While the ALA and VGA were dominant techniques, this review has highlighted the value of the combination of syntactic and experiential analyses. Importantly, VGA has been widely associated with the other syntactic techniques such as ALA and JPG, because perceptions are closely related to human movements and behaviours. That is, VGA should be a core method for investigating the socio-spatial experience of both architectural and urban spaces. Conversely, Hillier [72] promoted the combination of line and topological graphs, which were conventionally presented in ALA and CSA (or JPG), respectively. Despite this, only a few examples in urban research used such a hybrid method of axial and convex maps. This combination should be further explored in urban analytics, where convexity is rarely addressed. In this case, either Zhai's and Baran's research [43] or a stroke-based network [53] may provide an appropriate methodological precedent.

Interestingly, there are many syntactic studies using JPG or ALA only, where spatial analyses are largely designed to explore architectural theories or architects' intentions [78–82] and street structures or urban morphologies [32,38,83]. However, these approaches may also need VGA to accommodate the types of experiential and experimental approaches suggested in this review. For example, professional interviews [17], snapshot observations [19,20,52], and wayfinding experiments [29,31] would be very useful for this purpose, if samples are limited. Furthermore, this review has uncovered several wayfinding studies that brought a spatial-cognition-based approach to this domain [24,29,41]. Examining individual cognitive processes and behaviours, spatial cognition research can consider landmark or temporal and emotional dimensions that have been largely ignored in conventional space syntax research [24]. That is, this type of experimental research into wayfinding can provide important references for both architectural and urban design studies.

Many studies have already investigated the spatial configurations and morphologies of hundreds of cities and buildings. However, their multi-dimensional socio-spatial relationships should be further explored to examine cognitive and spatial experiences. Future research could also reveal the dynamic effects of syntactic properties on socio-spatial experience, examining the various β-values of the explanatory variables captured in this review, which would be able to overcome several fundamental limitations in the *simple* topological abstractions of space syntax research. In addition, the application of mathematics to space syntax approaches—for example, allometric equations [84], a Fibonacci sequence [17], and a modularity-based algorithm [85]—offer new research directions to test or improve their capabilities of predicting socio-spatial experiences. Likewise, additional ISA measures such as *circularity* [86] and *distance-weighted isovist area* [87] are also introduced in this field. In addition, a "weighted and directed JPG" has recently been developed for centrality measures [3,79,88]. These new centrality measures can be useful for more precisely examining architectural configurations as well as urban networks.

5. Conclusions

This research has addressed an important question, how has space syntax research characterised and predicted socio-spatial experience in the built environment, providing the first systematic scoping review of space syntax studies on socio-spatial experience. Syntactic techniques have been widely used in research to explore movement, behaviour, perception, and preference patterns in architectural, medical, and urban spaces, which synthetically form socio-spatial experience. One of the reasons this significant body of empirical research has never been synthesised in a comprehensive manner before relates to the complexity of socio-spatial relationships and the combination of heterogeneous syntactic and experiential measures.

Combining the PRISMA and PECO frameworks, this review rigorously synthesises different types of socio-spatial properties and examines their relations in the built environment, producing new integrated findings. This extension of the traditional scoping review ensures that this article not only classifies and reports past research but synthesises the findings of data in a new way. The proposed collective, predictive model in Figure 4, identifying the three key socio-spatial phenomena (movement, behaviour, and perception), can be used by researchers undertaking systematic reviews. The correlational and regression results of space syntax studies reported in this article also offers a model that can be expanded in a full systematic review to compare the available evidence.

Despite the rigorous use of these frameworks, this approach may not capture all relevant features and insights in the body of research being examined. The data selection process developed from the PRISMA standard will not uncover every theoretical, social, and cognitive study ever completed using space syntax. This is why the present article emphasises critical synthesis, presenting three discussions of results, along with key literature that can augment these findings and provide insights for future space syntax research. In this way, this article contributes to identifying important research areas for future research as well as clarifying key concepts and characteristics in the literature.

Significantly, the synthesis of the findings of the 38 articles has identified descriptive and correlational relationships, as well as a collective model of the predictive relationships between syntactic and experiential variables. Such findings are not only useful for the space syntax community, but also a valuable reference for urban analytics researchers examining various aspects of planning and design. Considering the complex nature of socio-spatial phenomena, a relational mapping such as the proposed collective model, although it could also be a multi-dimensional matrix of spatial and social theories, should be inevitable for a following study to identify the appropriate syntactic and experimental parameters and investigating the various effects of a design. This article contributes to the development of future syntactic, empirical studies to develop research problems and methodologies that positively develop socio-spatial experience in the built environment.

Author Contributions: Conceptualization, J.H.L. and M.J.O.; methodology, J.H.L. and M.J.O.; formal analysis, J.H.L. and L.Z.; investigation, J.H.L., M.J.O., and L.Z.; writing—original draft preparation, J.H.L. and L.Z.; writing—review and editing, J.H.L. and M.J.O.; visualization, J.H.L. and L.Z.; supervision, J.H.L. and M.J.O.; funding acquisition, J.H.L. All authors have read and agreed to the published version of the manuscript.

Funding: This research was supported by the Scientia program and ADA Fellowship at UNSW Sydney.

Data Availability Statement: The data presented in this study are available on request from the corresponding author. The data are not publicly available due to the policy of research projects.

Conflicts of Interest: The authors declare no conflict of interest.

References

1. Hillier, B.; Hanson, J. *The Social Logic of Space*; Cambridge University Press: Cambridge, MA, USA, 1984.
2. Lee, J.H.; Ostwald, M.J.; Lee, H. Measuring the spatial and social characteristics of the architectural plans of aged care facilities. *Front. Archit. Res.* **2017**, *6*, 431–441. [CrossRef]
3. Lee, J.H.; Ostwald, M.J. *Grammatical and Syntactical Approaches in Architecture: Emerging Research and Opportunities*; IGI Global: Hershey, PA, USA, 2020.
4. Hillier, B. The architecture of the urban object. *Ekistics* **1989**, *56*, 5–21.
5. Vaughan, L. Glossary of Space Syntax. In *Suburban Urbanities: Suburbs and the Life of the High Street*; Vaughan, L., Ed.; UCL Press: London, UK, 2015; pp. 307–312.
6. Haq, S.; Luo, Y. Space Syntax in Healthcare Facilities Research: A Review. *HERD Health Environ. Res. Des. J.* **2012**, *5*, 98–117. [CrossRef] [PubMed]
7. Sharmin, S.; Kamruzzaman, M. Meta-analysis of the relationships between space syntax measures and pedestrian movement. *Transp. Rev.* **2018**, *38*, 524–550. [CrossRef]
8. Yamu, C.; van Nes, A.; Garau, C. Bill Hillier's Legacy: Space Syntax—A Synopsis of Basic Concepts, Measures, and Empirical Application. *Sustainability* **2021**, *13*, 3394. [CrossRef]
9. Nes, A.V.; Yamu, C. *Introduction to Space Syntax in Urban Studies*; Springer: Cham, Switzerland, 2021. [CrossRef]
10. Moher, D.; Liberati, A.; Tetzlaff, J.; Altman, D.G.; The, P.G. Preferred Reporting Items for Systematic Reviews and Meta-Analyses: The PRISMA Statement. *PLoS Med.* **2009**, *6*, e1000097. [CrossRef] [PubMed]
11. Morgan, R.L.; Whaley, P.; Thayer, K.A.; Schünemann, H.J. Identifying the PECO: A framework for formulating good questions to explore the association of environmental and other exposures with health outcomes. *Environ. Int.* **2018**, *121*, 1027–1031. [CrossRef]
12. Lim, Y.; Edelenbos, J.; Gianoli, A. Identifying the results of smart city development: Findings from systematic literature review. *Cities* **2019**, *95*, 102397. [CrossRef]
13. Tura, N.; Ojanen, V. Sustainability-oriented innovations in smart cities: A systematic review and emerging themes. *Cities* **2022**, *126*, 103716. [CrossRef]
14. Wang, J.; Biljecki, F. Unsupervised machine learning in urban studies: A systematic review of applications. *Cities* **2022**, *129*, 103925. [CrossRef]

15. Munn, Z.; Peters, M.D.J.; Stern, C.; Tufanaru, C.; McArthur, A.; Aromataris, E. Systematic review or scoping review? Guidance for authors when choosing between a systematic or scoping review approach. *BMC Med. Res. Methodol.* **2018**, *18*, 143. [CrossRef]
16. Page, M.J.; McKenzie, J.E.; Bossuyt, P.M.; Boutron, I.; Hoffmann, T.C.; Mulrow, C.D.; Shamseer, L.; Tetzlaff, J.M.; Akl, E.A.; Brennan, S.E.; et al. The PRISMA 2020 statement: An updated guideline for reporting systematic reviews. *BMJ* **2021**, *372*, n71. [CrossRef]
17. Aknar, M.; Atun, R.A. Predicting movement in architectural space. *Archit. Sci. Rev.* **2017**, *60*, 78–95. [CrossRef]
18. Alalouch, C.; Aspinall, P. Spatial attributes of hospital multi-bed wards and preferences for privacy. *Facilities* **2007**, *25*, 345–362. [CrossRef]
19. Askarizad, R.; Safari, H. The influence of social interactions on the behavioral patterns of the people in urban spaces (case study: The pedestrian zone of Rasht Municipality Square, Iran). *Cities* **2020**, *101*, 102687. [CrossRef]
20. Can, I.; Heath, T. In-between spaces and social interaction: A morphological analysis of Izmir using space syntax. *J. Hous. Built Environ.* **2016**, *31*, 31–49. [CrossRef]
21. Chiang, Y.-C.; Li, D. Metric or topological proximity? The associations among proximity to parks, the frequency of residents' visits to parks, and perceived stress. *Urban For. Urban Green.* **2019**, *38*, 205–214. [CrossRef]
22. El-Hadedy, N.; El-Husseiny, M. Evidence-Based Design for Workplace Violence Prevention in Emergency Departments Utilizing CPTED and Space Syntax Analyses. *HERD Health Environ. Res. Des. J.* **2022**, *15*, 333–352. [CrossRef] [PubMed]
23. Elshater, A.; Abusaada, H.; Afifi, S. What makes livable cities of today alike? Revisiting the criterion of singularity through two case studies. *Cities* **2019**, *92*, 273–291. [CrossRef]
24. Esposito, D.; Santoro, S.; Camarda, D. Agent-Based Analysis of Urban Spaces Using Space Syntax and Spatial Cognition Approaches: A Case Study in Bari, Italy. *Sustainability* **2020**, *12*, 4626. [CrossRef]
25. Ferdous, F.; Moore, K.D. Field Observations into the Environmental Soul: Spatial Configuration and Social Life for People Experiencing Dementia. *Am. J. Alzheimer's Dis. Other Dement.* **2015**, *30*, 209–218. [CrossRef] [PubMed]
26. Geng, S.; Chau, H.-W.; Yan, S.; Zhang, W.; Zhang, C. Comparative analysis of hospital environments in Australia and China using the space syntax approach. *Int. J. Build. Pathol. Adapt.* **2021**, *39*, 525–546. [CrossRef]
27. Güngör, O.; Harman Aslan, E. Defining urban design strategies: An analysis of Iskenderun city center's imageability. *Open House Int.* **2020**, *45*, 407–425. [CrossRef]
28. Hidayati, I.; Yamu, C.; Tan, W. Realised pedestrian accessibility of an informal settlement in Jakarta, Indonesia. *J. Urban. Int. Res. Placemaking Urban Sustain.* **2021**, *14*, 434–456. [CrossRef]
29. Hölscher, C.; Brösamle, M.; Vrachliotis, G. Challenges in Multilevel Wayfinding: A Case Study with the Space Syntax Technique. *Environ. Plan. B Plan. Des.* **2012**, *39*, 63–82. [CrossRef]
30. Knöll, M.; Neuheuser, K.; Cleff, T.; Rudolph-Cleff, A. A tool to predict perceived urban stress in open public spaces. *Environ. Plan. B Urban Anal. City Sci.* **2018**, *45*, 797–813. [CrossRef]
31. Li, R.; Klippel, A. Wayfinding in Libraries: Can Problems Be Predicted? *J. Map Geogr. Libr.* **2012**, *8*, 21–38. [CrossRef]
32. Liu, P.; Xiao, X.; Zhang, J.; Wu, R.; Zhang, H. Spatial Configuration and Online Attention: A Space Syntax Perspective. *Sustainability* **2018**, *10*, 221. [CrossRef]
33. Mansouri, M.; Ujang, N. Space syntax analysis of tourists' movement patterns in the historical district of Kuala Lumpur, Malaysia. *J. Urban. Int. Res. Placemaking Urban Sustain.* **2017**, *10*, 163–180. [CrossRef]
34. Marquardt, G.; Johnston, D.; Black, B.S.; Morrison, A.; Rosenblatt, A.; Lyketsos, C.G.; Samus, Q.M. Association of the Spatial Layout of the Home and ADL Abilities Among Older Adults With Dementia. *Am. J. Alzheimer's Dis. Other Dement.* **2011**, *26*, 51–57. [CrossRef]
35. Mohamed, A.A.; Stanek, D. The influence of street network configuration on sexual harassment patterns in Cairo. *Cities* **2020**, *98*, 102583. [CrossRef]
36. Nubani, L.; Puryear, A.; Kellom, K. Measuring the Effect of Visual Exposure and Saliency of Museum Exhibits on Visitors' Level of Contact and Engagement. *Behav. Sci.* **2018**, *8*, 100. [CrossRef]
37. O'Hara, S.; Klar, R.T.; Patterson, E.S.; Morris, N.S.; Ascenzi, J.; Fackler, J.C.; Perry, D.J. Macrocognition in the Healthcare Built Environment (mHCBE): A Focused Ethnographic Study of "Neighborhoods" in a Pediatric Intensive Care Unit. *HERD Health Environ. Res. Des. J.* **2018**, *11*, 104–123. [CrossRef] [PubMed]
38. Omer, I.; Goldblatt, R. Using space syntax and Q-analysis for investigating movement patterns in buildings: The case of shopping malls. *Environ. Plan. B Urban Anal. City Sci.* **2017**, *44*, 504–530. [CrossRef]
39. Ozbil, A.; Yesiltepe, D.; Argin, G.; Rybarczyk, G. Children's Active School Travel: Examining the Combined Perceived and Objective Built-Environment Factors from Space Syntax. *Int. J. Environ. Res. Public Health* **2021**, *18*, 286. [CrossRef] [PubMed]
40. Rashid, M.; Boyle, D.K.; Crosser, M. Network of Spaces and Interaction-Related Behaviors in Adult Intensive Care Units. *Behav. Sci.* **2014**, *4*, 487. [CrossRef] [PubMed]
41. Tzeng, S.-Y.; Huang, J.-S. Spatial Forms and Signage in Wayfinding Decision Points for Hospital Outpatient Services. *J. Asian Archit. Build. Eng.* **2009**, *8*, 453–460. [CrossRef]
42. Zeng, M.; Wang, F.; Xiang, S.; Lin, B.; Gao, C.; Li, J. Inheritance or variation? Spatial regeneration and acculturation via implantation of cultural and creative industries in Beijing's traditional compounds. *Habitat Int.* **2020**, *95*, 102071. [CrossRef]
43. Zhai, Y.; Baran, P.K. Do configurational attributes matter in context of urban parks? Park pathway configurational attributes and senior walking. *Landsc. Urban Plan.* **2016**, *148*, 188–202. [CrossRef]

44. Domènech, A.; Gutiérrez, A.; Anton Clavé, S. Built environment and urban cruise tourists' mobility. *Ann. Tour. Res.* **2020**, *81*, 102889. [CrossRef]
45. Keszei, B.; Halász, B.; Losonczi, A.; Dúll, A. Space Syntax's Relation to Seating Choices from an Evolutionary Approach. *Period. Polytech. Archit.* **2019**, *50*, 115–123. [CrossRef]
46. Koohsari, M.J.; Owen, N.; Cerin, E.; Giles-Corti, B.; Sugiyama, T. Walkability and walking for transport: Characterizing the built environment using space syntax. *Int. J. Behav. Nutr. Phys. Act.* **2016**, *13*, 121. [CrossRef] [PubMed]
47. Neo, J.R.J.; Sagha-Zadeh, R. The influence of spatial configuration on the frequency of use of hand sanitizing stations in health care environments. *Am. J. Infect. Control* **2017**, *45*, 615–619. [CrossRef] [PubMed]
48. Pagkratidou, M.; Galati, A.; Avraamides, M.N. Do environmental characteristics predict spatial memory about unfamiliar environments? *Spat. Cogn. Comput.* **2020**, *20*, 1–32. [CrossRef]
49. Rashid, M.; Khan, N.; Jones, B. Physical and Visual Accessibilities in Intensive Care Units: A Comparative Study of Open-Plan and Racetrack Units. *Crit. Care Nurs. Q.* **2016**, *39*, 313–334. [CrossRef]
50. Shatu, F.; Yigitcanlar, T.; Bunker, J. Shortest path distance vs. least directional change: Empirical testing of space syntax and geographic theories concerning pedestrian route choice behaviour. *J. Transp. Geogr.* **2019**, *74*, 37–52. [CrossRef]
51. Sheng, Q.; Wan, D.; Yu, B. Effect of Space Configurational Attributes on Social Interactions in Urban Parks. *Sustainability* **2021**, *13*, 7805. [CrossRef]
52. Zerouati, W.; Bellal, T. Evaluating the impact of mass housings' in-between spaces' spatial configuration on users' social interaction. *Front. Archit. Res.* **2020**, *9*, 34–53. [CrossRef]
53. Zhai, Y.; Korça Baran, P.; Wu, C. Can trail spatial attributes predict trail use level in urban forest park? An examination integrating GPS data and space syntax theory. *Urban For. Urban Green.* **2018**, *29*, 171–182. [CrossRef]
54. Zhang, T.; Lian, Z.; Xu, Y. Combining GPS and space syntax analysis to improve understanding of visitor temporal–spatial behaviour: A case study of the Lion Grove in China. *Landsc. Res.* **2020**, *45*, 534–546. [CrossRef]
55. Turner, A. Depthmap: A program to perform visibility graph analysis. In Proceedings of the 3rd International Symposium on Space Syntax, Atlanta, GA, USA, 7–11 May 2004; Georgia Institute of Technology: Atlanta, GA, USA, 2001; pp. 31.31–31.39.
56. Turner, A.; Doxa, M.; O'Sullivan, D.; Penn, A. From Isovists to Visibility Graphs: A Methodology for the Analysis of Architectural Space. *Environ. Plan. B Plan. Des.* **2001**, *28*, 103–121. [CrossRef]
57. Hillier, B. *Space is the Machine: A Configurational Theory of Architecture*; Cambridge University Press: Cambridge, MA, USA, 1996.
58. Jiang, B. Ranking spaces for predicting human movement in an urban environment. *Int. J. Geogr. Inf. Sci.* **2009**, *23*, 823–837. [CrossRef]
59. Hillier, B.; Iida, S. Network and Psychological Effects in Urban Movement. In *Spatial Information Theory, Proceedings of the International Conference, COSIT 2005, Ellicottville, NY, USA, 14–18 September 2005*; Springer: Berlin/Heidelberg, Germany, 2005; pp. 475–490.
60. Mahmoud, A.H.; Omar, R.H. Planting design for urban parks: Space syntax as a landscape design assessment tool. *Front. Archit. Res.* **2015**, *4*, 35–45. [CrossRef]
61. Serra, M.; Hillier, B. Angular and Metric Distance in Road Network Analysis: A nationwide correlation study. *Comput. Environ. Urban Syst.* **2019**, *74*, 194–207. [CrossRef]
62. Hillier, W.; Yang, T.; Turner, A. Normalising least angle choice in Depthmap-and how it opens up new perspectives on the global and local analysis of city space. *J. Space Syntax.* **2012**, *3*, 155–193.
63. Hillier, B. Studying Cities to Learn about Minds: Some Possible Implications of Space Syntax for Spatial Cognition. *Environ. Plan. B Plan. Des.* **2012**, *39*, 12–32. [CrossRef]
64. Turner, A. From Axial to Road-Centre Lines: A New Representation for Space Syntax and a New Model of Route Choice for Transport Network Analysis. *Environ. Plan. B Plan. Des.* **2007**, *34*, 539–555. [CrossRef]
65. Beck, M.P.; Turkienicz, B. Visibility and Permeability Complementary Syntactical Attributes of Wayfinding. In Proceedings of the 7th International Space Syntax Symposium, Stockholm, Sweden, 8–11 June 2009; Koch, D., Marcus, L., Steen, E., Eds.; KTH: Stockholm, Sweden, 2009; pp. 009:001–007.
66. Hillier, B.; Hanson, J.; Graham, H. Ideas are in things: An application of the space syntax method to discovering house genotypes. *Environ. Plan. B Plan. Des.* **1987**, *14*, 363–385. [CrossRef]
67. Lu, Y.; Gou, Z.; Ye, Y.; Sheng, Q. Three-dimensional visibility graph analysis and its application. *Environ. Plan. B Urban Anal. City Sci.* **2019**, *46*, 948–962. [CrossRef]
68. Baradaran Rahimi, F.; Levy, R.M.; Boyd, J.E.; Dadkhahfard, S. Human behaviour and cognition of spatial experience; a model for enhancing the quality of spatial experiences in the built environment. *Int. J. Ind. Ergon.* **2018**, *68*, 245–255. [CrossRef]
69. Pektaş, Ş.T. A scientometric analysis and review of spatial cognition studies within the framework of neuroscience and architecture. *Archit. Sci. Rev.* **2021**, *64*, 374–382. [CrossRef]
70. Netto, V.M. 'What is space syntax not?' Reflections on space syntax as sociospatial theory. *Urban Des. Int.* **2016**, *21*, 25–40. [CrossRef]
71. Hillier, B.; Vaughan, L. The city as one thing. *Prog. Plan.* **2007**, *67*, 205–230.
72. Hillier, B. The Hidden Geometry of Deformed Grids: Or, Why Space Syntax Works, When it Looks as Though it Shouldn't. *Environ. Plan. B Plan. Des.* **1999**, *26*, 169–191. [CrossRef]
73. Hillier, B.; Penn, A. Rejoinder to Carlo Ratti. *Environ. Plan. B Plan. Des.* **2004**, *31*, 501–511. [CrossRef]

74. Penn, A. Space Syntax And Spatial Cognition:Or Why the Axial Line? *Environ. Behav.* **2003**, *35*, 30–65. [CrossRef]
75. Bafna, S. Space Syntax:A Brief Introduction to Its Logic and Analytical Techniques. *Environ. Behav.* **2003**, *35*, 17–29. [CrossRef]
76. Pafka, E.; Dovey, K.; Aschwanden, G.D.P.A. Limits of space syntax for urban design: Axiality, scale and sinuosity. *Environ. Plan. B Urban Anal. City Sci.* **2020**, *47*, 508–522. [CrossRef]
77. Ratti, C. Space Syntax: Some Inconsistencies. *Environ. Plan. B Plan. Des.* **2004**, *31*, 487–499. [CrossRef]
78. Dawes, M.J.; Ostwald, M.J.; Lee, J.H. Examining control, centrality and flexibility in Palladio's villa plans using space syntax measurements. *Front. Archit. Res.* **2021**, *10*, 467–482. [CrossRef]
79. Lee, J.H.; Ostwald, M.J.; Dawes, M.J. Examining Visitor-Inhabitant Relations in Palladian Villas. *Nexus Netw. J.* **2022**, *24*, 315–332. [CrossRef]
80. Lee, J.H.; Ostwald, M.J.; Gu, N. A Justified Plan Graph (JPG) grammar approach to identifying spatial design patterns in an architectural style. *Environ. Plan. B Urban Anal. City Sci.* **2016**, *45*, 67–89. [CrossRef]
81. Ostwald, M.J. The Mathematics of Spatial Configuration: Revisiting, Revising and Critiquing Justified Plan Graph Theory. *Nexus Netw. J.* **2011**, *13*, 445–470. [CrossRef]
82. Ostwald, M.J. A Justified Plan Graph Analysis of the Early Houses (1975–1982) of Glenn Murcutt. *Nexus Netw. J.* **2011**, *13*, 737–762. [CrossRef]
83. Omer, I.; Goldblatt, R. Spatial patterns of retail activity and street network structure in new and traditional Israeli cities. *Urban Geogr.* **2016**, *37*, 629–649. [CrossRef]
84. Shpuza, E. Allometry in the Syntax of Street Networks: Evolution of Adriatic and Ionian Coastal Cities 1800–2010. *Environ. Plan. B Plan. Des.* **2014**, *41*, 450–471. [CrossRef]
85. Law, S. Defining Street-based Local Area and measuring its effect on house price using a hedonic price approach: The case study of Metropolitan London. *Cities* **2017**, *60*, 166–179. [CrossRef]
86. Dawes, M.J.; Lee, J.H.; Ostwald, M.J. Intelligibility and cellularity in the villas of Palladio and Le Corbusier: Examining Rowe's observations. *Archnet-IJAR Int. J. Archit. Res.* **2022**. *ahead-of-print*. [CrossRef]
87. Kim, Y.; Jung, S.K. Distance-weighted isovist area: An isovist index representing spatial proximity. *Autom. Constr.* **2014**, *43*, 92–97. [CrossRef]
88. Dawes, M.J.; Lee, J.; Ostwald, M.J. 'Visual excitation' in Richard Neutra's residential architecture: An analysis using weighted graphs and centrality measures. *Front. Archit. Res.* **2022**, *11*, 1092–1103. [CrossRef]

 *buildings*

Article

A Grammar-Based Approach for Generating Spatial Layout Solutions for the Adaptive Reuse of *Sobrado* Buildings

Daniele M. S. Paulino *, Heather Ligler * and Rebecca Napolitano *

Department of Architectural Engineering, The Pennsylvania State University, 201 Old Main, University Park, PA 16801, USA
* Correspondence: dpaulino@psu.edu (D.M.S.P.); hml5521@psu.edu (H.L.); nap@psu.edu (R.N.)

Abstract: This research develops a shape grammar to generate solutions for the adaptive reuse of historic buildings. More precisely, it proposes a transformation grammar for *sobrado* buildings, a typology present in the historic center of São Luís, Brazil. The methodology defines a workflow for adapting *sobrado* buildings, once characterized for single-families, into multi-family apartments considering spatial and structural requirements. The grammar specifies a framework for repurposing historic buildings into social housing and considers the allocation of three types of apartments in the floor plan: studios, one-bedroom apartments, and two-bedroom apartments. The adopted strategy for spatial planning prioritizes access to natural daylight. The grammar supports different layout solutions for the same building and aims to accelerate the reuse of historic structures for contemporary housing needs. This paper describes the grammar rules and their application to three case study buildings.

Keywords: adaptive reuse; floor plan generation; shape grammar; generative design; space planning

Citation: Paulino, D.M.S.; Ligler, H.; Napolitano, R. A Grammar-Based Approach for Generating Spatial Layout Solutions for the Adaptive Reuse of *Sobrado* Buildings. *Buildings* **2023**, *13*, 722. https://doi.org/10.3390/buildings13030722

Academic Editors: Morten Gjerde, Michael J Ostwald and Ju Hyun Lee

Received: 24 January 2023
Revised: 15 February 2023
Accepted: 6 March 2023
Published: 9 March 2023

1. Introduction

New technologies have emerged with the industrial revolution, resulting in a rapid increase in population and industrial activities, expanding urban areas, and developing new forms of land use [1]. This movement of cities often caused the isolation of historic downtowns and a decrease in urban life around these areas, leading to an increase in vacant buildings in many historic downtown areas [2]. What is nowadays called a *historical center* or a *historic downtown* in many cities worldwide once represented the city itself, constituting an essential part of the urban context.

Theorists defend the non-abandonment of buildings, highlighting that one of the most efficient means of preserving them is through encouraging their occupation and use [3]. Vacancy directly affects the state of conservation of buildings but also has economic and social consequences since it often decreases safety and community engagement in a particular neighborhood [4]. Thus, revitalization of historic downtown areas concerns not only maintenance requirements but also engages community life around these districts.

In parallel, real estate speculation can cause an increase in property values, which low income families cannot afford, causing the development of irregular settlements in cities and suburban areas [5]. The affordable housing crisis is a global reality, affecting both developing and developed countries, especially during economic and political recessions [6]. In Brazil, for example, more than 6.9 million families need housing and around 6 million buildings are vacant [7].

In this scenario, adaptive reuse emerges as a sustainable solution for transforming existing buildings, revitalizing historic districts, and addressing new land use needs, especially for affordable housing [8–10]. In summary, adaptive reuse modifies a building's functional typology, giving it a new use [11]. In this process, various actions related to construction adaptability can be executed to modify a building's function. Spatial modifications are the most evident, but adaptive reuse can also encompass improvements related to the building's structural and/or energy performance.

Modifications to a building's function affect the spatial configuration of the existing building, which can directly impact the building's structural performance. Balancing spatial and structural requirements is one of the most challenging aspects of adaptive reuse. An adaptive reuse project must consider the allocation of rooms and elements in the existing building, and consider how these modifications can be economically and structurally feasible. Design exploration techniques can assist designers with spatial planning tasks, reducing conceptualization time and guaranteeing feasible solutions in terms of different requirements.

Developing workflows to support design decisions in adaptive reuse projects can lead to efficient layout solutions considering different aspects of building adaptation, such as maximizing material reuse and maintaining structural integrity. Modifications in the building's use type lead to a change in the relationship between spaces and circulation areas, which may require inserting new elements (i.e., walls, windows, etc.) and demolishing existing ones [12–14]. These modifications may lead to structural instability, especially considering the current damage state of existing structural components. Thus, determining spatial solutions that meet the structural requirements is a crucial step in adaptive reuse.

Currently, there is a limitation in tools to assist designers and professionals in implementing adaptive reuse solutions [15], especially in identifying how spatial modifications affect a building's structural integrity [16,17]. Usually, the implementation relies on professional knowledge to provide a more detailed analysis, such as optimizing the placement of openings and the addition of new elements, maximizing material reuse, and minimizing construction and demolition (C&D) costs [18]. In recent years, computational tools have introduced innovative form-finding techniques, revolutionizing architectural design and production. In computer-aided design (CAD), this process is defined as generative design (GD) [19]. GD systems produce designs using algorithms, offering the ability to explore design space solutions for complex design tasks.

Since necessary modifications affect the layout configuration of existing buildings, one can consider an adaptive reuse problem as a space planning problem. In this scenario, generative design techniques support the exploration of different arrangements for a predefined problem, assisting professionals during complex design tasks [19]. GD includes a wide range of techniques, for example, rule-based systems, evolutionary computation, and computer vision methods [20]. Rule-based systems (RBS) can support the strategic encoding of sequential steps to generate designs. This is particularly useful for defining computational workflows, since design is a highly complex task combining creativity, specialist knowledge, experience, and judgment concerning the objective and the esthetic aspects of a problem domain [21].

Thus, encoding the designer's knowledge is essential to produce feasible solutions. For example, when adapting an existing building, new walls must be constructed and some may be demolished. One logical strategy is to restrict new walls from intersecting existing elements, such as windows. In this context, shape grammar (SG) is a rule-based design analysis and generation approach [22] that comprises a set of transformation rules that can be applied recursively to an initial form to generate new forms [23–25]. SG is widely applied to understand the formal and syntactic information underlying plan layouts for spatial configuration. Studies have successfully developed SG to encode expert knowledge and reproduce design patterns of renowned architects or architectural styles [26–29]. However, few attempts have been made to produce reusable grammatical design systems for architectural design tasks, especially regarding the adaptation of buildings [30–33].

This project aims to understand how GD techniques can assist designers during the adaptive reuse process. More specifically, this work explores the potential of SG in generating floor plan layout solutions for the adaptive reuse of historic buildings, focusing on *sobrado* buildings in the historic center of São Luís in Brazil, a UNESCO World Cultural Heritage Site. The methodology specifies a framework for adapting buildings into multi-family apartments, considering the allocation of spatial and structural elements in the design space, aiming to maximize the reuse of existing elements while producing various solutions for a given design space. The strategy can generate a variety of design solutions, considering the allocation of studios, one-bedroom (1-BR), and two-bedroom (2-BR) apartments.

2. The Architectural Typology in the Historic Center of São Luís

This research explores the architectural typology of buildings in the historic center of São Luís (HCSL), recognized as a UNESCO World Cultural Heritage Site since 1997. This ensemble preserves a significant architectural, historical, and urban collection reminiscent of the XVIII and XIX centuries, a time of economic prosperity through the exportation of rice and cotton [34]. Efforts have arisen from both federal and state organizations to preserve and maintain the historic buildings in the HCSL, contributing to reviving the sense of community in the historic district.

In fact, transformations in the functional typology of these buildings are presently being recommended and implemented by the Social and Habitation Promotion Subprogram (SPSH in Portuguese: *Subprograma de Promoção Social e Habitação*) [35]. This program emerged aiming to stimulate the implementation of housing in the HCSL, expanding access to urban lands for low income families. Despite significant efforts, the number of buildings and families covered by the SPSH initiative does not meet the occupation requirements in the HCSL, especially when analyzing the need for affordable housing in São Luís [35].

The HCSL demonstrates strong traces of the *Pombalino* architectural style [36]. This constructive system emerged in Lisbon after the devastating 1755 earthquake in the region known as *Baixa Pombalina*. This style is named after Sebastião José de Carvalho e Melo, the first Marquês de Pombal, responsible for supervising the reconstruction plan [37]. One of the most remarkable characteristics of buildings in São Luís is the presence of ceramic tiles in their facades (Figure 1a,b). Besides decorative purposes, the tiles provided impermeability and thermal isolation during the heavy rainy and dry seasons, respectively [34]. Additionally, buildings usually have internal patios with a shutter system, allowing ventilation and light into the buildings (Figure 1c).

Figure 1. Characteristics of the architectural typology in the historic center of São Luís: (**a**) facade of buildings at Rua Portugal, one of the most famous streets in the historic center; (**b**) different patterns of the facade tiles; (**c**) internal patios; (**d**) wall evidencing the Santo André cross structure.

The design of *Pombalino* buildings included earthquake safety standards and sewage channeling, benefiting from modular construction methods [37]. Structurally, Pombalino's buildings presented a three-dimensional interlocking system, also known as *gaiola pombalina*, formed externally by the main walls in stone masonry and internally by the frontal walls [36]. The frontal walls are characterized by masonry walls reinforced with a wood frame, formed by vertical, horizontal, and diagonal pieces, known as a Santo André cross [37]. This system allowed building lighter and more flexible elements in the event of an earthquake [38]. The same structural arrangement can be found in many buildings in the HCSL, although there was no risk of earthquakes in the region (Figure 1d).

In most buildings, structural masonry walls are constructed in stone and lime and are plastered with clay, lime, and sand mortar. These walls are usually located externally, and their thickness varies from 60 cm to 100 cm [37]. Additionally, they can be found internally on the ground floor, in contrast to upper floors, which mainly include frontal walls as internal structural elements and subdivision elements between spaces. Their thicknesses may vary along the building's height, at between 60 cm and 90 cm in most cases [39].

One of the most prominent differences between the *Pombalino* buildings and the typology in the HCSL is the volume distribution within blocks. *Pombalino* buildings followed a consistent distribution per block, resulting in a uniform volume, with usually five-story facilities. In contrast, buildings in São Luís demonstrate a variation in the number of floors in the same block distribution [34]. In addition, these buildings can present variations in their overall mass within the urban block. They can be projected in an "L", "C", "U", or "O" shape (Figure 2). This variability allows the incorporation of patios as described previously, facilitating ventilation and daylight requirements. Similar to most constructions in Brazilian urban centers of this period, the civic architecture of the HCSL is characterized as a contiguous block due to the absence of frontal or lateral setbacks. Thus, buildings are attached to each other, separated by gable walls.

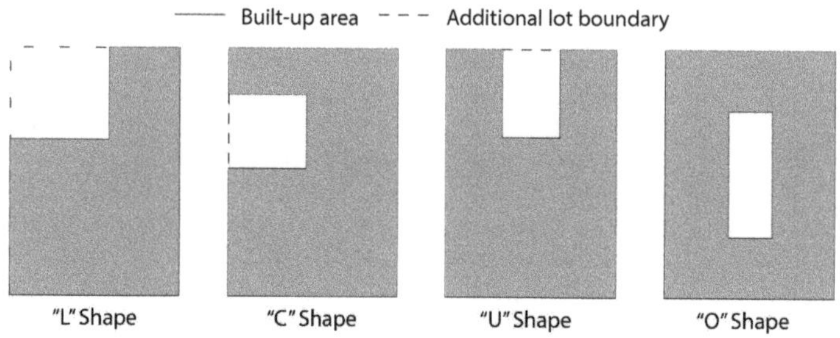

Figure 2. *Sobrado* massing types on the buildable urban lot.

Therefore, considering the volume and composition of the façade elements, the buildings in the historic center of São Luís are classified as *solar*, *sobrado*, and *casa térrea*. *Solar* is a house type intended for noble families. In São Luís they were built with refinement by the rural elite of the XVIII and XIX centuries, with an essentially residential function, to accommodate the family of the planters and the cotton and sugar producers in the capital. This layout typology is composed of the largest and most imposing buildings in the HCSL. The *sobrado* house type consists of buildings of two to four stories in height, with the ground floor reserved for commercial activities and the upper floor for residential areas. *Sobrado* buildings were occupied by merchants, also part of noble families. They are similar to the *solar* house type; however, the level of detail and refinement are lower when compared to *solar* buildings. *Casas terreas*, which can be translated as single-story houses, are present in a smaller area distribution than *solar* and *sobrado* buildings, intended for less wealthy population classes.

This study will focus on the *sobrado* house type, the most prevalent layout typology of buildings in the HCSL, especially in the federal and UNESCO areas of protection [37]. *Sobrado* buildings can have different shapes when analyzing the the buildable urban lot implementation. Considering the dimensions of such structures and high maintenance costs, single family use is not feasible in modern urban contexts [35]. The adaptation of *sobrado* buildings into multi-family apartments is a common solution today. Thus, this work defines a grammar-based methodology for assisting the adaptive reuse of *sobrado* buildings into multi-family apartments. The grammar presents a general approach for dealing with the high variability of *sobrado* building shapes by conceptualizing strategic transformations to modify a typical upper floor plan from single to multi-family use.

For formally defining rules that can be general for different building lot shapes while analyzing the *sobrado* house type, this study focused on three case study buildings (Figure 3). All buildings are part of the SPSH rehabilitation program, which defines mixed-use projects, with public offices on the ground floor and multi-family apartments on the upper floors. These buildings range from three classes of building lot implementation. Building 1 is located on Rua do Giz, no. 445, classified as a "C" shape. Building 2 is also located on Rua do Giz, no. 66; however, it is classified as a "U" shape. Lastly, building 3 is located on Rua da Palma, no. 336, and it is classified as an "L" shape.

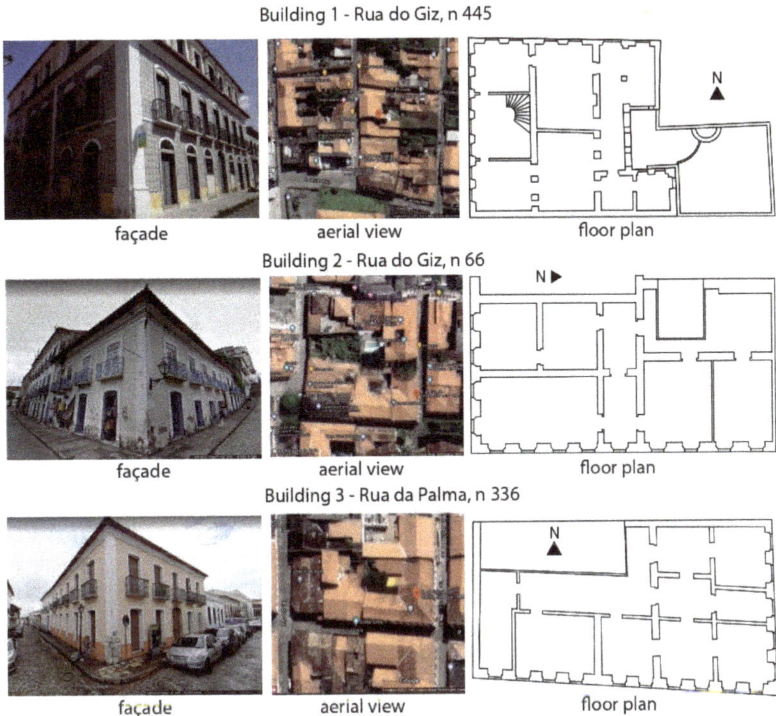

Figure 3. Case study buildings.

3. *Reviver* Grammar

This work explores the potential of SG in generating floor plan layout solutions for the adaptive reuse of historic buildings in the historic center of São Luís. The proposed grammar is entitled *Reviver*, a word from Portuguese meaning 'to revive'. The name is based on a rehabilitation program created in the late 1980s to preserve the city's heritage identity. Presently, locals refer to the historic center using this term. Grammar rules are defined to establish a strategic approach for allocating spaces (more specifically, apartments for multi-family use purposes and consecutively rooms within apartments) and architectural elements, such as walls, openings, corridors, etc. This process considers the base information of existing building elements as the initial input and encodes design constraints as shape rule sets, such as daylight requirements (for example, living rooms and bedrooms must contain a window) and dimension limits considering different apartment configurations (studios, 1-BR, and 2-BR apartments). Grammar aims to develop a formalism for repurposing historic buildings into social housing, in accordance with existing rehabilitation programs in the HCSL.

Although this paper focuses on the grammar-based methodology, this work is part of a broader context hybrid method framework for generating various layout solutions during adaptive reuse projects. It investigates how GD techniques can automate the adaptive reuse of historic buildings into multi-family apartments. The framework offers the potential to advance design workflows by combining human and machine knowledge to reduce conceptualization time and generate feasible designs considering the target objectives. More specifically, it investigates how the spatial requirements (number and type of apartments) in future work can be integrated with structural performance conditions, targeting solutions that maximize the reuse of existing elements and minimize C&D costs.

Since this methodology explores layout solutions for the adaptive reuse of historic buildings, the process focuses on floor plan drawings. Any floor plan configuration is abstracted as two datasets (Figure 4): (a) a list of points defining the precise boundaries of the building and (b) geometric information of existing elements such as walls, windows, stairs, patios, etc.

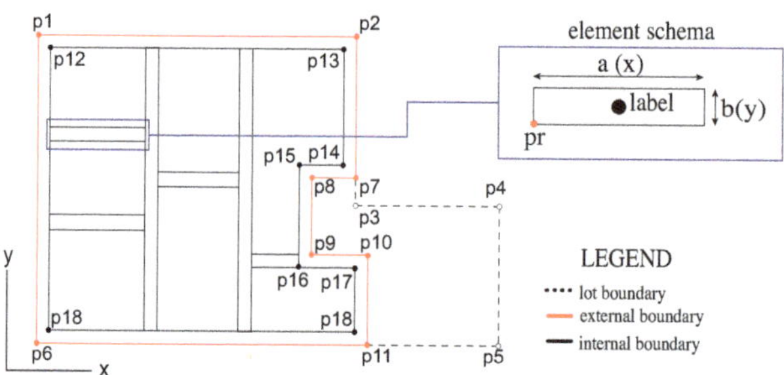

Figure 4. Abstract schema for defining the data input for the floor plan configuration (adapted from [34]).

First, three boundaries are defined as relevant for the adaptive reuse process: the urban lot boundary (lot) and the external (ext) and internal (int) boundaries for the built-up region (inside external walls). The data input for representing these curves are the coordinates (x, y) of the points defining each boundary. Second, information on each element on the floor plan must be provided. Floor plan elements are abstracted as a rectangle, as detailed in the sample internal element in Figure 4. To determine the element configuration, each element is abstracted as the following input variables: reference point defined by coordinates in x and y direction (p_r), dimension of the element in the x-direction (a), dimension of the element in the y-direction (b), and a label for each element (label). The label information refers to the type of element, such as walls, windows, stairs, etc. Although this process can be general for any floor plan, the label list can be more specific for incorporating details on the analyzed architectural typology.

The strategy considers the allocation of three types of apartments in the floor plan: studios, one-bedroom (1-BR), and two-bedroom (2-BR) apartments. It is important to consider that multi-family apartments are intended to be allocated only to upper floors, in accordance with the SPSH rehabilitation program guidelines. The transformation process considers the original built-up area (within exterior walls) as the design space. During the space allocation, façade elements will be used to guide the apartment allocation process, maximizing the use of natural daylight in apartments. In addition, the process will target the reuse of existing elements, aiming to minimize the C&D costs.

Although the three case study buildings are essential in the rule definition process, this section will initially consider only building 1 to exemplify the grammar stages and derivation process in detail. Then, the results on the other two buildings will be presented subsequently to validate and demonstrate the application of the *Reviver* grammar to different building types. In summary, the analysis can be divided into five stages (Figure 5). The first stage defines the initial floor plan configuration for the grammar-based space allocation process. In this stage, the numeric data input is converted into a labeled floor plan, where relevant elements are differentiated using colors, aiming to visually encode distinct shapes [40]. The second stage consists of defining a grid system for subdividing the design space into smaller areas. This step is essential for later classifying these areas according to daylight requirements (proximity to windows), which is part of the third stage. The fourth stage concerns the space allocation process, where apartments and connectivity elements (such as corridors and stairs) are allocated to the design space based on the subdivision areas. Lastly, the fifth stage includes the internal subdivision of individual apartments (units), also allocating openings to connect spaces and defining the final adaptive reuse layout. This process is implemented using Python and rhino script syntax. Although the computational implementation is an essential step in this project, this paper will mainly focus on describing the ruleset for the five stages (Figure 6) involved in the grammar-based methodology for facilitating the adaptive reuse process of the *sobrado* house type.

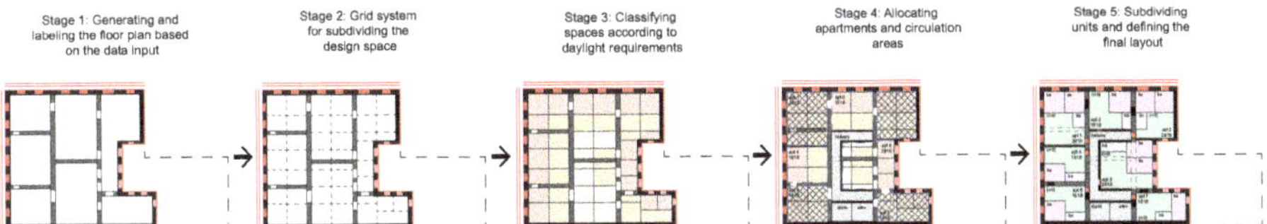

Figure 5. Summary of the *Reviver* grammar process.

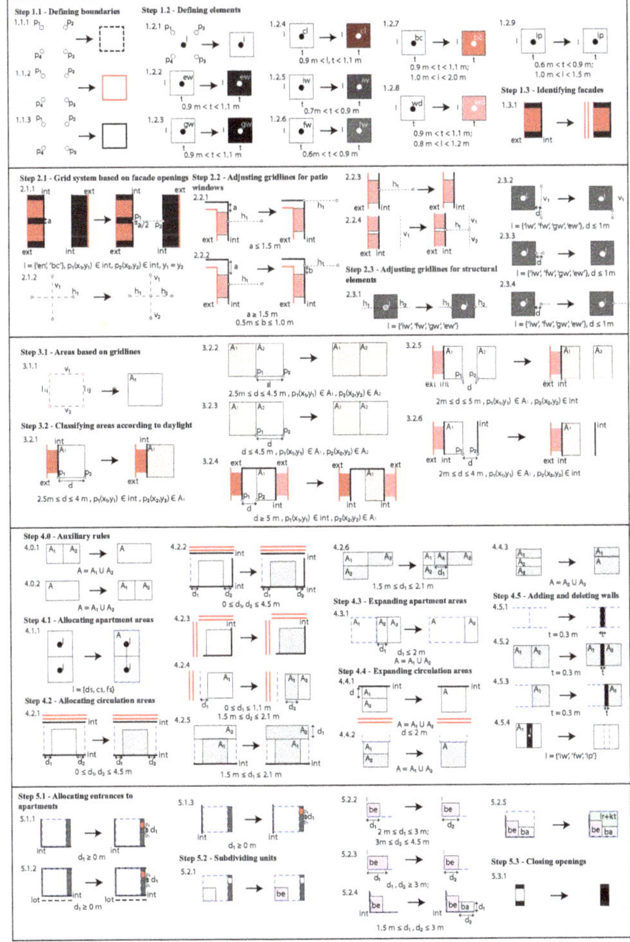

Figure 6. Ruleset for the *Reviver* grammar.

3.1. Stage 1: Defining the Existing Floor Plan Layout for the Analysis Process

Initially, grammar rules are defined to convert numeric data into visual components (steps 1.1 and 1.2 in Figure 6). The two datasets, defined in the previous section, can be stored in a comma-separated value file (csv) file, or any readable file, and used as data input for running the analysis. Step 1.1 consists of defining the three types of boundary (lot, external, and internal), based on the list of points representing the vertex of the shape to be drawn. Each boundary is visually differentiated. Rule 1.1.1 defines the lot boundary as a dashed line, while rules 1.1.2 and 1.1.3 defines the external and internal boundaries as red and black lines, respectively. It is important to stress that for simplification purposes, shapes in these rules are shown as basic rectangles to covey the overall concept. However, since *sobrado* buildings vary in urban lot implementation shapes, depending on the number of points given as input, these boundaries may assume a high variability of shapes.

The second step focuses on generating and labeling elements in the floor plan. Each element is abstracted as a rectangle, according to Figure 4. Thus, based on the given input, four vertex points and the label of each element can be used to draw a labeled rectangle (rule 1.2.1). Then, a color code is defined to visually differentiate each relevant element for the analysis (rules 1.2.2 through 1.2.9). These rules only include elements to be preserved during the adaptive reuse process. Considering the Luso-Brazilian architecture context, the list of element labels includes:

- Structural elements: The main structural components are masonry walls, classified into external masonry walls (ew), gable walls (gw), internal masonry walls (iw), and frontal walls (fw). Furthermore, some buildings may present stone columns (cl), especially on the ground floor.
- Opening elements: Mainly, one can classify the opening elements as facade elements, which includes the main entrance (en) and the facade windows, commonly containing balconies (bc). In addition, secondary windows (wd) are defined, encompassing windows located in the region facing internal patios and internal passages (ip).

It is important to note that existing non-structural partition walls and original stair locations are disregarded in this analysis. This is due to our focus on maximizing the reuse capacity of the base building structure and updating circulation components, especially stairs, to satisfy modern standards for egress in multi-family residential buildings. The last step in stage 1 differentiates façades regions (rule 1.3.1). When rule 1.3.1 is applied twice, it means the building is located in a corner lot. This is particularly important to identify, since this preliminary study will focus only on corner lot buildings. Figure 7 illustrates the derivation process of stage 1 for building 1, considering the numeric data (points and label information) as the data input. First, information on points (coordinates) is used to generate each boundary region, labeled according to ruleset 1.1 (dashed for the lot, red for the external, and black for the internal boundaries). Then, information on each element (points and labels) is used to draw labeled polygons by applying rule 1.2.1. Subsequently, each element is labeled by applying rules 1.2.2 through 1.2.9. Lastly, façade regions are differentiated by applying rule 1.3.1.

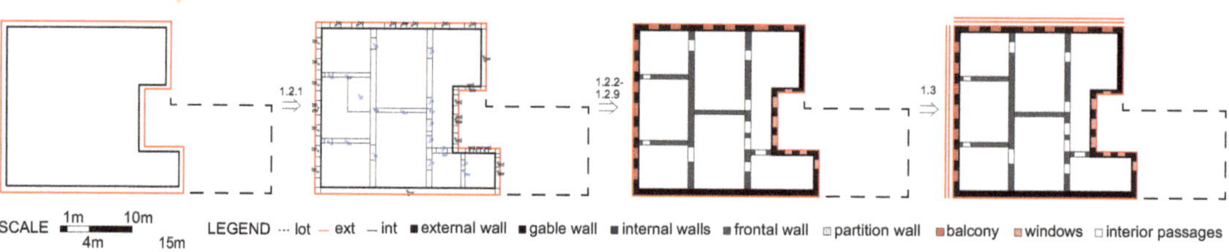

Figure 7. Derivation process considering steps for stage 1 for the case study building 1.

3.2. Stage 2: Grid System for Subdividing the Design Space

The second stage consists of defining a grid system to subdivide the design space area (steps 2.1 through 2.3 in Figure 6). The first step in this process consists of rules for allocating grid lines (step 2.1). Grid lines are generated based on the mid distance between two consecutive openings in the façade (balconies). It starts from the center point between two façade elements and it is extended until it meets the opposite internal boundary (rule 2.1.1), forming a 90° angle. In sequence, each grid line is subdivided into smaller components considering the points where vertical and horizontal lines connect with each other (rule 2.1.2).

When applying the grid division as defined in step 2.1, some undesirable patterns may arise due to the configuration of the existing building, such as grid lines coalescing with secondary windows or intersecting with structural elements (such as external and internal walls). Step 2.2 presents rules for adjusting grid lines to consider patio windows, while step 2.3 presents rules for modifying grid lines around structural elements.

In step 2.2, rules 2.2.1 and 2.2.2 are applied for adjusting grid lines that may intersect corner secondary windows, differing in the distance d from the window to the corner of the internal boundary (int). These rules may be applied simultaneously. In sequence, rules 2.2.3 and 2.2.4 are applied to intermediate secondary windows and should be applied in order. The first is applied to remove grid lines that may intersect these windows, while the latter is applied for defining a grid line between two windows.

In step 2.3, the first rule (2.3.1) is applied to split grid lines that intersect structural elements. In sequence, rules 2.3.2 and 2.3.3 can be applied simultaneously to remove grid lines that are located within a distance d (smaller than 1 m) to structural elements. Lastly, rule 2.3.4 is applied to extend any grid line that may be within a distance d from the structural element, a pattern that may arise due to the application of previous rules. These rules are critical while preparing the floor plan since they restrict intersection with existing structural elements to satisfy one of the primary goals of the grammar: to preserve the majority of the existing structure.

Figure 8 illustrates the derivation process for stage 2, considering the building 1 configuration. Starting with the labeled floor plan from stage 1, the first step is to apply rules 2.1.1 and 2.1.2 to generate the grid lines. Subsequently, adjustments to grid lines will be performed according to steps 2.2 and 2.3. Each grid line that will be modified is highlighted in blue. Thus, the rules from step 2.2 are applied to adjust the grid lines near the regions of patio windows, and rules from step 2.3 are applied to adjust the grid lines near structural elements.

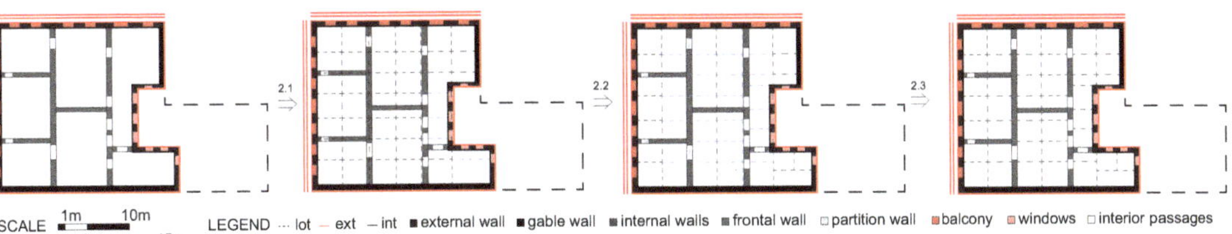

Figure 8. Derivation process considering steps for stage 2 for the case study building 1.

3.3. Stage 3: Classifying the Design Space According to Daylight

Based on the final grid configuration, stage 3 can specify areas to transform the existing building layout for a multi-family apartment floor plan (steps 3.1 and 3.2 in Figure 6). Step 3.1 presents a rule for generating areas based on the grid lines (3.1.1). In sequence, step 3.2 focuses on labeling areas according to their proximity to openings. These areas are classified as: (a) areas exposed to direct sunlight, "ds", labeled as a light orange shade; (b) areas without openings but still relatively close to spaces with openings, "cs", labeled as a light yellow shade; and (c) areas far from sunlight, "fs", labeled as a greyish-yellow shade.

First, rules 3.2.1 through 3.2.3 are applied in sequence to define "ds", "cs", and "fs" regions considering facade windows/balconies as references. Since external patios may have different shapes, which can affect the area distribution for secondary windows, rules 3.2.4 through 3.2.6 are used to label areas near secondary windows. Rule 3.2.4 is applied when an area with an adjacent secondary window is within a distance of $d \geq 5$ m to a facade window. It is important to note that this rule can only be applied to corner windows (when two or more sides of the apartment boundary intersect the internal boundary). Rule 3.2.5 is applied when an area is within a distance $2 \text{ m} \leq d \leq 5 \text{ m}$ from an "fs" region (defined based on facade windows). Lastly, rule 3.2.6 is applied when an area with an adjacent secondary window is within a distance of $2 \text{ m} \leq d \leq 4 \text{ m}$ from the internal boundary. Once these rules are applied, rules 3.2.2 and 3.2.3 can be applied again to modify the areas near to "ds" regions, defined based on the secondary windows.

Figure 9 illustrates the derivation process for stage 3, considering the building 1 layout. Starting from the grid system defined in stage 2, the first step is to apply rules 3.1.1 to define areas based on grid lines. Then, each area is classified according to its proximity to openings. First, rules 3.2.1 through 3.2.3 can be applied in sequence to classify areas near facade windows (balconies). Next, areas near patio windows are classified. Once rules 3.2.4, 3.2.5, and 3.2.6 are applied, rules 3.2.2 and 3.2.3 can be applied again to modify regions "cs" and "fs" based on the "ds" regions for patio windows.

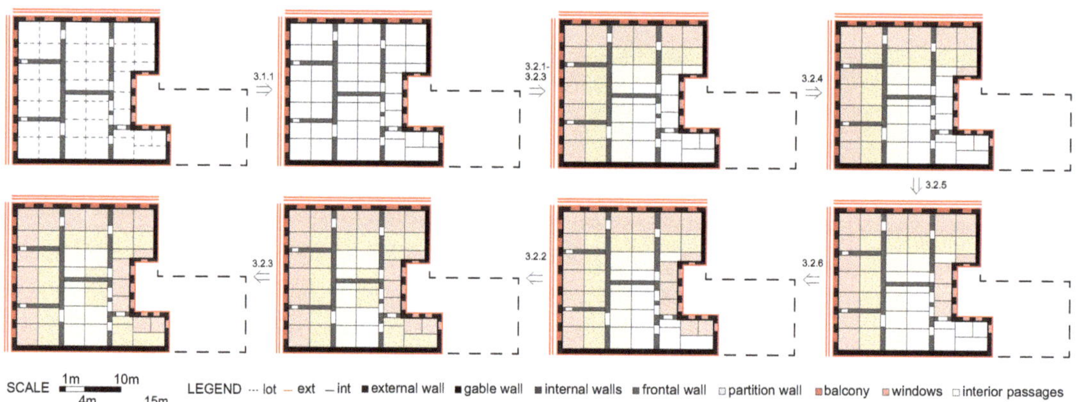

Figure 9. Derivation process considering steps for stage 3 for the case study building 1.

3.4. Stage 4: Allocating Apartments and Circulation Areas

This stage defines a general approach for allocating areas into apartments and circulation spaces, summarized in five steps (steps 4.1 through 4.5 in Figure 6). In addition, step 4.0 defines auxiliary rules, which may be applied concurrently with other rules during stages 4 and 5. Rule 4.0.1 is defined to merge areas, while rule 4.0.2 is defined to split a single area into two new areas.

Before describing the rules for the allocation process, it is essential to define the minimum requirements for each apartment configuration considered in the analysis (studio, 1-BR, and 2-BR), as summarized in Table 1. Studios must have at least two areas with direct sunlight, with a minimum area of 30 m^2, while 1-BR apartments must have at least two areas with direct sunlight and a minimum area of 40 m^2 and a 2-BR apartment must have at least three areas with direct sunlight and a minimum area of 50 m^2.

Based on this classification, the first step is allocating areas for apartments and circulation purposes. Rule 4.1.1 concerns the process of grouping areas into apartments, considering the minimum requirements in Table 1. While grouping areas, "ds" areas are mandatory, and to meet area requirements, priority will be given to "cs" regions over "fs" regions. Once areas are grouped, respecting the minimum requirements for the three types of apartments, the apartment boundary is differentiated as illustrated in rule 4.1.1.

Table 1. Minimum apartment requirements.

Apartment Type	n of Windows	Minimum Area (m^2)
Studio	2	30
1-BR	2	40
2-BR	3	50

Then, the second step is to allocate circulation areas. These rules modify "fs", regions since they are not preferably used in apartments. Rules 4.2.1 through 4.2.3 are defined to allocate potential vertical circulation areas (elevator and stairs). They can be applied simultaneously when possible and consider facade walls as a reference. They are applied to regions adjacent to internal boundaries opposite to facade walls. Rule 4.2.1 applies when an "fs" region is between two apartments. Similarly, rule 4.2.2 applies when an "fs" region is located between one apartment and one internal boundary. Lastly, rule 4.2.3 applies to "fs" regions on opposite sides of facade walls.

Rule 4.2.4 is applied to define horizontal circulation regions (corridors/hallways). It maintains a minimum distance, d, from an apartment boundary and splits an existing area into two new regions. One subarea defines a feasible circulation region, following a predefined range for the corridor width (d_1). Similarly, rule 4.2.5 is applied to determine corridor regions but considers vertical circulation regions as a reference instead of apartment boundaries. Then, rule 4.2.6 can be used to connect horizontal circulation areas that may be unattached if necessary. By applying these rules, several regions can be allocated. Thus, rule 4.0.1 can be used to combine areas into a single region for horizontal circulation purposes.

Then, the third step of stage 4 focuses on adjusting apartment areas. When an "fs" area is located between an apartment and a circulation area, rule 4.3.1 can be applied to merge the "fs" area into the apartment configuration. Similarly, stage 4 focuses on adjusting corridor areas. Basically, rules 4.4.1 through 4.4.3 can be applied to transform any remaining "fs" area left into circulation areas, focusing on horizontal circulation.

Once this process is finalized, the fifth step focuses on adding and removing walls. Rules 4.5.1 through 4.5.3 define new structural walls around apartment boundaries and vertical circulation regions. It is important to emphasize that these rules should be only applied in spaces where existing structural walls are not delimiting these regions. Lastly, rule 4.5.4 defines a strategy for removing existing walls and internal openings inside an area. In this stage, only walls inside the horizontal circulation region must be demolished. However, this rule can be used in later steps.

Figure 10 illustrates the derivation process for stage 4 considering the case study building 1. Starting from the final design in stage 3 (Figure 9), rule 4.1.1 is applied to combine areas into apartment units. Then, rule 4.2.1 applies to allocate the vertical circulation regions. In sequence, rules 4.2.4 and 4.2.5 are applied to define horizontal circulation (corridor) regions. Then, rule 4.0.1 is applied to merge all corridor regions. Later, rule 4.3.1 is applied to expand apartment regions. In this building, stage 4 is not necessary. Then, stage 5 rules are applied. First, rules 4.5.1 through 4.5.3 are applied to add new walls, which later can be used as structural elements. Then, rule 4.5.4 is applied to remove wall portions inside corridor regions.

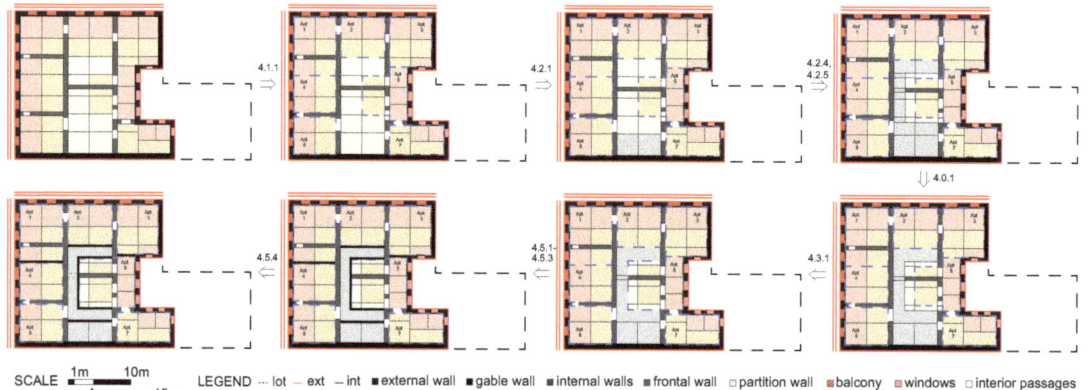

SCALE 1m 10m 4m 15m   LEGEND ··· lot — ext — int ■external wall ■gable wall ■internal walls ■frontal wall ▫partition wall ■balcony ▫windows ▫interior passages

Figure 10. Derivation process considering steps for stage 4 for the case study building 1.

3.5. Stage 5: Defining the Final Layout Considering the Adaptive Reuse Project

This stage concerns the final arrangements for generating a floor plan layout considering the adaptive reuse strategy. It is divided into three steps, defined as steps 5.1 through 5.3 in Figure 6. The first step of stage 5 concerns the allocation of openings representing the entrance of apartments (units). Rules 5.1.1 and 5.1.2 are applied to corner units (when two or more sides of the apartment boundary intersect the internal boundary). The first one is applied to apartments where two sides of its boundary intersect the internal building boundary. It defines the allocation of openings on opposite sides from the internal boundary location. The latter is applied to apartments where three sides of its boundary intersect the internal building boundary. It defines the allocation of openings on opposite sides of the lot boundary location. Lastly, rule 5.1.3 is applied to intermediate units. During this process, if an existing interior opening ("ip") is positioned near the apartment boundary, then entrance elements must be allocated to maximize the reuse of such circulation elements.

Then, step 5.1 defines rules for subdividing units internally. Three spaces are defined: bedroom (be), bathroom (ba), and a open kitchen space, with a living room shared with the kitchen (lr + kt). Rule 5.2.1 starts by allocating a bedroom in a "ds" area. In the case of a 2-BR apartment, rule 5.2.1 is applied again to define a second bedroom. When a bedroom has one dimension in the range of 2 m $\leq d_1 \leq$ 3 m, rule 5.2.2 can be used to expand the bedroom area. In addition, rule 5.2.3 can be applied to reduce the bedroom area. These rules must respect minimum dimension ranges, aiming to guarantee the usability of spaces.

After bedrooms are allocated, the remainder areas can be grouped using rule 4.0.1. This process must respect the daylight classification. For example, all "ds" areas can be merged. The same is valid for "cs" and "fs" regions. Then, walls that are internally distributed within an apartment region can be deleted, according to rule 4.5.4. Subsequently, rule 5.2.4 is applied to allocate the bathroom, while rule 5.2.5 transforms the remaining area into the living room and kitchen space. Lastly, step three focuses on closing existing openings that are not necessary for circulation. It is important to note that when applying rules 5.2.2 through 5.2.5, rule 4.0.2 is applied concurrently to split regions and allow the subdivision process.

Figure 11 illustrates this process for the case study building 1. Starting from the final configuration from stage 4, rules from step 5.1 are applied to allocate entrances to each unit. Next, rule 5.2.1 is applied to allocate bedrooms. Then, rule 5.2.2 and rule 4.0.2 are applied to expand the bedroom areas for apartments 5 and 7. Next, rule 4.0.1 is used to merge regions labeled with the same classification according to its proximity to openings, and rule 4.5.4 is applied to delete walls that are located internally to apartments and cannot be used to subdivide spaces. Then, rule 5.2.4 is applied to allocate bathrooms, and subsequently rule 5.2.5 is applied to allocate living room and kitchen spaces to any remainder regions inside apartment units. Lastly, rule 5.3.1 is applied to close any existing opening that may not be used as a circulation element in the transformation process.

Figure 11. Derivation process considering steps for stage 5 for the case study building 1.

3.6. Validating the Reviver Grammar

Using the three case study buildings, it is possible to define a grammar that is likely to suit the three studied types of projected shapes on the buildable urban lot: "C", "U", and "L". This section demonstrates the applicability of the *Reviver* grammar to case study buildings 2 and 3, classified as "U" and "L" shapes, respectively. Figures 12 and 13 demonstrate the simplified derivation process for buildings 2 and 3.

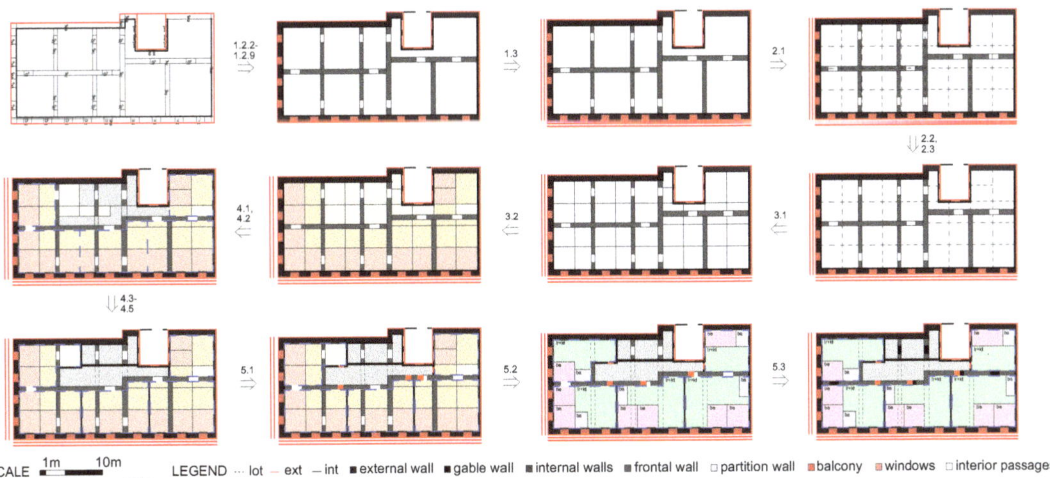

Figure 12. Derivation process for the case study building 2.

By verifying the applicability of rules to the three buildings, each rule can be classified according to its existence: mandatory, exception, and optional. A mandatory rule is applied regardless of the building configuration. An exception rule is applied if it exists in the floor plan configuration, meaning that in some building configurations, it may not be possible to apply it. Lastly, an optional rule allows the generation of design variations. Although they can be applied in a specific building configuration, they are not necessarily applied.

Figure 13. Derivation process for the case study building 3.

In stage 1, rules from steps 1.1 through 1.3 are mandatory, except for rules 1.2.4 and 1.2.5, since these elements may only sometimes be present in the floor plan. In stage 2, rules in step 2.1 are mandatory, while rules from steps 2.2 and 2.3 are classified as exception rules. Analyzing the two buildings, from step 2.2, only rule 2.2.4 is applied for building 2, while 2.2.3 and 2.2.4 are applied for building 3. Although rules from step 2.3 may be exceptions, this is not true for buildings 2 and 3, since all four rules were necessary. In stage 3, rules 3.1.1 and 3.2.1 through 3.2.3 are mandatory, while 3.2.4 through 3.2.6 are exceptions. From rules 3.2.4 through 3.2.6, only rule 3.2.4 is applied for both buildings 2 and 3. In addition, it is important to stress that if rules 3.2.4 through 3.2.6 are applied, rules 3.2.2 and 3.2.3 must be applied in sequence since they modify areas considering their proximity to openings.

In stage 4, step 4.1 specifies the apartment allocation process with a single rule. Thus, this rule is mandatory during the derivation process. In step 4.2, rules 4.2.1 through 4.2.3 concern the allocation of vertical circulation elements; sometimes only one of these rules is applied. For building 2, only rule 4.2.2 is applied, while for building 3 both rules 4.2.1 and 4.2.3 are applied. Then, when verifying rules for allocating horizontal circulation, rule 4.2.4 is always applied (mandatory), while 4.2.5 and 4.2.6 are only applied when needed (exception). For building 2, only rules 4.2.4 and 4.2.6 are applied, while for building 3, only rule 4.2.5 is applied. Then, rules from step 4.3 and 4.4 are classified as exception rules. This can be verified by analyzing the case study buildings. Rule 4.3.1 is only applied to building 2. Then, for step 4.4, rule 4.4.3 is applied to building 2 and rules 4.4.1 and 4.4.2 are applied to building 3. Lastly, all rules in step 4.5 are mandatory.

In stage 5, rules 5.1.1 and 5.1.3 are classified as mandatory, while 5.1.2 is classified as an exception. This can be verified for the case study buildings, since there will always be corner and intermediate apartments for any building type. However, a corner apartment that has three sides coinciding with the internal building boundary does not exist in the building 3 layout. Then, rules 5.2.1 and 5.2.2 are classified as mandatory, while rule 5.2.3 is classified as optional. Reducing the dimensions of bedrooms is an optional process included in the ruleset to generate variations in design. Finally, rules 5.2.4, 5.2.5, and 5.3.1 are mandatory. By the end of the derivation process, building 2 was split into five apartments (two 1-BR and three 2-BR) and building 3 was split into six apartments (two studios, one 1-BR, and three 2-BR).

Table 2 summarizes the *Reviver* grammar, defining the steps and corresponding actions. The rules in each step are classified according to their application during the derivation process.

Table 2. Summary of steps and classification of rules according to their application during the derivation process.

Step	Action	Rule Classification
Step 1.1	creating and labeling boundaries	mandatory
Step 1.2	creating elements (1.2.1)	mandatory
	labeling elements(1.2.2–1.2.9)	mandatory
Step 1.3	identifying façade walls	mandatory
Step 2.1	defining grid system	mandatory
Step 2.2	adjusting grid lines for patio windows	exception
Step 2.3	adjusting grid lines for structural elements	exception
Step 3.1	defining areas based on grid lines	mandatory
Step 3.2	classifying areas according to daylight	mandatory (3.2.1–3.2.3)
		exception (3.2.4–3.2.6)
Step 4.1	allocating apartment areas	mandatory
Step 4.2	allocating circulation areas	exception
Step 4.3	expanding apartment areas	exception
Step 4.4	expanding circulation areas	exception
Step 4.5	adding and deleting walls	mandatory
Step 5.1	allocating entrance to apartment units	mandatory (5.1.1,5.1.3)
		exception (5.1.2)
Step 5.2	subdividing units	all mandatory except rule
		5.2.3 (optional)
Step 5.3	closing openings	mandatory

4. Variations in Multi-Family Apartment Solutions Using the *Reviver* Grammar

This section produces different floor plan layouts for each case study building to verify the proposed grammar's ability to achieve various design solutions. For building 1, three combinations of seven apartments are explored during stage 4. The first and third variation achieved one studio, three 1-BR apartments, and three 2-BR apartments, while the second variation achieved four 1-BR apartments and three 2-BR apartments. By applying the subdivision process of apartments in stage 5, each variation from stage 4 can have sub-variations. Figure 14 shows six different layouts produced for building 1 using the *Reviver* grammar.

For buildings 2 and 3, four variations are presented (Figures 15 and 16). During stage four, two combinations are explored. For building 2, the first variation resulted in five units (two 1-BR and three 2-BR apartments), while the second variation produced six apartments (one studio, four 1-BR apartments, and one 2-BR apartment). For building 3, the first variation produced six units (two studios, one 1-BR apartment, and three 2-BR apartments), while the second variation produced seven apartments (two studios, four 1-BR apartments, and one 2-BR apartment). Then, for each variation from stage 4, two sub-variations are illustrated for the subdivision process of apartment units in stage 5.

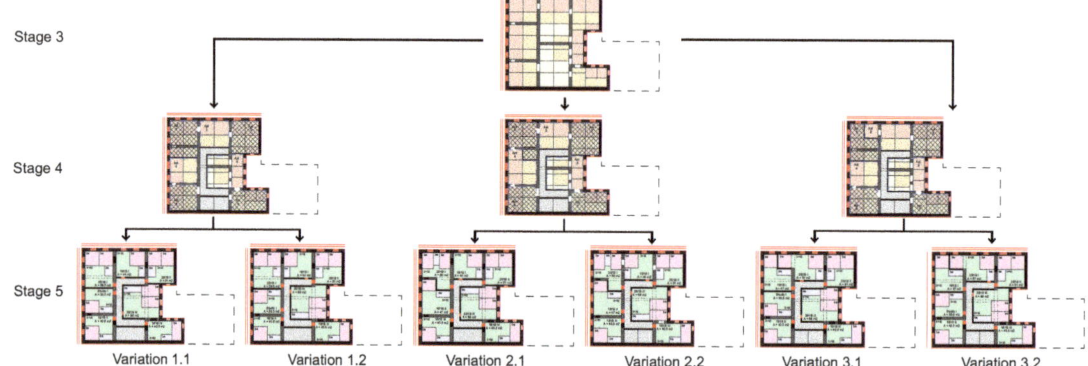

Figure 14. Variation process for the case study building 1.

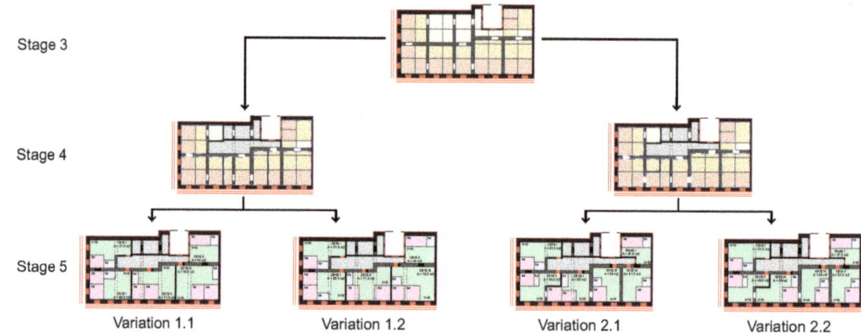

Figure 15. Variation process for the case study building 2.

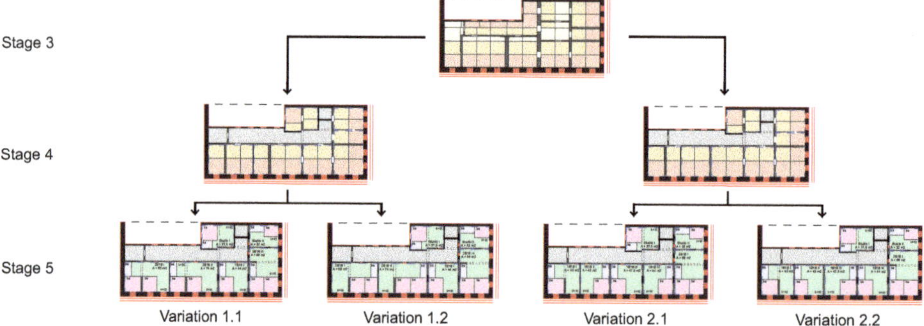

Figure 16. Variation process for the case study building 3.

5. Conclusions and Future Work

Adaptive reuse is a challenging task that involves understanding a building's configuration and repurposing it for a new use. Thus, this research investigates how the SG formalism can automate the space allocation process for adaptive reuse projects, focusing on historic buildings. It explores a building typology named *sobrado*, the predominant building type in the historic center of São Luís, Brazil, a UNESCO World Cultural Heritage Site. The described framework contributes to the exploration of design space by formalizing the steps for allocating multi-family apartments in the floor plans of typical *sobrado* buildings. The grammar, entitled *Reviver*, presents five stages that assist in transforming an existing floor plan configuration, originally designed for single-family use, into a new layout reconfigured as units subdivided as housing units for multiple families. It effectively produces variations in designs considering three different apartment types: studios, 1-BR apartments, and 2-BR apartments. The *Reviver* grammar targets the maximization of reuse of existing elements, not only considering daylight but also considering structural (walls) and circulation elements (interior openings). The strategy encodes a design logic that strategically allocates apartments into regions that profit from existing façade openings and windows, supporting daylight requirements.

Since many historic downtown areas, especially in Europe and European colonized countries, present similarities in how buildings interact among blocks, the proposed strategy can be further investigated with different case studies, verifying the transferability of a rule-based framework for adaptive reuse. Despite differences in architectural styles and functional typology, often historic downtown areas have low-rise buildings distributed in the same block, prevalently from the same architectural style, lined up shoulder-to-shoulder and sharing one or both walls, which can be defined as *row buildings*. Thus, the *reviver* grammar can be investigated as a general approach for row buildings in historic downtown areas. As described in this work, the HCSL presents similarities with the *Baixa Pombalina* region in Lisbon, which can be considered in future work as a transfer case study to extrapolate the results in this paper.

The steps in the *reviver* grammar are a general approach for analyzing an existing building and allocating spaces to plan for the future use of urban infrastructure. When studying the transferability of this approach to other case studies, some adaptation may be necessary. In step one, the labeled colors are defined based on the constructive system encountered in the HCSL. Due to a substantial similarity with the *Pombalino* style, the ruleset can be maintained when analyzing buildings in the historical region of Lisbon. However, new rules may have to be defined in this step to accommodate other architectural styles depending on how different the constructive system is compared to the buildings in the HCSL.

In steps 2 and 3, the general idea can be translated to other building types since the strategy investigates the subdivision of the interior spatial layout based on existing façade openings and windows. However, new rules may be necessary, especially in adjusting gridlines (step 2.2 and 2.3) and in classifying areas according to the distance to openings, especially rules 3.2.4 through 3.2.6. These rules were defined based on the massing types present in the urban context in São Luís (Figure 2). Thus, it is essential to investigate how they can be transferred to other urban configurations.

Similarly, the logic for allocating apartments (step 4.1) can be easily extrapolated to other building types. Modifications to apartment requirements (such as total area and the number of windows) can also be included by modifying the specifications in Table 1. However, steps 4.2 through 4.5 could be more challenging, depending on the existing floor plan layout. Finally, rules in step 5 would also be easily adapted to different building typologies since they deal with the subdivision of apartment units.

This project is part of a broader scope of mixed methods to study how generative design can support adaptive reuse in terms of spatial and structural problems. Future work will integrate the *reviver* grammar with genetic algorithms (GAs) and generative adversarial networks (GANs) to support the combination of human and machine expertise for the adaptive reuse of historic buildings. Some advantages of combining GD methods into the adaptive reuse process include: (1) facilitating a broader exploration of the design space to evaluate different scenarios based on objective functions and design requirements; (2) formalizing human design knowledge, experience, and expertise in SG rules to automate feasible designs in an accelerated and adaptable framework; and (3) developing more robust, repeatable frameworks, which may be further integrated with participatory design to promote feedback and interaction from the perspective of the designer and the end-user.

Author Contributions: Conceptualization, D.M.S.P., H.L. and R.N.; methodology, D.M.S.P.; software, D.M.S.P.; validation, D.M.S.P., H.L. and R.N.; formal analysis, D.M.S.P.; investigation, D.M.S.P.; resources, H.L. and R.N.; data curation, D.M.S.P.; writing—original draft preparation, D.M.S.P.; writing—review and editing, H.L. and R.N.; visualization, D.M.S.P.; supervision, H.L. and R.N.; project administration, H.L. and R.N; funding acquisition, H.L. and R.N. All authors have read and agreed to the published version of the manuscript.

Funding: This material is based upon work supported by the National Science Foundation under grant IIS-2123343. Any opinions, findings, and conclusions or recommendations expressed in this material are those of the author(s) and do not necessarily reflect the views of the National Science Foundation. This work was partially supported by the Maranhão Research Foundation (FAPEMA), under grant BD-09053/22. This paper was partially supported by The H. Campbell and Eleanor R. Stuckeman Fund for Design Computing at The Pennsylvania State University and by the Gordon D. Kissinger Graduate Research Fellowship from the Architectural Engineering Department at Pennsylvania State University.

Institutional Review Board Statement: Not applicable.

Informed Consent Statement: Not applicable.

Data Availability Statement: The data presented in this study are available on request from the corresponding author.

Acknowledgments: We thank José Pinto Duarte, Thomas E. Boothby, and Lisa Iulo for fruitful discussions during committee meetings, and Cooper Knapp for their assistance during undergraduate research activities related to Shape Grammars. We also would like to thank IPHAN-MA and Margareth Figueiredo for their assistance during data collection. We acknowledge FAPEMA, SECTI and the Maranhão State Government for providing partial funds for this work. We also acknowledge the Architectural Engineering department and the Stuckeman Center for Design Computing at The Pennsylvania State University for all the support during the execution of this work.

Conflicts of Interest: The authors declare no conflict of interest.

References

1. Grabowski, M. The 18th And 19th Century Industrialization Process as The Main Aspect of City Creation and Its Impact On Contemporary City Structures: The Case of Lodz. *IOP Conf. Ser.* **2019**, *471*, 82046. [CrossRef]
2. Hollander, J.B.; Hartt, M.D.; Wiley, A.; Vavra, S. Vacancy in shrinking downtowns: A comparative study of Québec, Ontario, and New England. *J. Hous. Built Environ.* **2018**, *33*, 591–613. [CrossRef]
3. Conejos, S.; Langston, C.; Smith, J. Improving the implementation of adaptive reuse strategies for historic buildings. In *The IX International Forum of Studies: S.A.V.E. Heritage*; Bond University: Naples, Italy, 2011.
4. Robertson, K.A. Downtown Redevelopment Strategies in the United States: An End-of-the-Century Assessment. *J. Am. Plan. Assoc.* **1995**, *61*, 429–437. [CrossRef]
5. King, R.; Orloff, M.; Virsilas, T.; Pande, T. *Confronting the Urban Housing Crisis in the Global South: Adequate, Secure, and Affordable Housing*; World Resources Institute: Washington, DC, USA, 2019; p. 40.
6. Coupe, T. How global is the affordable housing crisis? *Int. J. Hous. Mark. Anal.* **2020**, *14*, 429–445. [CrossRef]
7. Odilla, F.; Passarinho, N.; Barrucho, L. Brasil tem 6,9 Milhões de Famílias sem Casa e 6 Milhões de Imóveis Vazios, diz Urbanista. BBC News Brasil, 2018. Available online: https://www.bbc.com/portuguese/brasil-44028774 (accessed on 18 July 2022).
8. Foster, G.; Kreinin, H. A review of environmental impact indicators of cultural heritage buildings: A circular economy perspective. *Environ. Res. Lett.* **2020**, *15*, 1–14. [CrossRef]
9. Mısırlısoy, D.; Günçe, K. Adaptive reuse strategies for heritage buildings: A holistic approach. *Sustain. Cities Soc.* **2016**, *26*, 91–98. [CrossRef]
10. Yung, E.H.K.; Chan, E.H.W.; Xu, Y. Community-Initiated Adaptive Reuse of Historic Buildings and Sustainable Development in the Inner City of Shanghai. *J. Urban Plan. Dev.* **2014**, *140*, 05014003. [CrossRef]
11. Shahi, S.; Esnaashary Esfahani, M.; Bachmann, C.; Haas, C. A definition framework for building adaptation projects. *Sustain. Cities Soc.* **2020**, *63*, 102345. [CrossRef] [PubMed]
12. Bullen, P.A. Adaptive reuse and sustainability of commercial buildings. *Facilities* **2007**, *25*, 20–31. . [CrossRef]
13. Brooker, G.; Stone, S. *Rereadings: Interior Architecture and the Design Principles of Remodelling Existing Buildings*; RIBA Enterprises: London, UK, 2004.
14. Plevoets, B.; Cleempoel, K.V. Adaptive Reuse as a Strategy towards Conservation of Cultural Heritage: A Literature Review. In *Structural Studies, Repairs and Maintenance of Heritage Architecture XII*; Brebbia, C., Binda, L., Eds.; WIT Transactions on The Built Environment: London, UK, 2011. [CrossRef]

15. Fisher-Gewirtzman, D. Adaptive Reuse Architecture Documentation and Analysis. *J. Archit. Eng. Technol.* **2016**, *5*, 3. [CrossRef]
16. Davila Delgado, J.; Hofmeyer, H. Automated generation of structural solutions based on spatial designs. *Autom. Constr.* **2013**, *35*, 528–541. .: 10.1016/j.autcon.2013.06.008. [CrossRef]
17. Kim, S.; Ryu, H.; Kim, J. Automated and qualitative structural evaluation of floor plans for remodeling of apartment housing. *J. Comput. Des. Eng.* **2021**, *8*, 376–391. [CrossRef]
18. Sanchez, B.; Haas, C. A novel selective disassembly sequence planning method for adaptive reuse of buildings. *J. Clean. Prod.* **2018**, *183*, 998–1010. [CrossRef]
19. Agkathidis, A. Generative Design Methods Implementing Computational Techniques in Undergraduate Architectural Education. In Proceedings of the 33rd eCAADe Conference, Vienna, Austria, 16–18 September 2015; pp. 47–55. [CrossRef]
20. Vatandoost, M.; Yazdanfar, S.; Ekhlassi, A. Computer-Aided Design: Classification of Design problems in Architecture. *J. Inf. Comput. Sci.* **2019**, *9*, 605–625.
21. Mountstephens, J.; Teo, J. Progress and Challenges in Generative Product Design: A Review of Systems. *Computers* **2020**, *9*, 80. [CrossRef]
22. Sönmez, N.O. A review of the use of examples for automating architectural design tasks. *Comput. Aided Des.* **2018**, *96*, 13–30. [CrossRef]
23. Stiny, G.; Mitchell, W. The Palladian Grammar. *Environ. Plan. B* **1978**, *5*, 5–18. [CrossRef]
24. Stiny, G. *Shape: Talking about Seeing and Doing*; The MIT Press: Cambridge, MA, USA, 2006; ISBN 978-0-262-19531-7.
25. March, L. Forty Years of Shape and Shape Grammars, 1971–2011. *Nexus Netw. J.* **2011**, *13*, 5–13. [CrossRef]
26. Ligler, H. Reconfiguring atrium hotels: Generating hybrid designs with visual computations in Shape Machine. *Autom. Constr.* **2021**, *132*, 103923. [CrossRef]
27. Granadeiro, V.; Duarte, J.P.; Palensky, P. Building envelope shape design using a shape grammar-based parametric design system integrating energy simulation. In Proceedings of the IEEE Africon'11, Victoria Falls, Zambia, 13–15 September 2011; IEEE: Livingstone, Zambia, 2011; pp. 1–6. [CrossRef]
28. Benrós, D.; Duarte, J.; Hanna, S. A New Palladian Shape Grammar. *Int. J. Archit. Comput.* **2012**, *10*, 521–540. [CrossRef]
29. Lee, J.H.; Ostwald, M.J.; Gu, N. A Combined Plan Graph and Massing Grammar Approach to Frank Lloyd Wright's Prairie Architecture. *Nexus Netw. J.* **2017**, *19*, 279–299. [CrossRef]
30. Eloy, S.; Duarte, J. A transformation-grammar-based methodology for the adaptation of existing housetypes: The case of the rabo-de-bacalhau. *Environ. Plan. B* **2015**, *42*, 775–800. [CrossRef]
31. Eloy, S.; Guerreiro, R. Transforming housing typologies. Space syntax evaluation and shape grammar generation. *Arq. Urb* **2016**, *15*, 86–114.
32. Guerritore, C.; Duarte, J. *Rule-Based Systems in Adaptation Processes: A Methodological Framework for the Adaptation of Office Buildings into Housing*; Springer: Cham, Switzerland, 2019; pp. 499–517. [CrossRef]
33. Belčič, A.; Eloy, S. Architecture for Community-Based Ageing—A Shape Grammar for Transforming Typical Single-Family Houses Into Old People's Cohousing. *Buildings* **2023**, *13*, 453. [CrossRef]
34. Paulino, D.M.S.; Knapp, C.; Ligler, H.; Napolitano, R. The Reviver Grammar: Transforming the historic center of São Luís through social housing. In Proceedings of the XXVI International Conference of the Iberoamerican Society of Digital Graphics, Lima, Peru, 9–11 November 2022; pp. 347–358. Available online: https://conf.dap.tuwien.ac.at/papers/sigradi/2022/sigradi2022_94.pdf (accessed on 22 January 2023).
35. da Conceição Silva, I. Politicas Habitacionais No Centro Historico Ludovicense. In Proceedings of the IX International Seminar of Public Policies, São Luís, Brazil, 20–23 August 2019; pp. 1–12. Available online: http://www.joinpp.ufma.br/jornadas/joinpp201 9/images/trabalhos/trabalho_submissaoId_545_5455ca5106ce4781.pdf (accessed on 22 July 2022).
36. Lopes, M.; Meireles, H.; Cattari, S.; Bento, R.; Lagomarsino, S. Pombalino Constructions: Description and Seismic Assessment. In *Structural Rehabilitation of Old Buildings*; Costa, A., Guedes, J.M., Varum, H., Eds.; Building Pathology and Rehabilitation, Springer: Berlin/Heidelberg, Germany, 2014; pp. 187–233. [CrossRef]
37. Figueiredo, M.G. Influência pombalina na morfologia urbana de São Luís do Maranhão. *Converg. Lus.* **2014**, *25*, 168–180.
38. Mateus, J.M. (Ed.) *Baixa Pombalina: Bases Para uma Intervenção de Salvaguarda/Baixa Pombalina: Bases for Intervention and Preservation*; Lisbon City Council: Lisbon, Portugal, 2005.
39. Lourenço, P.B.; Vasconcelos, G.; Poletti, E. Edifícios Pombalinos: Comportamento e Reforço. *Semin. Intervir Construções Exist. Madeira* **2014**, *I*, 103–112.
40. Knight, T. Color grammars: Designing with lines and colors. *Environ. Plan. B* **1989**, *16*, 417–449. [CrossRef]

 *buildings*

Article

Experience Grammar: Creative Space Planning with Generative Graph and Shape for Early Design Stage

Rizal Muslimin

School of Architecture, Design, and Planning, The University of Sydney, Sydney, NSW 2008, Australia; rizal.muslimin@sydney.edu.au

Abstract: This paper presents a method to synthesise functional relationships and spatial configuration simultaneously using shape and graph computation from shape grammar and space syntax theories. The study revisits seminal works and summarises the compatibilities between shape and graph computation as a set of rules. The rule computation is demonstrated in two cases from hospitality and retail, where current applications, opportunities, and limitations are discussed. The results from the study show that incorporating graph and shape rules allows sequences of functions and spatial arrangements to be developed in parallel. The method could help the designer anticipate the impact on the users' flow of activities more explicitly during the early design process and could also assist in generating new functional configurations to provide alternative spatial strategies in broader applications.

Keywords: shape; graph; shape grammar; space syntax; rule-based design; hotel spatial experience; retail transformation; flows of activities

1. Introduction

Building occupancy modes have been rapidly shifting in the past decade as users adapt to trends in technology and global issues (e.g., the COVID-19 pandemic and IoT). As users reinvent their businesses, functional configuration in a building sometimes becomes contested, and design strategies relying on aesthetic and form articulation do not always provide satisfactory answers [1,2]. While form-centric design remains commonly practised, diverse logic on how a building adapts to functional changes—and how a function adjusts to a building—is becoming more necessary today. This includes the logic of how users transform the space to fit new functions or modify their functions to suit existing spaces. A new-looking building might be operating for a conventional function, while an old structure could be reused for a new, trending function.

As such, it is necessary to distinguish spatial from functional innovations. For instance, some mid-to-high-rise office buildings today maintain a similar configuration to the first office building with an elevator during the Gilded Age in the US, i.e., bound by the building envelope, utility core, and structure [3]. What is usually considered an innovation in office design is mostly related to an incremental improvement, e.g., taller and wider-spanning structures with more cores and atriums [4]. Yet, in a more subtle area not immediately visible, the office occupancy mode has grown through a series of phases, from a single tenant to multiple tenants, from a single function to multiple functions (e.g., mixed-use with retail and residential) and from a single tenant per floor to multiple tenants per desk with hourly based tenancy (e.g., *WeWork*).

Nevertheless, progress in functional evolution is not always compatible with spatial innovation. While functional growth is commonly anticipated by providing wide-open space in a building, the function can evolve beyond the spatial boundary, not necessarily by capacity but by increasing the complexity or simplicity. When a function requires flexibility that a building cannot provide, e.g., work from home, wide-open space becomes irrelevant,

Citation: Muslimin, R. Experience Grammar: Creative Space Planning with Generative Graph and Shape for Early Design Stage. *Buildings* **2023**, *13*, 869. https://doi.org/10.3390/buildings13040869

Academic Editors: Michael J Ostwald and Ju Hyun Lee

Received: 27 February 2023
Revised: 23 March 2023
Accepted: 24 March 2023
Published: 26 March 2023

and the function may evolve elsewhere—such as by adjusting to a residential unit's spatial flexibility (e.g., a home office) or by generating a new typology, such as the emerging shared communal office [5–7].

The above example illustrates how a spatial configuration in a building can be vulnerable to functional evolution. A lack of functional adaptation can make a building space obsolete, especially if it hosts virtualisable or portable elements. Based on this observation, this study maintains that functional configuration needs to be developed simultaneously with spatial articulation in building design, and, to pair them effectively, it is necessary to understand the logic of how a function develops. The method outlined in this paper aims to map functional development in conjunction with the spatial arrangement. The study starts by reviewing foundational studies on functional analysis and spatial transformation, evaluating the compatibility between methods, and formulating a set of logics to analyse how function and space evolve and adapt. Based on this logic, this study then investigates how a function transforms, to what degree a function determines a space, and how a space adapts to functions in two cases: hospitality and retail. The research outputs are applicable for functional and spatial analysis and synthesis at the early design stage, namely the pre-design stage and feasibility study [8,9]. The investigation is presented with diagrammatic plans and corresponds to the ideation process during these stages.

2. Materials and Methods

This study incorporates the representation techniques used in space syntax (SS) and shape grammar (SG) to map the sequences of functions and their relationships in building spaces. The two methods use different types of representation, which complement each other in this research. SS explains a spatial relationship with a graph to measure the quality of space according to its relationship (e.g., space adjacency, connectivity, and depth of hierarchy) [10]. In architecture, SS application involves assessing the evolution of the functional relationship of a building type and mapping the users' movement and interaction within their corresponding spaces [11–14]. SG analyses and synthesises the shapes and spatial relationships in design with the use of rules (e.g., transformation, substitution, and subdivision rules) [15]. Haakonsen et al. found that most SG studies since 1985 were focused on the floor-plan generation and primarily on residential cases, such as vernacular houses and apartments [16–22].

Nevertheless, graph calculation, such as graph grammar and graph rewriting, as developed in [23,24], has been increasingly integrated with SG for the defining of design languages [25]. Compared with other generative design methods outlined in [26], the rule-based method offers design-friendly algorithms which are explicit enough for design analysis and synthesis. It is not as opaque as probabilistic modelling or too explicit, as in calculus modelling [27]. Shape and graph combinations in SS and SG have been approached on different grounds. As seen in the study by Heitor, Duarte, and Pinto on the SS and SG combination, the shape rules are described with a symbolic description, such as that of the house's site condition (context), required functions (typology), morphology, connection/adjacency (topology), and costs [28]. The SS graph is used to describe the topological description. As their rule computes, the iteration generates house design and topological descriptions. SS is then used to identify degrees of privacy and connectivity (e.g., a room with a dead end or circulating access). Studies by Eloy and Guerreiro take this approach. SS was used to evaluate floor plans in Lisbon housing built between the 18th century and the 20th century (e.g., depth and integration analysis), and SG generated variations of the housing design corpus that comply with a justified plan graph (JPG) from SS analysis [29]. Some SS studies started by generating a JPG, then followed this with SG iteration to generate shapes (e.g., walls and roofs), as seen in the studies of Lee et al., which evaluate the houses designed by Glenn Murcutt and Frank Lloyd Wright (FLW) [30–32]. Others start with a workflow scheme graph (e.g., patient flow in a hospital) to define shape rules, generate the designs, then evaluate the results in SS [33]. In a different application, Grasl and Economou use graphs to recognise the shape topology [34]. In their method, the spatial relationship between shapes, from points to lines to faces, is represented through nodes and edges in a graph—such as a 'plan graph' for point connections (e.g., end points and intersection points), an 'edge-graph' for a line connection, and a face connection graph. Their rules incorporate this graph to identify a shape by matching its graph structure and its topological patterns, such as point order, parallelism, and line length. The SS and SG methods have been widely studied, and they have been digitally automated as software plugins for space planning or shape matching (e.g., SG interpreter) [35–38]. Additionally, the graph has been used to navigate rules, rule iterations, and design outputs in design space [39–41].

To pursue the study's aim, it is necessary to integrate graph and shape computation to analyse functional evolution and spatial configuration in parallel. The paper first revisits the foundational studies on how shape and graph are computed in design. For brevity, the technical terminology is listed in Table 1.

Table 1. Glossary. Readers can find further information on this terminology in [15,42,43] for SG and [10,44,45] for SS.

Terminology	Definition
Shape	A set shape of i dimensions arranged in j-dimensional space within an algebra U_{ij}, where j- refers to 0-, 1-, 2- or 3- dimensional space, and i- refers to 0 for points, 1 for lines, 2 for planes, and 3 for solids.
Label	A mark to define shape U_{ij} orientation and position by using a shape in algebra V_{ij} (e.g., points (V_{0j}), lines (V_{1j}), or planes (V_{2j})) or symbols (e.g., with a letter or number).
Weight	A graphical property associated with shape U_{ij} in algebra W_{ij}; for instance, a point (U_{0j}) can be weighted with a plane (W_{2j}), a line (U_{1j}) weighted with thickness (W_{1j}), or a plane (U_{2j}) weighted with tones (W_{2j}). In the figures, plane weight is shown in a grey tone and line weight with a black colour.
Shape rule	A mapping function (R) to transform shape A into shape B in format R: A → B. Given a shape C, a formula (C-t(A)) + t(B) applies if there is a transformation of shape A or t(A) in shape C.
Rule computation	A step-by-step shape rule iteration application whenever a shape matches the left-hand shape of a rule, noted by a double arrow, e.g., $C_0 \Rightarrow C_1 \Rightarrow C_2 \ldots \Rightarrow C_n$, where shapes C_0 to C_n mark the output shapes from the previous step as the input shape for the next step. The three dots [...] indicate multiple rules applied in one step (nested between steps).
Rule schema	An assignment function in format x → y that assigns a predicate shape as the values for variable x and y with assignment g, such that g(x) → g(y) would compose a shape rule A → B.
Shape description	A set of descriptions (D) to describe a context for applying shape rules (G) (e.g., a spatial, functional context), assigned with a description function h: L_G → D.
Graph	A set of nodes (V), such as nodes {a, b, c, d}, connected by a set of edges (E), such as ((a, b), (b, c), (c, d), (d, e)), such that graph G = (V, E). In SS, nodes in the graph usually denote spaces (e.g., streets, rooms, or corridors), and the lines represent the connection or intersection between pairs of spaces. To distinguish the edge from a line in U_{1j}, the edges are coloured cyan.

It is important to acknowledge the principal differences between SS and SG in calculating a shape and graph to understand the limits and opportunities of their integration. When a shape is used in SS, it serves as a proxy for a graph for measuring spatial and functional relationships. SS considers nodes and edges in a graph numerically as discrete units and calculates them combinatorially (e.g., by counting numbers of rooms to measure spatial integration). Hillier states that it is *'by expressing these pattern properties in a numerical way that we can find clear relations between space patterns and how collections of people use them'* [45] (p. 22). In contrast, SG considers shapes visually as discrete and continuous elements and calculates them by first finding embedded shapes in a design. As such, SS graph calculation is recognised in SG as *'calculating with shapes when embedding is identity—in the special case where shapes behave like symbols [...] The rules in set grammars use identity to check embedding as rules do in all algebras where i is zero'* [42] (p. 275). While label and weight can be presented with high-dimensional geometry to augment graph visualisation, as seen in [46] (pp. 124, 148), they remain an attribute to a shape or, in Stiny's terms, a *'visual analogy'*. A point in U_0 will remain a point regardless of being accompanied by label V_{0+i} and weight W_{0+i} (e.g., a point appears solid). Similarly, while appearing as a line, an edge in the graph stays as a label in V_{1j}—not as a shape—as it indicates the point's orientation and connection. To further discuss shape and graph computation, the following section induces rules from the seminal studies by March and Steadman and the early SS studies by Hillier.

2.1. Early Shape-to-Graph Computation

Not long after the graph was introduced for formal functional analysis in *Notes on the Synthesis of Form* [47], the relationship between graph and shape was thoroughly demonstrated by March and Steadman in *The Geometry of Environment* [48]. One of their examples analyses three FLW house plans, stating, *'If each functional space is mapped onto a point and if, when two spaces interconnect, a line is drawn between their representative points, we produce a mapping known as a graph'* [48] (p. 28). As highlighted in *The Geometry*, the three houses shared the same graph, despite their unique forms and different construction periods. The similarity implies that sequences of how occupants experience the functions are *'topologically equivalent'*. For instance, the family room always has direct access to terraces, entrances, and dining rooms; the entrance is always connected to the office and the family room; and the dining room always buffers the family room from the office.

Figure 1 shows an interpretation of how FLW plans are converted into a graph focusing on family, lounge, and dining relationships in the Vigo Sundt house. The middle row in the figure visualises their statements into rule iteration. The upper parts of the iteration show induced rules from each iterative step [49]. For instance, rule R1 extracts a room's node; R2 links two nodes when their representative rooms share a boundary; and R3 removes the shapes' attributes to leave only nodes and edges as a graph. The inverse versions of these rules, from A $\rightarrow$ B to B $\rightarrow$ A format, convert the graph back to the shape. The lower part of the figure shows the identity rules to indicate whether the shapes are represented in algebra U_{0+1} to analyse the spatial condition with visual calculation or in U_0 to analyse the functional relationship with the symbolic calculation. The accompanying description of the algebra of U_{ij}, V_{ij}, and W_{ij} clarifies the shapes' condition.

Step one represents the statement *'If each functional space is mapped onto a point'*, where the shapes are analysed in U_{12} to recognise the rooms' boundaries. In step two, the shape is reduced into symbols in U_{02} (points), where part of the attention is shifted from a spatial condition to a functional condition. The point is accompanied by weight W_{22} (tone) to delineate the initial shape boundary, as expressed with the description 'U_{12} became $U_{02} + W_{22}$'. In step three, after the statement *'when two spaces interconnect, a line is drawn between their representative points'*, the focus is on the room's connectivity, where two points are connected with an edge (line) if their shapes share a boundary or are overlapping, $A \cap B$, or if one is inside the other, $A \subset B$ or $A \supset B$. The edges are represented with the label V_{12}, as described by $U_{02} + W_{22} + V_{12}$. Note that with the weight that remains, the spatial condition and functional relationship are visible and coexist in this step. The last step confirms *'a mapping known as a graph'*, where the weights are removed, leaving only points and labels as nodes and edges of a graph. Without weight, the attention at this step is purely invested in a functional relationship. The recurring shape U_{02} in this iteration confirms that the shape-to-graph conversion operates with a symbolic calculation, hence a set grammar.

2.2. Graph by Shape Decomposition

To apply visual computation, a shape should be considered in algebra $U_{0+i,\,j}$, where different compositions can be explored, as can be seen in Figure 2. Five different compositions from the Sundt house living area were assembled from triangles *a*, *b*, *c*, *d*, *e*. Rule R0 adds a label to a shape, while rules R4b and R4c recognise two shapes, where one is inside the other. Applying the rules to these different compositions produces various graphs, including a null graph if no shared boundaries are found. When the number of shapes is determined, the algebra goes back to U_{0j}, e.g., seeing two triangles means seeing two nodes.

As such, the transformation rules may produce different configurations, but they do not always change the graph configuration if the shapes are regarded as points. To vary the graph with transformation rules, the operation works better if it shifts back and forth with algebra $U_{0+i,j}$.

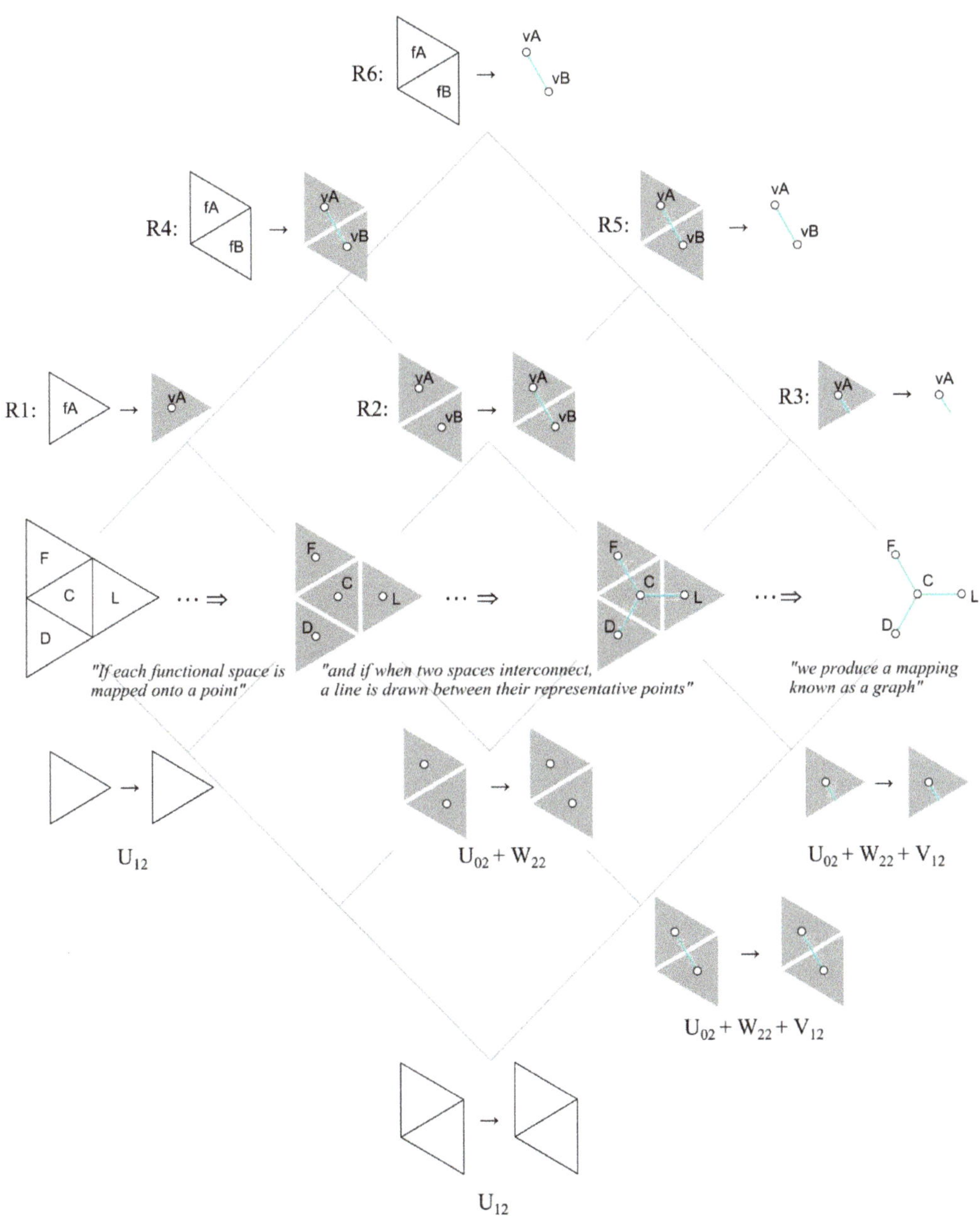

Figure 1. Shape rules induced from March and Steadman statements for the Sundt house plan. The initial shape represents the family (F), dining (D), and living room (L) zones, with the circulation (C) in between.

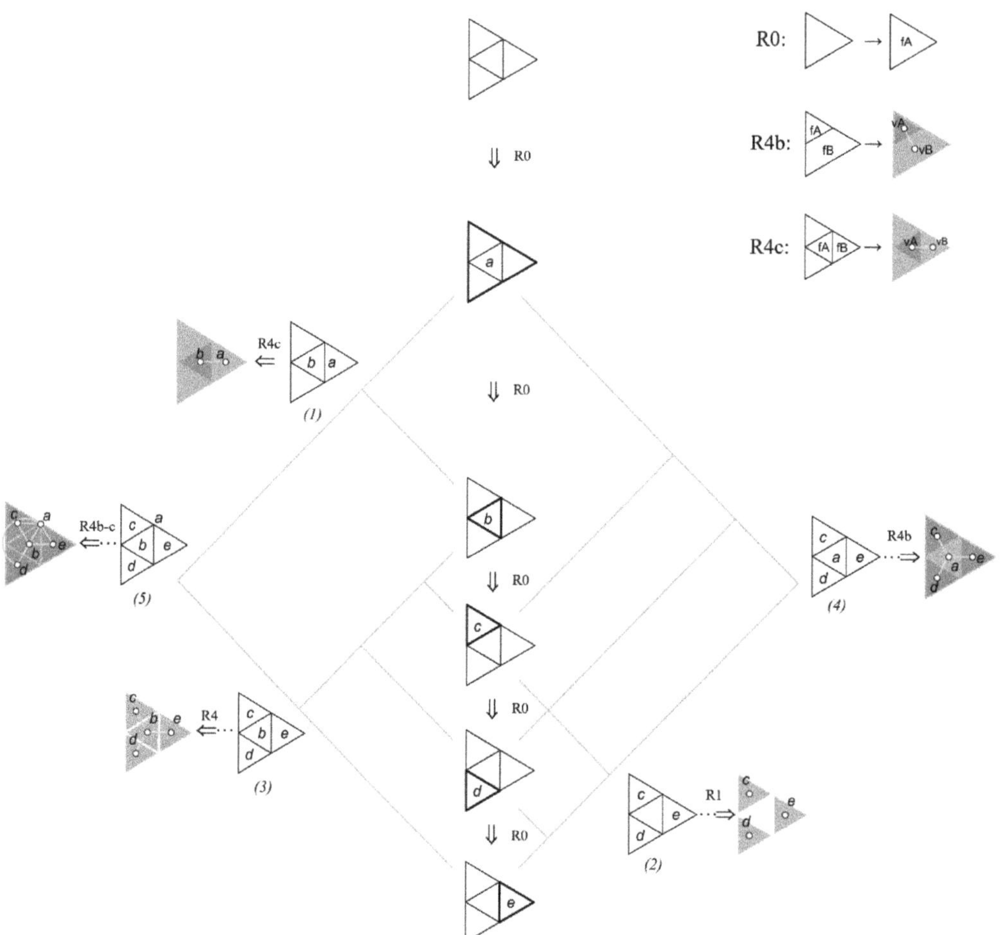

Figure 2. Different graphs are derived from one shape by decomposition.

Figure 3 shows how the translation rule R7 works with symbolic and visual calculation by varying the distance parameter (y). R7 is applied on the small triangle b in composition 1 and simultaneously on the three triangles c, d, and e in composition 2. Additionally, rule R1′ returns the triangles and removes their weight, and rule R0′ removes the labels added by rule R0. As can be seen, despite producing different compositions by varying y values, the graphs do not change much in algebra $U_{02} + W_{22}$. Yet, when the weight W_{22} is converted back to shape U_{12}, more triangles are found in the compositions, each producing different graphs after applying rules R0 and R1.

2.3. Graph by Rule Schema

By associating different plans with the same graph, March and Steadman suggest that FLW uses a *'range of "grammars"'* as *'the controlling geometric unit, which ordered the plan and pervaded the details'* [48] (p. 28). This study considers FLW's grammar as an integration of shape rules and functional configurations, and they can be analysed by associating their rules and their graph schemas (see more discussion on SG rule schemas in [50,51] and SS graph schemas in [30–32,52]). Figure 4 shows an interpretation of the shape rules, rule schemas, and graph iteration of Vigo Sundt's house and Ralph Jester's house floor plans (note that the interpreted iteration refers to design thinking rather than the drawing process). To produce shape rules, the transformation, additive, and subdivision schemas are assigned with a triangle and a circle, which correspond to each house's unique patterns (Figure 4a). The schemas are associated with the graphs, e.g., the rules from the additive and subdivision schemas produce two node graphs. The description with the variables D_{mn} replaces labels *a* and *b* in the shape rules (Figure 4d). In the iteration, steps one to five show a similar graph derived from the two plans to distribute living, utility, and sleeping zones before they diverge to generate more detailed rooms (Figure 4b,c). The only anomaly is in the last step, where the circle is substituted by a rectangle in the Jester house.

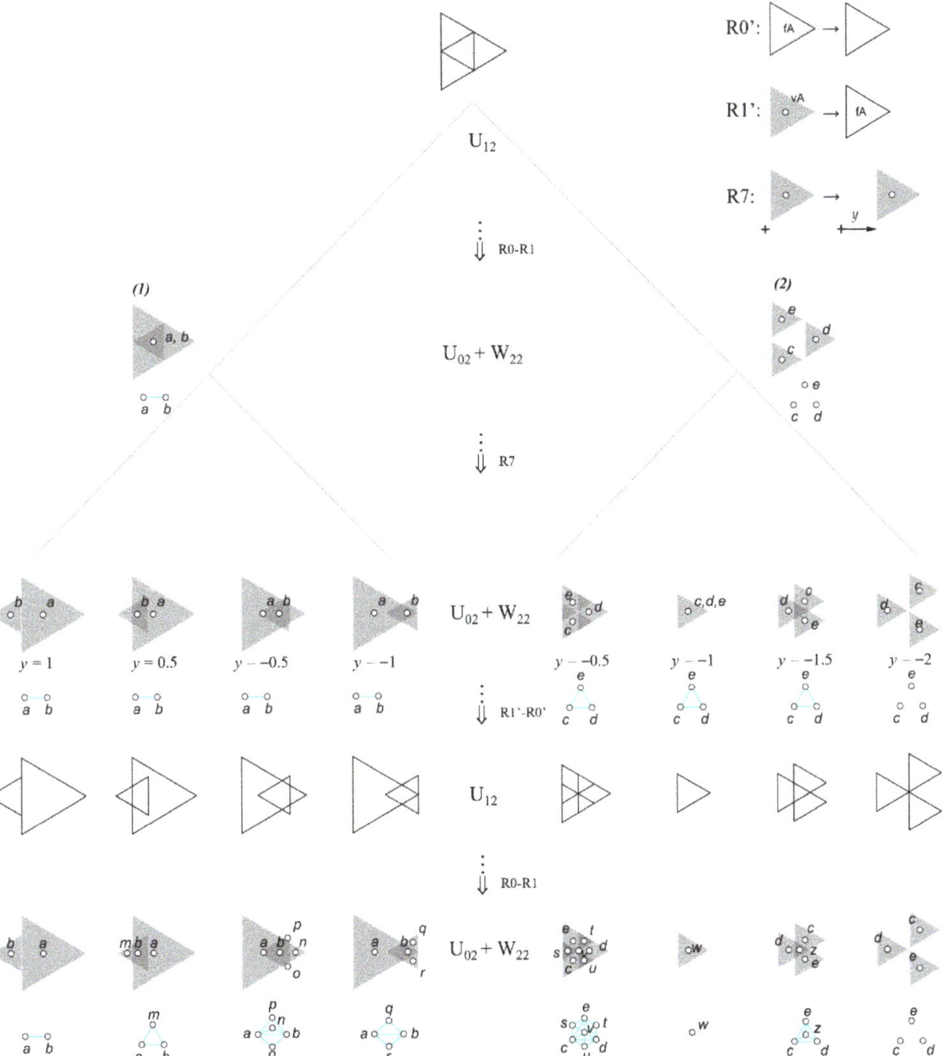

Figure 3. Comparison between graphs derived from discrete shape transformation (middle) and embedding on continuous shapes (bottom). The description indicates their operating algebras.

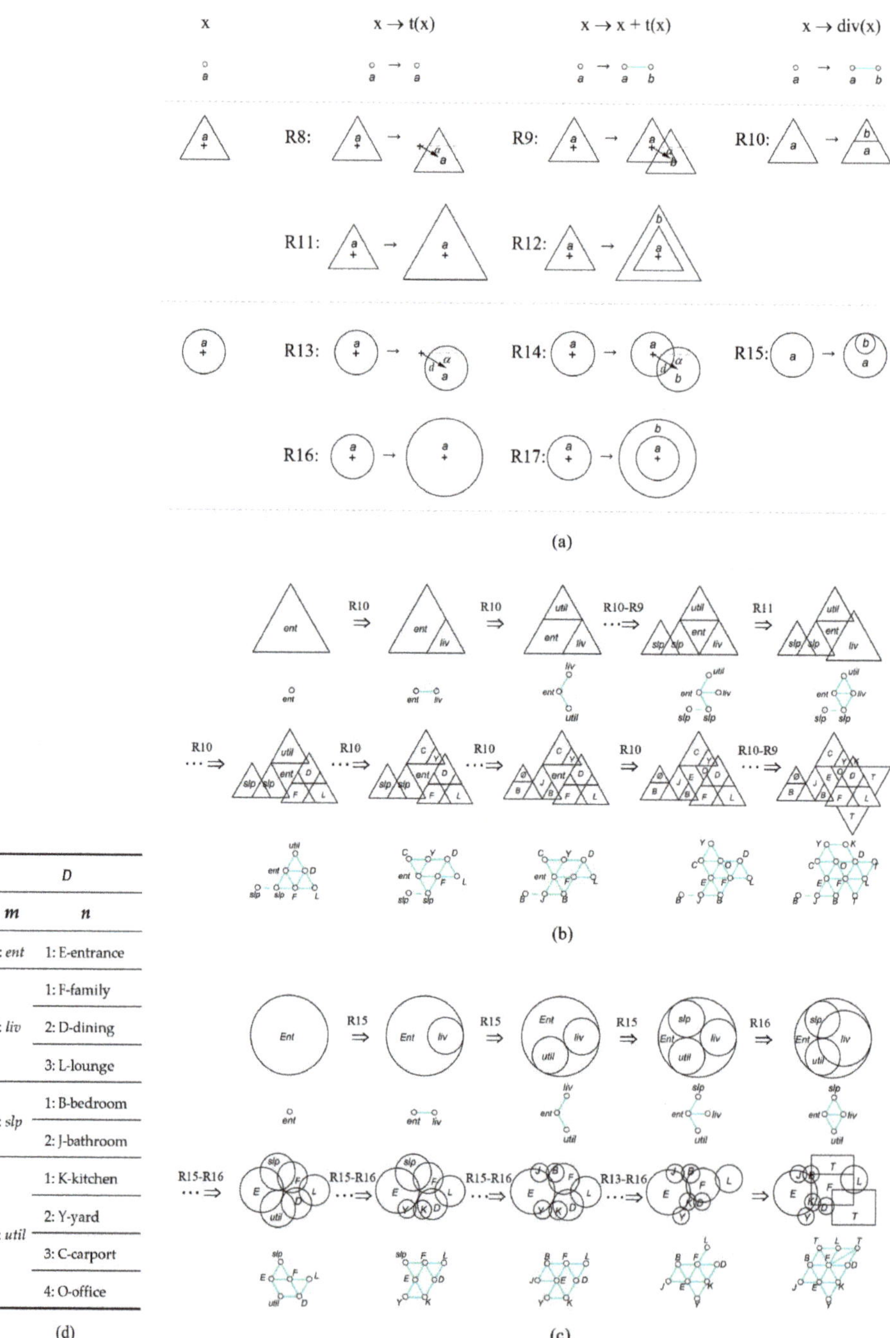

(a)

(b)

(c)

D	
m	**n**
1: *ent*	1: E-entrance
2: *liv*	1: F-family
	2: D-dining
	3: L-lounge
3: *slp*	1: B-bedroom
	2: J-bathroom
4: *util*	1: K-kitchen
	2: Y-yard
	3: C-carport
	4: O-office

(d)

Figure 4. Rule schema, adjacency graph rules, and shape rules (**a**), with their rule iteration on regenerating plans of the Vigo Sundt house (**b**) and the Ralph Jester house (**c**). Labels *a* and *b* are replaced by names outlined in the list of descriptions D_{mn} (**d**).

2.4. Shape and Graph Generative Rules

The revisit to March and Steadman's studies indicates that while graph calculation is inherently symbolic, it can be paired with visual calculation. In addition to using shape decomposition and transformation rules to produce graphs, a graph can be generative, as in graph rewriting. It can use the operators typically used in rule-based design (e.g., addition and subtraction). This section discusses the rule-based nature of SS by revisiting Hillier's rules in SG format for his *Laws of construction of space* (beady ring dwelling generation), *Laws from society to space* (graph generation), and *Laws from space to society* (axial map generation). The generative feature of SS is implied in Hillier's early dissemination, where he considered informal settlement spaces as analogous to '*irregular beads on a string*' [10]. In his 'beady rings' model, closed- and open-space relationships are represented with black and grey squares, and the settlement is assumed to be an aggregate of these beads, generated from bottom-up rules within a grid. One rule states: '*Each new unit must be joined by its open space with a full facewise join onto an open space already in the aggregation*'; another states: '*While full facewise joins of dwellings are allowed where they occur randomly, the joining of dwelling by their vertices is not allowed*' [44] (p. 170). In other words, the first rule adds a dwelling unit onto an aggregate of an existing open space (a grid) as the initial condition, and the second rule restricts overlapping between dwelling units. By applying these rules, a beady ring configuration can be produced with a graph consisting of square beads linking open and closed spaces within a grid. Figure 5 visualises Hillier's rules and their iterations in a 3×3 point grid as the initial shape. To avoid the overlapping of dwellings, rule R18 labels an open space as circulation (*c*) and only adds a dwelling unit on a point labelled with zero (0), oriented towards another point labelled 0 or *c*. The unit is represented by its node, weighted with a tone, and labelled with a number/letter and line. A counter variable *n* is used to label the unit's number in the iteration. As seen in iteration one, the rules managed to convert open space into beady ring dwellings, H1 (Figure 5-1, far right).

Nevertheless, the resulting beady rings only generated a weighted point configuration. To show the dwelling shapes, rules R19 and R20 replace the points with a set of lines that represent square walls with openings. The dwelling's opening position follows the line label orientation, while the open space openings are applied to all the walls. Iteration two in Figure 5-2 demonstrates these rules to generate the corresponding walls, where the final plan is clearer. To represent a connectivity graph of rooms from the design, Hiller's technique works similarly to that of March and Steadman: '*represent each space in the house by a small circle and each relation of direct access by a line*' [44] (p. 170). Rule R21 applies this technique by removing the weights of the dwelling nodes. The numbering labels and line labels are left to identify the room and the graph's edges. Additionally, rule R22 adds an edge between two *c* nodes to define access, and rule R24 extends a line to indicate access to the outside grid area. Iteration three in Figure 5-3 shows an iteration that converts a shape into a graph. Furthermore, the graph edges can be converted into lines to analyse access and axial integration. The notion of an axial map that Hillier refers to as '*the least set of straight lines that cover the open space of the area*' recalls a 'maximal line' in SG [44] (p. 174). The discrete lines converted from the graph's edges with rule R23 can be considered minimal lines and are reduced (or fused) into a set maximal line with this reduction rule: '*if lines I and I' are collinear and discrete, and share an endpoint, then replace both lines with the line I" fixed by the remaining endpoints for I and I'*' [42] (p. 187). The fourth iteration in Figure 5 shows the conversion from a connectivity graph into an axial map. The resulting maximal lines are weighted with thickness with rule R25, based on their number of initial minimal lines (e.g., a maximal line's thickness from four minimal lines is 0.4 mm). The line parameter from a weighted graph can be linked to the building's shape transformation. For example, in adjusting the corridor dimensions or surface articulation that responds to the users' traffic patterns, a thicker line may suggest a wider corridor and/or a more elaborate wall.

Note that the iteration from the Hillier study is constrained by the grid as an initial input. See more off-grid iteration in [46] (p. 153).

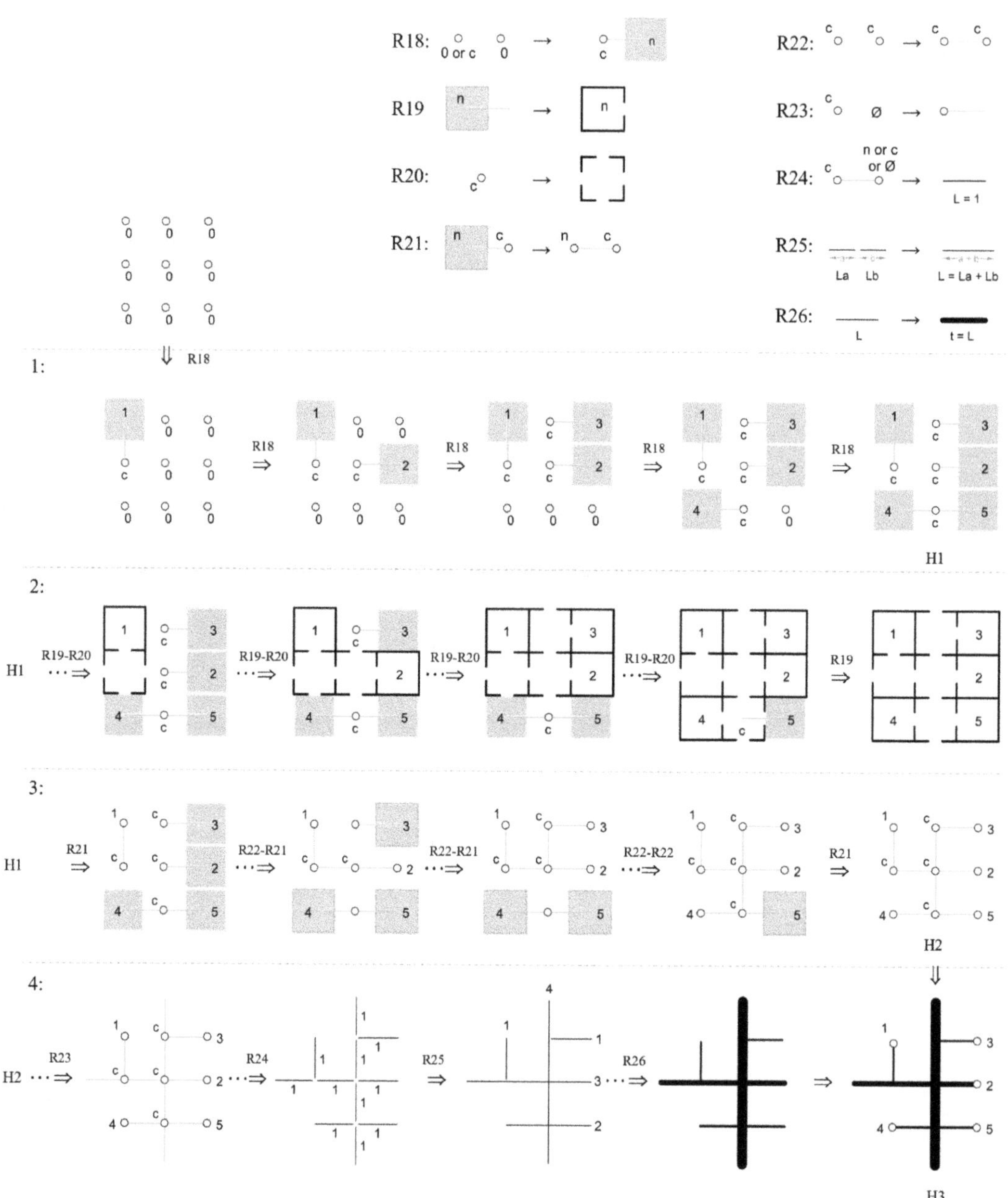

Figure 5. Shape rules (R18–R25) and their iterations to regenerate Hillier's laws for the construction of spatial orders (H1) in iteration one (**1**), laws from society to space (H2) in iteration three (**3**), and laws from space to society (H3) in iteration four (combined with graph H2) (**4**). Iteration two converts graph H1 into a plan (**2**).

2.5. Revisit Synopsis

The review of graph and shape computation studies highlights areas where graph and shape computation are compatible. The first is the interplay between graphs, shapes, and their attributes. While SS uses graphs mainly to measure how spaces are positioned against one another, the nodes and edges in the graph can be visually augmented with labels and weights as the attributes of symbolic shapes—for instance, weighted by a triangle or circle in FLW's plan or as a square in Hillier's beady ring.

The annotation distinguishes graphs from shapes so as not to confuse them for symbolic or visual calculations. Shifting shapes and their attributes back and forth helps to view spatial and functional relationships concurrently, and changing their algebras helps to reciprocate symbolic and visual calculations to analyse and synthesise the design. The latter brings to light a second compatibility: the generative rules for a graph. While generative rules are not always presented in SS studies, the revisited study exhibits rules for generating spatial relationship graphs and converting shapes and graphs. Their generative functions enable shape and graph rules to run in parallel. The rules derived from the revisited study are outlined in Figure 6.

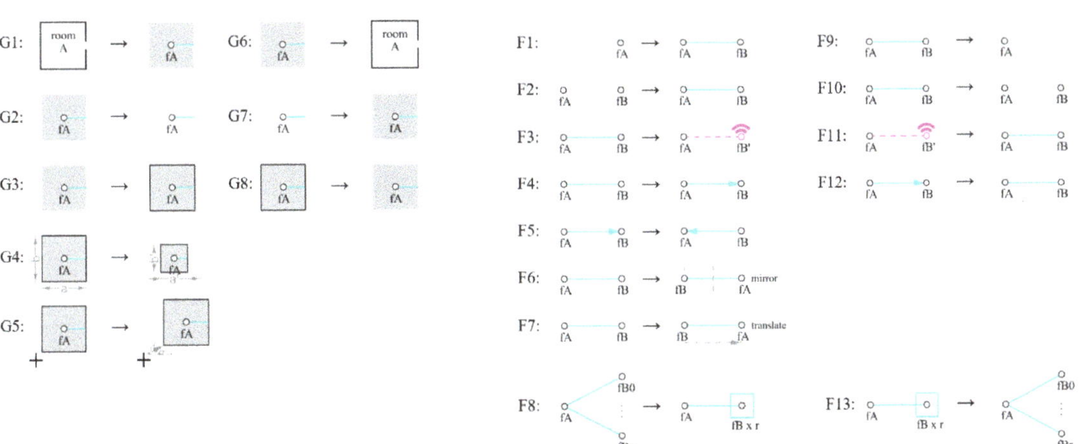

Figure 6. A set of shape and graph rules from revisiting studies.

2.6. Shape-to-Graph Rules

Rules G1 to G8 convert a shape into a graph and vice versa. Rule G1 returns a room's function node, represented by a point, weighted by a plane (tone) representing the room's boundaries, and labelled by edges to indicate access position. The function name corresponds to the room's name to indicate the user's activity (e.g., bedroom for sleeping). Rule G3 returns the weighted node into a shape to enable shape transformation by rule G4 (scaling) or G5 (translation and rotation). When the shape boundary changes, the weight follows. Rule G2 removes the weight and leaves only nodes and edges as a function graph without hinting at the spatial condition. In graph mode, modifications can only be made through functional reconfiguration and their sequence orders (as explained in the subsequent section). Rules G6, G7, and G8 are the inverse of G1, G2, and G3 to convert the graph back to a shape. Rule G7 returns the weight. Rule G8 removes the shape boundary. Rule G6 removes the weight and returns a function node to a room's shape with its corresponding name. The room's opening is positioned based on the line label orientation.

2.7. Graph-to-Graph Rules

Rules F1 and F9 add and remove an activity node in a graph, while rules F2 and F10 add and remove connectivity between nodes. To incorporate virtualisation trends, rule F3, marked with a Wi-Fi symbol, replaces a physical activity with a digital one (e.g., an online check-in), while its inverse (rule F11) returns the digital to a physical activity (e.g., from an online to an in-person meeting). Direction rules F4 and F12 add and remove direction to an edge, while rule F5 reverses the direction. The transformation rules F6 and F7 rearrange the position of the nodes in the graph through translation and reflection (mirroring by node or edge) for the exploration of different functional configurations. Annotation rule F8 is added for graph organisation by grouping the same functions into one (e.g., groups of bedrooms or classrooms), and rule F13 ungroups them. These rules are labelled by the number of elements (r) in a group (F × r). Grouping could be useful in the early design stage, such as in configuring private, public, and utility zones.

3. Case Study

The chosen case studies–hotel and trading spaces–represent a typology that is prone to functional changes. The iteration shows how the functions evolved and were followed by building plan transformation and how alternative building plans can be exercised while preserving functional relations. The case studies are presented in diagrammatic resolution to clarify the functional graph and shape arrangements.

3.1. Case 1: Hotel Lobby

Over the years, the strategies for accommodating travellers have evolved considerably. One significant transformation can be seen in the hotel lobby, which bridges public and private spaces, often coupled with various amenities [53–57]. The lobby features ambassadorial roles for the hotel, ranging from basic functions such as reservation and waiting to perceptual roles for simulating a welcoming place to attract guests. Figure 7 demonstrates how graph and shape rules are computed to examine how a lobby is transformed through amenities and technology. The horizontal iterations show shape-to-graph and graph-to-shape iterations, while the vertical iteration shows a graph-to-graph iteration applied to the first step at each row from Figure 7a–e. The horizontal iteration modifies spatial experience, while the vertical iteration alters the functional sequences.

3.1.1. Shape to Graph—Graph to Shape

Figure 7a shows a shape-to-graph conversion for a simple hotel layout consisting of a waiting lounge, a receptionist desk with an office, and several bedrooms. Given the room and access layout in step 1, labelled with names, rule G1 returns the room's nodes with the shape weight and edges between the nodes, labelled with activity names in step 2. At this step, both functional relationships and spatial conditions are visible. In step 3, rules G3, G4, and G5 normalise the weight into identical sizes (i.e., squares) and rearrange them to simplify their arrangement. These steps result in a graph, visualised with normalised weights. Step 4 removes the weight with rule G2, leaving only the nodes and edges as a graph. Rule F8 groups nodes of the same activities, and rule F4 adds direction to the edges according to a hotel guest's flow of activity. The iteration in Figure 7b reverses the graph from Figure 7a (step 4), back to a shape and into a different design. From step one to step two, rule F12 removes the directions of the edges; rule F13 ungroups the bedroom nodes; and rule G7 returns the shape weight for each node. Next, rules G3, G4, and G5 rearrange the nodes by distancing the bedroom nodes from the lobby. Rules G8 and G6 return the boundary shape of the weights. The plan in step 4 is different from the initial plan, yet the graph remains the same. The new arrangement suggests a similar functional experience for the user with a different spatial experience (e.g., a more private bedroom zone).

3.1.2. Graph to Graph—Graph to Shape

The iteration in Figure 7c shows how the sequences of functions are reconfigured. While the resulting plans in step 4 between Figure 7b,c look similar, the functional experience in the lobby is profoundly different. As seen from step 1 in Figure 7b to step 1 in Figure 7c, rule F6 mirrors the check-in and work nodes with the walking (foyer) node as the mirroring axis, resulting in a new graph, where the sitting and check-in nodes have merged, with a work node as their child node. The graph indicates that the area for sitting will also be used for check-in, and the receptionist will approach the sitting guests. The resulting design in step 4 reflects these consequences, where the reception and offices become part of the sitting lounge.

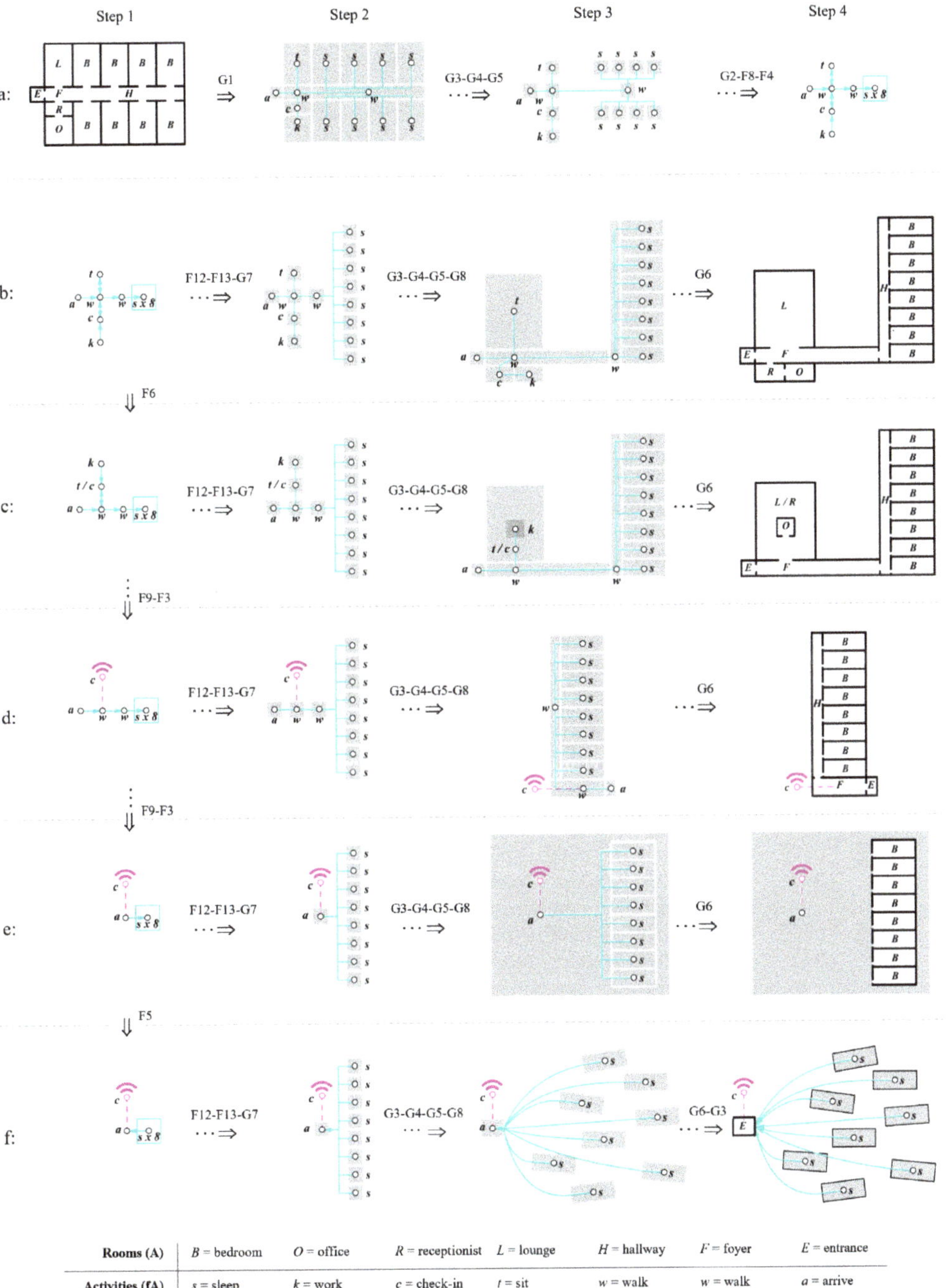

Figure 7. Graph and shape rule iterations for a hotel lobby (**a–f**). Graph-to-graph iteration is shown in the vertical direction (left-hand column) and graph-to-shape iteration in the horizontal direction.

The graph transformation from Figure 7c,d demonstrates a functional and spatial reduction. Rule F9 removes the sitting node, and rule F3 replaces the working and check-in nodes with a virtual node, which practically removes the sitting lounge and the receptionist from the plan. Instead, a guest would check in virtually (e.g., through a sensor, a vending machine, or a mobile device). The resulting graph has two walking nodes, as indicated by a foyer and a corridor in the plan in step 4.

The subsequent two iterations exemplify how a final design and an ongoing design process can coexist in the plan. In the iteration in Figure 7e, rule F9 continues to remove the walking node from the graph in Figure 7d, implying that a dedicated room for walking is no longer necessary. Rule F3 mirrors the virtual nodes with an edge between the arrival and walking nodes. As seen in the output plan in step 4, a corridor is no longer connected to the bedrooms. Instead, while the sleeping nodes have been finalised into bedroom shapes with rule G6, the arrival node remains as a graph with weight. It suggests that the bedrooms' design is fixed but the arrival position and context are flexible, such as in a capsule hotel in an airport or mall. In step 1 in Figure 7f, the difference with the graph in Figure 7e is subtle, with only the arrow reversing direction by rule F5. Yet, the reversal creates a substantial difference in functional and spatial experience. The arrow suggests that instead of a guest walking into a sleeping node from the arrival node, the sleeping node approaches the guest in the arrival node. After transforming the sleeping node's position and orientation with rules G3, G4, and G5, the resulting plan in step 4 produces a fixed entrance. Yet, the size and position of the bedroom remain in flexible positions, as indicated by the weighted nodes and directed edges (e.g., floating hotel).

As shown above, the graph iterations not only capture a functional evolution and alternative building plan they also embed different states of a design process in the final operation. Subtracting rules are predominantly used in this case and gradually shrink the graph. Other rule operations can expand the graph for different results to provide spaces for more functions (amenities) with addition rules (Figure 4) or translation rules (Figure 3) to create an overlapping space. Additionally, the number of graph nodes and similar shapes can inform preferences for construction types (e.g., in situ or off-site production). For instance, repetitive functions with identical shapes, such as bedrooms or corridors, can be considered for a prefabricated module, and non-repetitive functions, such as a hotel lounge, might be more suitable for in situ construction.

3.2. Case 2: Trading Space

Spatial configuration for trading space has experienced substantial disruption in recent decades, such as in retail, where customers browse and buy products [58–60]. Improving customer experiences has been the main goal in the retail industry [61]. This case demonstrates the graph and shape rules to examine function evolution and alternative spatial configuration in trading spaces by generating trade flows based on three basic nodes: the customer, product, and seller. The customer node represents the space where they are present; the product node represents the space for product display; and the seller node represents the space where the transaction occurs. Figure 8 outlined shape rules to generate spatial arrangements for trade flow functional evolution in Figure 9a. The flow transformation is arranged similarly to that of Figure 7, where the vertical iteration (Figure 9a) demonstrates trade flow transformation, and the horizontal iteration in Figure 9b shows the consequences of spatial transformation, as calculated with the rules explained in Figure 8.

3.2.1. Graph to Graph

Moving vertically from step one to step two in Figure 9a, rules F1 and F4 add a seller node to the product node. In step three, a customer node is added and linked to the seller node with the same rules. This graph can describe a travelling salesman or a street vendor's trade flows, where the seller brings the product to the customer's house or public areas in a city. In step four, rule F5 changes the direction of the edge, where the customer node goes to a seller node, which itself goes to the product node. This graph illustrates traditional retail, where the customer visits the seller to buy a product displayed behind or below the counter. Step five uses rule F6 to mirror the product and seller nodes' positions, allowing the customer to browse the product display directly and then bring the products to the seller for transaction. This swap can illustrate the shift from traditional to modern retail, where customers can browse products directly on display (with RFID tags attached for safety reasons) [62].

The subsequent steps involve rule F3, in which virtual activity replaces physical activity. This rule is applied in step 6a to virtualise the product node into, for instance, a screen (online) display. The customer interacts with the product virtually with the products located in a remote location (e.g., with pick-up or delivery options) or in close proximity to a customer (e.g., a restaurant that uses an app-based menu) [63]. Step 6b adds a virtual function to a physical store to combine digital and physical experiences (e.g., with immersive technology) [64]. Step 6c converts the seller node into a virtual node. It allows a customer to browse the product display physically and perform the transaction online without interacting with the seller, e.g., via a sensor or a self-checkout machine. Step seven virtualises both the product and the seller nodes. This graph illustrates the case of non-physical products (e.g., music or movies), where the seller can sell them online and the customer can purchase them online via a physical outlet (e.g., digital player). In the final step (eight), all the nodes become virtual. This graph can describe a condition where customers and sellers delegate their trading activities digitally for a virtual product (e.g., an automated or algorithmic trading system for shares and crypto) where no physical space is required [65].

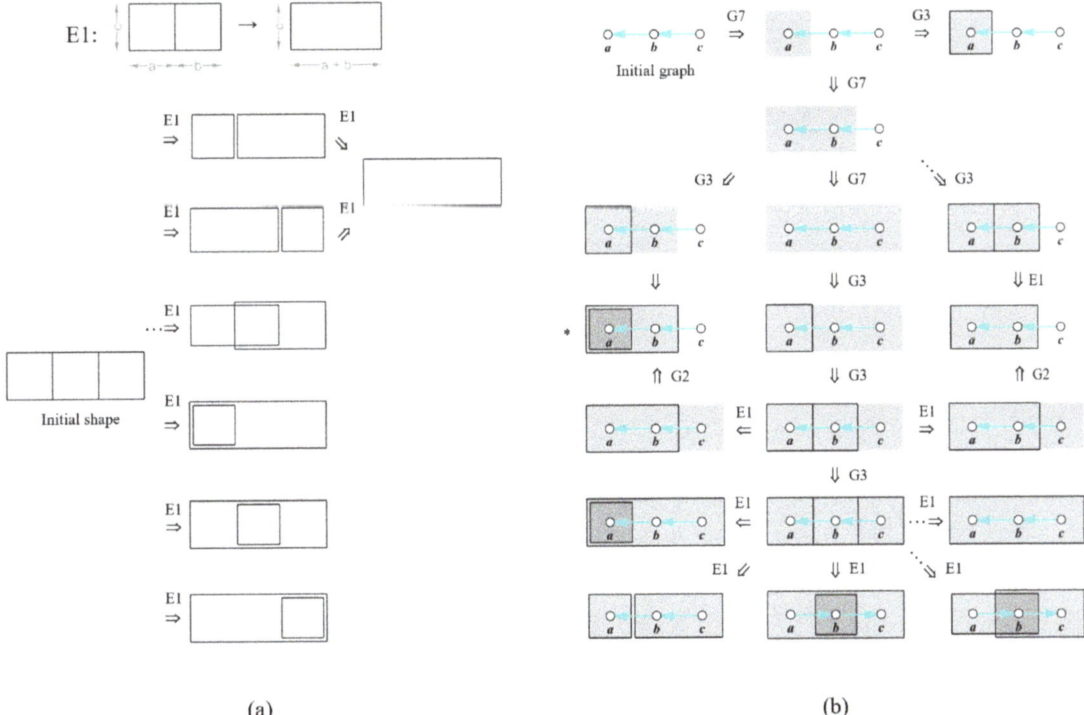

(a) (b)

Figure 8. Reduction rules for adjacent rectangles and example iterations (**a**) with their application graph-to-shape rules (**b**).

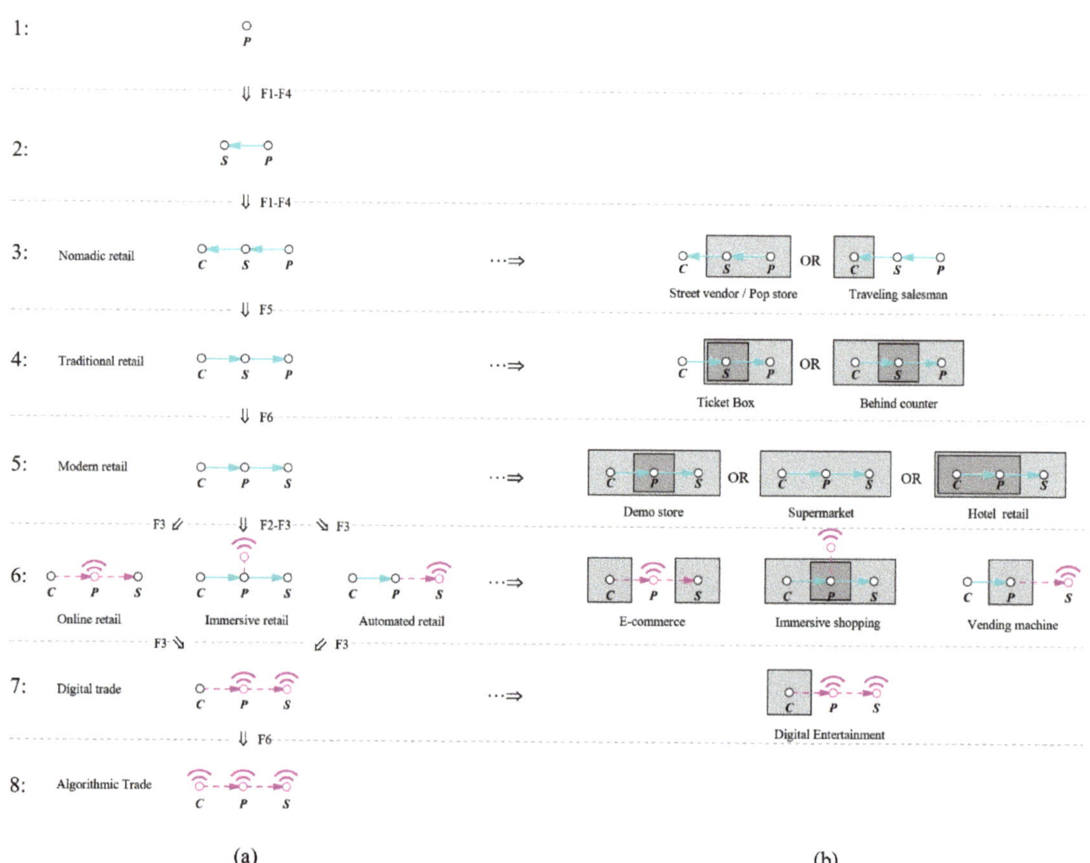

Figure 9. Different trade flow and retailing types produced from graph-to-graph iteration (**a**) and examples of spatial arrangement presented with diagrammatic sections (**b**). Nodes P = product, S = seller, and C = customer. The wi-fi symbols indicate a virtual function.

3.2.2. Graph to Shape—Shape to Shape

Most of the examples in the cases so far show spatial configuration arranged for a symbolic shape. To demonstrate shape embedding, this case adds rule E1, which combines two adjoining rectangles into one (each can have various lengths and yet have the same height). This rule utilises SG reduction rules to recognise sub-shapes embedded within lines along the rectangles' boundaries [42,66]. By recognising these lines as sub-shapes, the rule could produce various spatial compositions for a graph with a minimum of transformation rules.

With SG embedding, rule E1 iteration on two adjoining rectangles can generate six rectangular arrangements, two of which can further produce a single rectangle (Figure 8a). With such versatility, rule E1 can be applied to rectangular compositions generated from graph-to-shape rules to exercise possible spatial arrangements for a functional configuration. Figure 8b shows at least 18 possible spatial arrangements for a three-node graph, including a combination of their outputs (*). Some of them indicate functions with separated spaces, while others show functions with overlapped and confined spaces. Additionally, function nodes can be bound by definitive shapes (e.g., walls or partitions) or indicative space marked by weight (e.g., furniture arrangement) or they can be not bound by anything (e.g., outdoor space). The diagrammatic section could inform the decision-making process at a later stage, such as in choosing the structural configuration and envelope system. For instance, shapes with boundaries that face outdoor spaces require treatment to weather the building; shapes that cover more nodes may require a longer span, and a confined shape may be less constrained in the building structure and can be freely articulated.

In a retail context, the rule application is demonstrated in several examples of trading space arrangement for the same trade flow graph. After rules G2, G3, and G7 are applied to the graph in Figure 9a, each node, except for the virtual nodes, has a rectangular weight and shape boundary to represent a diagrammatic section of the space. These rectangles are then applied with rule E1 to generate the different compositions in Figure 9b. For instance, the graph from step three is transformed into two compositions. One is where the seller and the product share the same rectangle, separated from the customer's, which may illustrate a seller with a non-permanent space (e.g., pop-up stores or street vendors, such as food trucks). The other shows all the customers, sellers, and products sharing the same space (e.g., a salesman selling a product door to door). The graph from step four is transformed into overlapping rectangles, one with a square confined within a rectangle and a customer node on the outside (e.g., a ticket box at a cinema or stadium), another with the seller space separating the space between the customer and the product (e.g., as in a traditional behind-the-counter store). Composition one from step five's graph shows a product square confined within a rectangle shared by the customer and the seller (e.g., an enclosed space in a demo store, where a customer can experience the product). Composition two from the same graph shows a rectangle shared by all three nodes (e.g., a supermarket), and composition three shows an overlap between customer space and seller space (e.g., a retail store within a hotel, where the customer can experience the seller's products during their stay). In step six, one composition shows two squares separated by a virtual node, indicating they are physically distant (e.g., e-commerce); one composition mimics a demo store with a virtual node (e.g., AR/VR experience in-store); and the last composition shows a customer accesses the product physically while the seller is virtually connected (e.g., a vending machine). The last output from step seven only shows one rectangle, representing a space where only the customer is present physically (e.g., digital entertainment).

4. Results

The outputs from the above cases show several highlights from the interplay between the graph and shape rule, including the roles of shape, label, and weight.

4.1. Shape-Driven: New Spatial Experience for the Same Function

First, the experiments show how the iterations for graph and shape rules can work in tandem in a design process. The initial iterations with a simple hotel plan show different designs generated from the same graph, echoing the lessons from the revisited study. The dual presence of graphs and weighted shapes can assist in simultaneously evaluating early spatial and functional conditions. The use of graph transformation rules helps to quickly assemble different spatial arrangements (e.g., within the site). Additionally, the directed edges in the graph can ensure that the resulting room layout does not compromise the required access and sequences, and the weighted nodes indicate the intended sizes and proximity between rooms. Generating a functional–spatial graph schematically could clarify functional evolution and how it transforms building plans, as well as exercise alternative building plans while preserving functional relations.

4.2. Graph-Driven: New Functional Experience with Similar Shapes

Second, the iterations from the cases show that creative spatial solutions can start at functional iteration before shape articulations. The hotel lobby case demonstrated how graph rules generate various functional sequences and configurations. Some rules, such as the erasing and virtualisation rules, produce the expected results with disappearing functions and their respective spaces. Other rules produced rather unexpected effects, such as the mirroring rules, which could merge functions, and the reversing of a rule, which creates profound experiential changes by swapping the sequence direction. While it appears simple, a change in flow direction could help improve users' experiences, e.g., by providing a space designed for staff to approach guests for check-in. Some iteration in the cases also suggests that differences in user experience can be achieved by modifying the order of rules in the iteration and that the final design can be set without having all the shapes fixed. Leaving some shapes at their weighted graph state could provide alternative means for the users' flow of activities.

4.3. Graph- and Shape-Driven: New Functional and Spatial Experience

Third, shape embedding can help to generate a profound spatial impact. As demonstrated in the second case, rule iteration with shape embedding can produce different spatial configurations to accommodate the same trade flow activities, as seen in the trading space case.

4.4. Rule Characteristics in Graph-to-Shape Iteration

The study also notices different rule characteristics in generating graphs to shapes pertinent to the building design. The substitution rule is more deterministic as it initiates a shape from a node and an edge into a component (e.g., opening), and its geometry could be dominating in a building's design language. The subdivision rule is constraining as the iteration typically goes inward and ensures that spaces for functions are multiplied inside an initial shape (e.g., a site boundary or a building envelope). Some additive rules are expansive and less sensitive to boundaries when populating a shape unless they are accompanied by overlapping rules [67–70]. The identity rule is intriguing as it invites opportunities to explore what shapes and rules can do in graph-to-shape translation. For instance, it allows the same shape to be implemented across stages to maintain a design language, namely from the functional to the structural configuration stages. In Sundt house, the triangle for functional arrangement is also applied for columns profiles and column positions, while in the Jester house, the circles extruded into the walls and columns. This also suggests that consistency throughout function, design, and construction can be achieved if graph-to-shape iteration is aligned with, or generated over, structural grids, e.g., the triangular and rectilinear–circular grids.

5. Discussion

This study revisited early studies on graph and shape in computation design to examine their compatibility and to interpret a set of rules to map a function graph from design and generate a design from a function graph that mediates spatial and functional relationships. Two cases demonstrated the lessons learned from the revisited studies on how graph and shape rules can create sequences of functions in pairs with spatial development.

The experiments shown are predominantly run with the combinatorial operation, presented with schematic resolutions, and have yet to utilise a benchmarking system to evaluate the output performance in detail. There are other features in SG visual calculation and SS evaluation systems that have yet to be tested, such as isovist and depth measurement with JPG. Incorporating these other features could help increase design variety and separate the useful and useless results. Further studies could include resolution enhancement of the diagrammatic outputs with tools such as those mentioned in [36,38,71,72], and function descriptions can be organised into an ontological structure in semantic modelling [73–75].

Nevertheless, the main principles of operating shape and graph calculations have been outlined and demonstrated in this paper. The generative rules can help to analyse functional evolution as well as to synthesise different functional sequences and their alternative building plans to venture into different typologies, which, in Lionel March's words, *'minimises prejudice and maximises choice'*. Shape and graph integration can show how a function within a building, such as a hotel or a retail establishment, evolves and shrinks, as well as what alternatives exist for creating new experiences or new business models.

Funding: This research received no external funding.

Data Availability Statement: Not applicable.

Conflicts of Interest: The author declares no conflict of interest.

References

1. Boddy, N. Australia COVID: Office occupancy rates go backwards for the first time in six months. *Financial Review*, 11 August 2022.
2. Zettl, M. Office buildings are still less than 50% occupied. Who should worry? *Forbes*, 29 November 2022.
3. Duffy, F. Office buildings and organisational change. In *Buildings and Society*; Routledge: London, UK, 2003; pp. 147–163.
4. WAF. World Architecture Festival 2022 Winners Homepage. Available online: https://www.worldarchitecturefestival.com/waf2023/en/page/waf-2022 (accessed on 5 December 2022).
5. Amoils, J. The evolving workplace: Where to next in a post-pandemic world? *CRE Real Estate Issues* **2021**, *45*, 1–10.
6. Duchene, P. The evolving workplace. *Forbes*, 12 March 2019.
7. Sweeney, S. Post-Pandemic Utilization of Office to Residential Adaptive Reuse Strategies in Cities. Master's Thesis, University of North Carolina, Chapel Hill, NC, USA, 6 April 2021. [CrossRef]
8. RIBA. *RIBA Plan of Work 2020 Overview*; RIBA Publications: London, UK, 2020.
9. American Institute of Architects, *The Architect's Handbook of Professional Practice*; John Wiley & Sons, Incorporated: New York, NY, USA, 2013.
10. Hillier, B.; Leaman, A.; Stansall, P.; Bedford, M. Space syntax. *Environ. Plan. B Plan. Des.* **1976**, *3*, 147–185. [CrossRef]
11. Kim, Y.O.; Kim, J.Y.; Yum, H.Y.; Lee, J.K. A study on mega-shelter layout planning based on user behavior. *Buildings* **2022**, *12*, 1630. [CrossRef]
12. Zhang, X.; Cui, T. Evolution process of scientific space: Spatial analysis of three groups of laboratories in history (16th–20th century). *Buildings* **2022**, *12*, 1909. [CrossRef]
13. Lee, J.H.; Kim, Y.S. Rethinking art museum spaces and investigating how auxiliary paths work differently. *Buildings* **2022**, *12*, 248. [CrossRef]
14. Azizi, V.; Usman, M.; Sohn, S.S.; Schwartz, M.; Moon, S.; Faloutsos, P.; Kapadia, M. The role of latent representations for design space exploration of floorplans. *Simulation* **2022**, 375497221115734. [CrossRef]
15. Stiny, G.; Gips, J. Shape grammars and the generative specification of painting and sculpture. In Proceedings of the IFIP Congress 71, Ljubljana, Yugoslavia, 23–28 August 1971; pp. 125–135.
16. Belčič, A.; Eloy, S. Architecture for community-based ageing—A shape grammar for transforming typical single-family houses into older people's cohousing in Slovenia. *Buildings* **2023**, *13*, 453. [CrossRef]
17. Flemming, U. More than the sum of parts: The grammar of Queen Anne houses. *Environ. Plan. B Plan. Des.* **1987**, *14*, 323–350. [CrossRef]
18. Cagdas, G. A shape grammar: The language of traditional Turkish houses. *Environ. Plan. B Plan. Des.* **1996**, *23*, 443–464. [CrossRef]
19. Colakoglu, B. Design by grammar: An interpretation and generation of vernacular Hayat houses in contemporary context. *Environ. Plan. B Plan. Des.* **2005**, *32*, 141–149. [CrossRef]
20. Yousefniapasha, M.; Teeling, C.; Rollo, J.; Bunt, D. Shape grammar, culture, and generation of vernacular houses (a practice on the villages adjacent to rice fields of Mazandaran, in the north of Iran). *Environ. Plan. B Urban Anal. City Sci.* **2021**, *48*, 94–114. [CrossRef]
21. Erem, Ö.; Ermiyagil, M.S.A. Adapted design generation for Turkish vernacular housing grammar. *Environ. Plan. B Plan. Des.* **2016**, *43*, 893–919. [CrossRef]
22. Haakonsen, S.; Rønnquist, A.; Labonnote, N. Fifty years of shape grammars: A systematic mapping of its application in engineering and architecture. *Int. J. Arch. Comput.* **2022**, 14780771221089882. [CrossRef]
23. Ehrig, H.; Pfender, M.; Schneider, H.J. Graph-grammars: An algebraic approach. In *14th Annual Symposium on Switching and Automata Theory (swat 1973)*; IEEE: Piscataway, NJ, USA, 1973; pp. 167–180.
24. Rozenberg, G. *Handbook of Graph Grammars and Computing by Graph Transformation*; World Scientific: Singapore, 1997; Volume 1.
25. Voss, C.; Petzold, F.; Rudolph, S. Graph transformation in engineering design: An overview of the last decade. *Artif. Intell. Eng. Des. Anal. Manuf.* **2023**, *37*, e5. [CrossRef]
26. Weber, R.E.; Mueller, C.; Reinhart, C. Automated floorplan generation in architectural design: A review of methods and applications. *Autom. Constr.* **2022**, *140*, 104385. [CrossRef]
27. Nourian, P.; Azadi, S.; Oval, R. Generative design in architecture: From mathematical optimization to grammatical customization. In *Computational Design and Digital Manufacturing*; Springer: Berlin/Heidelberg, Germany, 2023; pp. 1–43. [CrossRef]
28. Heitor, T.V.; Duarte, J.P.; Pinto, R.M. Combing grammars and space syntax: Formulating, generating and evaluating designs. *Int. J. Arch. Comput.* **2004**, *2*, 491–515. [CrossRef]
29. Eloy, S.; Guerreiro, R. Transforming housing typologies. Space syntax evaluation and shape grammar generation. *Arq. Urb.* **2016**, *15*, 86–114.
30. Lee, J.H.; Ostwald, M.J.; Gu, N. A Justified Plan Graph (JPG) grammar approach to identifying spatial design patterns in an architectural style. *Environ. Plan. B Urban Anal. City Sci.* **2016**, *45*, 67–89. [CrossRef]

31. Lee, J.H.; Ostwald, M.J.; Gu, N. A Combined plan graph and massing grammar approach to frank lloyd wright's prairie architecture. *Nexus Netw. J.* **2017**, *19*, 279–299. [CrossRef]
32. Lee, J.H.; Ostwald, M.; Gu, N. A syntactical and grammatical approach to architectural configuration, analysis and generation. *Arch. Sci. Rev.* **2015**, *58*, 189–204. [CrossRef]
33. Etintaş, M.F.; Erem, N.Ö. Outpatient clinic design through rule based design methods. *Eskişeh Tech. Univ. J. Sci. Technol.-Appl. Sci. Eng.* **2022**, *23*, 156–171.
34. Grasl, T.; Economou, A. From topologies to shapes: Parametric shape grammars implemented by graphs. *Environ. Plan. B: Plan. Des.* **2013**, *40*, 905–922. [CrossRef]
35. Lima, F.; Brown, N.C.; Duarte, J.P. Optimizing urban grid layouts using proximity metrics. In *Artificial Intelligence in Urban Planning and Design*; Elsevier: Amsterdam, The Netherlands, 2022; pp. 181–200.
36. Hong, T.-C.K.; Economou, A. Implementation of shape embedding in 2D CAD systems. *Autom. Constr.* **2023**, *146*, 104640. [CrossRef]
37. Schaffranek, R.; Vasku, M. Space syntax for generative design: On the application of a new tool. In Proceedings of the Ninth International Space Syntax Symposium, Seoul, Republic of Korea, 31 October–3 November 2013.
38. Stouffs, R. A Multi-formalism Shape Grammar Interpreter. In *Computer-Aided Architectural Design. Design Imperatives Future Now*; Gerber, D., Pantazis, E., Bogosian, B., Nahmad, A., Miltiadis, C., Eds.; Springer: Singapore, 2022; Volume 1465, pp. 268–287. [CrossRef]
39. Lambe, N.R.; Dongre, A.R. A shape grammar approach to contextual design: A case study of the Pol houses of Ahmedabad, India. *Environ. Plan. B Urban Anal. City Sci.* **2019**, *46*, 845–861. [CrossRef]
40. Veloso, P.; Celani, G.; Scheeren, R. From the generation of layouts to the production of construction documents: An application in the customization of apartment plans. *Autom. Constr.* **2018**, *96*, 224–235. [CrossRef]
41. Vardouli, T. Skeletons, Shapes, and the Shift from Surface to Structure in Architectural Geometry. *Nexus Netw. J.* **2020**, *22*, 487–505. [CrossRef]
42. Stiny, G. *Shape: Talking about Seeing and Doing*; The MIT Press: Cambridge, MA, USA, 2006.
43. Stiny, G. *Shapes of Imagination: Calculating in Coleridge's Magical Realm*; MIT Press: Cambridge, MA, USA, 2022.
44. Hillier, B. The nature of the artificial: The contingent and the necessary in spatial form in architecture. *Geoforum* **1985**, *16*, 163–178. [CrossRef]
45. Hillier, B. *Space Is the Machine: A Configurational Theory of Architecture*; Cambridge University Press: Cambridge, MA, USA, 1996; p. 463.
46. Coates, P. *Programming Architecture*, 1st ed.; Routledge: London, UK, 2010.
47. Alexander, C. *Notes on the Synthesis of Form*; Harvard University Press: Cambridge, MA, USA, 1964; Volume 5.
48. March, L.; Steadman, P. *The Geometry of Environment. An Introduction to Spatial Organization in Design*; MIT Press: Cambridge, MA, USA, 1974.
49. Muslimin, R. Ethnocomputation: An Inductive Shape Grammar on Toraja Glpyh. In *Computer-Aided Architectural Design. Future Trajectories*; Çağdaş, G., Özkar, M., Gül, L.F., Gürer, E., Eds.; Springer: Singapore, 2017; Volume 724, pp. 329–347. [CrossRef]
50. Stiny, G. What rule(s) should I use? *Nexus Netw. J.* **2011**, *13*, 15–47. [CrossRef]
51. Kotsopoulos, S.D. Design without rigid rules. In *Design Computing and Cognition'20*; Springer: Berlin/Heidelberg, Germany, 2022; pp. 107–128.
52. Lee, J.H.; Ostwald, M.J.; Gu, N. Combining space syntax and shape grammar to investigate architectural style. In Proceedings of the Ninth International Space Syntax Symposium, Seoul, Republic of Korea, 31 October–3 November 2013; Sejong University: Seoul, Republic of Korea, 2013.
53. Bowie, D. Innovation and 19th century hotel industry evolution. *Tour. Manag.* **2018**, *64*, 314–323. [CrossRef]
54. Li, L.; Lu, L.; Xu, Y.; Sun, X. The spatiotemporal evolution and influencing factors of hotel industry in the metropolitan area: An empirical study based on China. *PLoS ONE* **2020**, *15*, e0231438. [CrossRef]
55. Murphy, J.; Schegg, R.; Olaru, D. Investigating the evolution of hotel internet adoption. *Inf. Technol. Tour.* **2006**, *8*, 161–177. [CrossRef]
56. James, K.J.; Sandoval-Strausz, A.K.; Maudlin, D.; Peleggi, M.; Humair, C.; Berger, M.W. The hotel in history: Evolving perspectives. *J. Tour Hist.* **2017**, *9*, 92–111. [CrossRef]
57. MacDonald, R. Urban hotel: Evolution of a hybrid typology. *Built Environ.* **1978**, *2000*, 142–151.
58. Gauri, D.K.; Jindal, R.P.; Ratchford, B.; Fox, E.; Bhatnagar, A.; Pandey, A.; Navallo, J.R.; Fogarty, J.; Carr, S.; Howerton, E. Evolution of retail formats: Past, present, and future. *J. Retail.* **2021**, *97*, 42–61. [CrossRef]
59. Miotto, A.P.; Parente, J.G. Retail evolution model in emerging markets: Apparel store formats in Brazil. *Int. J. Retail. Distrib. Manag.* **2015**, *43*, 242–260. [CrossRef]
60. Quartier, K. Retail design: What's in the name? In *Retail Design*; Routledge: London, UK, 2016; pp. 39–56.
61. Bonfrer, A.; Chintagunta, P.; Dhar, S. Retail store formats, competition and shopper behavior: A Systematic review. *J. Retail.* **2022**, *98*, 71–91. [CrossRef]
62. Stobart, J. A history of shopping: The missing link between retail and consumer revolutions. *J. Hist. Res. Mark.* **2010**, *2*, 342–349. [CrossRef]

63. Stieninger, M.; Gasperlmair, J.; Plasch, M.; Kellermayr-Scheucher, M. Identification of innovative technologies for store-based retailing—An evaluation of the status quo and of future retail practices. *Procedia Comput. Sci.* **2021**, *181*, 84–92. [CrossRef]
64. Hilken, T.; Chylinski, M.; Keeling, D.I.; Heller, J.; Ruyter, K.; Mahr, D. How to strategically choose or combine augmented and virtual reality for improved online experiential retailing. *Psychol. Mark.* **2021**, *39*, 495–507. [CrossRef]
65. Treleaven, P.; Galas, M.; Lalchand, V. Algorithmic trading review. *Commun. ACM* **2013**, *56*, 76–85. [CrossRef]
66. Kristic, D. Diagonal decompositions of shapes and their algebras. *Artif. Intell. Eng. Des. Anal. Manuf.* **2022**, *36*, e10. [CrossRef]
67. Caneparo, L. Semantic knowledge in generation of 3D layouts for decision-making. *Autom. Constr.* **2022**, *134*, 104012. [CrossRef]
68. Grasl, T. Transformational palladians. *Environ. Plan. B Plan. Des.* **2012**, *39*, 83–95. [CrossRef]
69. Duarte, J.P. Customizing Mass Housing: A Discursive Grammar for Siza's Malagueira Houses. Ph.D. Dissertation, Massachusetts Institute of Technology, Cambridge, MA, USA, 2001.
70. Stiny, G. Ice-ray: A note on the generation of Chinese lattice designs. *Environ. Plan. B Plan. Des.* **1977**, *4*, 89–98. [CrossRef]
71. Okhoya, V.W.; Bernal, M.; Economou, A.; Saha, N.; Vaivodiss, R.; Hong, T.-C.K.; Haymaker, J. Generative workplace and space planning in architectural practice. *Int. J. Arch. Comput.* **2022**, *20*, 645–672. [CrossRef]
72. Al-Jokhadar, A.; Jabi, W. Spatial reasoning as a syntactic method for programming socio-spatial parametric grammar for vertical residential buildings. *Archit. Sci. Rev* **2020**, *63*, 135–153. [CrossRef]
73. Fernandes, P.A. Space syntax with prolog. In Proceedings of the 13th Space Syntax Symposium, Bergen, Norway, 20–24 June 2022.
74. Andaroodi, E.; Andres, F.; Einifar, A.; Lebigre, P.; Kando, N. Ontology-based shape-grammar schema for classification of caravanserais: A specific corpus of Iranian Safavid and Ghajar open, on-route samples. *J. Cult. Heritage* **2006**, *7*, 312–328. [CrossRef]
75. Grobler, F.; Aksamija, A.; Kim, H.; Krishnamurti, R.; Yue, K.; Hickerson, C. Ontologies and shape grammars: Communication between knowledge-based and generative systems. In *Design Computing and Cognition '08*; Gero, J.S., Goel, A.K., Eds.; Springer: Dordrecht, The Netherlands, 2008; pp. 23–40.

MDPI

St. Alban-Anlage 66

4052 Basel

Switzerland

www.mdpi.com

Buildings Editorial Office

E-mail: buildings@mdpi.com

www.mdpi.com/journal/buildings